THE CHANGING STRUCTURE OF AMERICAN INDUSTRY AND ENERGY USE PATTERNS

Issues, Scenarios, and Forecasting Models

Edited by

AHMAD FARUQUI
Battelle Palo Alto

and

JOHN BROEHL
Battelle Columbus Division

CLARK W. GELLINGS
EPRI Program Manager

BATTELLE PRESS
Columbus • Richland

EPRI
Electric Power
Research Institute

3412 Hillview Avenue
Palo Alto, California 94304

Distributed by
WESTVIEW PRESS
Frederick A. Praeger, Publisher
5500 Central Avenue
Boulder, Colorado 80301

EPRI Report EM-5075-SR

TOPICS
Industrial load forecasting
Economic impact
Demand-side planning
Structural change
Indepth
Industrial sector

Distributed by

Battelle Press
505 King Avenue
Columbus, Ohio 43201
614-424-6393

Library of Congress Cataloging-in-Publication Data

EPRI Seminar (1985 : Arlington, Va.)
The changing structure of American industry and energy use patterns.

1. Electric utilities—United States—Forecasting— Congresses. 2. Electric power consumption—United States—Forecasting—Congresses. 3. United States—Industries—Energy consumption—Forecasting—Congresses. 4. Economic forecasting—United States—Congresses. 5. United States—Economic conditions—1981- —Congresses. I. Faruqui, Ahmad. II. Broehl, John, 1938- . III. Electric Power Research Institute. IV. Title.
HD9685.U5E68 1986 333.79'32 86-17488
ISBN 0-935470-33-6

ABSTRACT

This volume is the proceedings of a seminar sponsored by the Electric Power Research Institute (EPRI) in Fall 1985 on "The Changing Structure of American Industry and Energy Use Patterns: Issues, Scenarios, and Forecasting Models." The seminar brought together for the first time on a national level industrial analysts concerned with the future of American industry and utility planners concerned with the future of American utilities.

Participants represented a wide variety of organizations, including industry, utilities, government agencies, universities, research organizations, and consulting firms. Over a period of 3 days, more than 80 attendees heard 22 presentations representing a multiplicity of disciplinary, institutional, and regional perspectives.

Seminar highlights included discussions of the following topics:

- U.S. economic outlook--long-term perspective.
- Industrial electricity consumption and economic conditions.
- International competitiveness of U.S. industries.
- Outlook for energy-intensive industries.
- Results from INDEPTH industrial sector forecasting model.

From the seminar discussions, a deeper understanding has emerged of the forces that determine industrial structure and the role electric utilities can play in improving customer profitability through the development and marketing of selected electrotechnologies.

CONTENTS

1

EDITORS' OVERVIEW

Ahmad Faruqui
Electric Power Research Institute

John Broehl
Battelle Columbus Division

This volume is the proceedings of a seminar sponsored by the Electric Power Research Institute (EPRI) in Fall 1985 on "Industrial Structural Change and Future Electric Sales: Issues, Scenarios, and End-Use Models." The seminar brought together for the first time on a national level industrial analysts concerned with the future of American industry **and** utility planners concerned with the future of American utilities.

Participants represented a wide variety of organizations, including industry, utilities, government agencies, universities, research organizations, and consulting firms. Over a period of three days, more than 80 attendees heard 22 presentations representing a multiplicity of disciplinary, institutional, and regional perspectives.

Seminar highlights included a discussion of the following topics:

- U.S. economic outlook—long-term perspective
- Industrial electricity consumption and economic conditions
- International competitiveness of U.S. industries
- Outlook for energy-intensive industries
- Results from INDEPTH industrial sector forecasting model

Specific questions examined by the speakers included:

- What constitutes structural change? Has there been any? Do we expect any?
- What are the implications of changing energy prices on future industrial structure? How will industrial structure changes affect electricity use?
- What specific new technologies are emerging? How will these technology changes affect industrial structure and electricity use? (For example, how might electricity-intensive/cost-saving technologies be received by firms?)
- What are the implications of international competition on industrial structure and electricity use? What are the determinants of international competitiveness?

From the seminar discussions, a deeper understanding has emerged of the forces that determine industrial structure and the role electric utilities can play in improving customer profitability through the development and marketing of selected electrotechnologies.

This volume contains all the essential discussions that occurred at the seminar, including the questions and answers associated with each speaker. This section seeks to convey to the reader some of the flavor of the discussions that occurred at the seminar. It does not, however, summarize the entire range of discussions that occurred.

BACKGROUND

As developed in the paper by Ahmad Faruqui of EPRI, electricity sales in the industrial sector are influenced by the level of output, the product mix, and the technological processes utilized in the production of that output: see Figure 1-1. A major driving force behind product mix changes is increased international competition, which is causing structural shifts with the U.S. economy. In one scenario, the U.S. economy is expected to move toward advanced manufacturing and away from the traditional energy-intensive sectors, resulting in a decline in industrial electricity sales. This could have an adverse impact not only on electric utilities' sales but also on their financial performance, since industrial sales are made at a high load factor. These points are also discussed in the paper by Tom Moore of Tampa Electric.

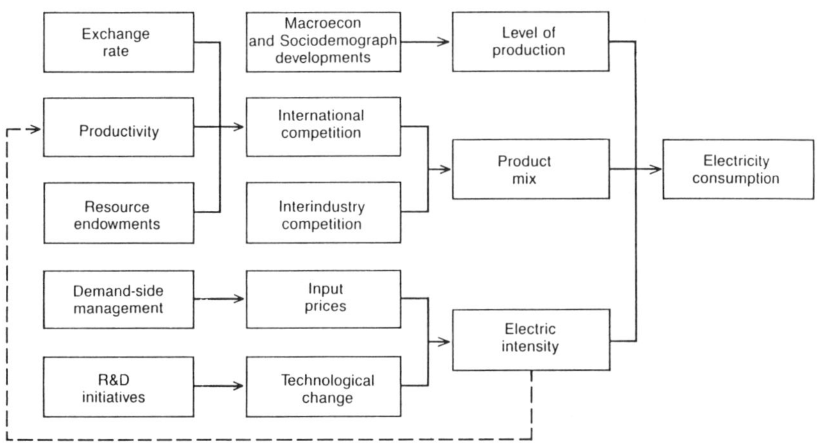

FIGURE 1-1 Structure of Industrial Electricity Demand

At the same time, several new technologies that are being rapidly developed and commercialized in the new competitive environment offer the electric utility industry many new opportunities, including the increased marketability of electricity for industrial applications. Electricity's form value, i.e., workside efficiency, controllability, and precision of application, make possible increases in business productivity. Recognition of this fact by industrial consumers may lead to increased market penetration of new electric-powered processes such as robotics, laser beam welding, and electrolytic coatings. How fast these technologies enter the market place will depend upon the financial health of the various user industries, as well as the rate of change in product mix.

While this is only one of many scenarios currently under consideration, industrial analysts expect structural as well as technological changes in the economy to be the critical determinants of industrial electricity sales in the future. The volume explores the historical impact of structural change in the U.S. economy on industrial electricity sales and suggests several scenarios of future structural change and its likely impact on electricity sales. The paper by William Huss of Battelle presents a summary of the scenarios. Opportunities as well as threats to electric utilities resulting from structural change are identified and discussed. In addition, EPRI's new industrial forecasting and market planning system for utilities, INDEPTH, is described.

ORGANIZATION

The volume is structured into four parts:

Energy Planning and Structural Change

- Utility planning and industrial structural change
- Industrial electricity consumption and economic structure
- Scenarios in utility planning

Structural Change—Sectoral Perspective

- Macroeconomic outlook
- Role of final demand changes
- Trade, technology, and management factors

Structural Change in Five Key Industries

- Metals industries
- Chemical/petrochemical industries
- Pulp and paper industry
- Transportation and assembly industries
- "High technology" industries

Model-Based Energy Forecasting
- National models
- Forecasting at the service area level
- Structural change and regional growth

SELECTED FINDINGS

Eight major themes emerge from the discussions in this volume. First, the long-term macroeconomic outlook for the U.S. industrial sector is quite sensitive to macroeconomic variables, especially the interest rate and the dollar exchange rate. This point is brought out, first of all, by Kurt Karl of Wharton, and followed up in several of the other papers. In the broad context of the overall level of production, high interest rates and dollar exchange rates lead to lower economic growth.

Second, there is a pretty strong consensus that considerable structural change has occurred within the U.S. industrial sector over the past two decades: details are contained in the paper by Barry Bluestone of Boston College. The future outlook for the industrial sector is characterized by a rather wide distribution of growth rates across the industries. Certain industries where the U.S. economy has comparative disadvantages—textiles, apparel, footwear, commodity parts of the chemicals industry, most of the steel industry and the auto industry—are low-growth industries across the cross-section of scenarios presented by Kurt Karl, Stan Feldman of Data Resources, Inc., and others. Another category of industries where U.S. manufacturing has a comparative advantage—high-technology industries, such as specialty chemicals, computing equipment, aerospace, electronic components, telecommunications equipment, and scientific instruments—consistently appear as high-growth industries.

There is no expectation of a substantial change in the sectoral mix of the economy as a whole, with manufacturing output continuing to account for 25 percent of the GNP through 2000 and manufacturing employment for about 18 percent of national employment. This is consistent with the Bureau of Labor Statistics projections (1).

Third, structural change has considerable implications for electric energy consumption. This point is brought out in the papers by Bob Marley and Paul Werbos of the U.S. Department of Energy, Bruce Humphrey of the Edison Electric Institute, and Barry Bluestone. These authors indicate that structural effects have dominated price effects, and much of what has been regarded as efficiency improvements at an aggregate level is really structural change at a more disaggregated level. In addition, an important finding about the character of structural change, particularly based on Barry Bluestone's paper, is that the impact of structural change on employment can often be quite different from the impact on energy consumption, especially electricity consumption. For many industries, the employment intensity of production has gone down dramatically (labor productivity has increased), but that has not been accompanied by dramatic reductions in either the electricity or the energy intensity.

Fourth, two major sets of factors determine structural change: factors within the domestic economy and factors in the international economy.

Fifth, within the domestic category, Stan Feldman of DRI and several of the industry representatives identified two major factors. One is the saturation of end-use markets for several products as the economy grows. For example, there is less need for basic materials because less infrastructure building needs to occur. These end products typically have low-income elasticities of demand (food, clothing) or are affected by demographic developments (automobiles, housing).

The second factor associated with domestic structural change is product substitution. For example, as indicated by Jim Collins of the American Iron and Steel Institute, Al Sobey of General Motors, and Bill Shephard of Battelle-Columbus, graphic composites and engineering plastics are replacing metals in several markets. Product substitution is determined by a combination of technological and market factors, whereas the first are more or less market determined.

Sixth, in terms of the international forces, there are two major schools of thought: those that emphasize macroeconomic factors such as the exchange rate, interest rates, and budget deficits (such as Robert Lawrence of Brookings), and those that emphasize microeconomic factors involving the costs of production and the industrial policies of trading nations such as dumping and subsidization, quotas, and tariffs (see Gupta-Faruqui of EPRI and Stewart of the National Science Foundation, **2**).

Seventh, future trends point toward increasing energy efficiency in most industries. However, in many cases, increased electricity use is expected per unit of output. Several of the industry authors point toward the expanded use of electrotechnologies to substitute for fossil fuels as a means of improving overall energy efficiency and boosting business productivity. Al Sobey of General Motors cites the use of induction-based processes to replace fossil-fuel processes, and Stan Lancey of the American Paper Institute cites the rapid penetration of thermomechanical pulping in the paper industry.

This trend toward energy efficiency and electrification is evident in several of the model-based simulations reported by Laurel Andrews of Synergic Resources Corporation, who uses econometric models at the two-digit level, Hill Huntington of Stanford University, who reports on the Energy Modeling Forum's comparative evaluations of end-use and econometric models, and John Broehl of Battelle's Columbus Division, who uses process models.

Sam Sugiyama of the Bonneville Power Administration presents that agency's experience with the pulp and paper process model, one of the ten included in EPRI's INDEPTH system covered in John Broehl's paper. Sugiyama's paper touches on both the strengths and weaknesses of process models and provides a candid evaluation of process models for electric utilities.

Eighth, the utility industry can exercise considerable leverage in shaping the evolution of future structural change. This works through the fact that electricity costs are a small share of production costs, with labor and material accounting for the bulk of production costs. However, little empirical evidence is yet available on how this leverage can be exploited in practice. This remains a topic for future research.

REFERENCES

1. Valerie A. Personick. "A Second Look at Industry Output and Employment Trends Through 1995." Monthly Labor Review, November 1985.

2. Lester C. Thurow. The Zero Sum Solution. New York: Simon & Schuster, 1985.

2

UTILITY PLANNING AND INDUSTRIAL STRUCTURAL CHANGE

Ahmad Faruqui
Electric Power Research Institute

This paper explores the relationship between industrial structural change and electric utility planning. Specifically, it tries to answer three basic questions: What is utility planning? What is industrial structural change? And what are the implications of industrial structural change for utility planning? Individuals involved with utility planning have typically not analyzed developments in the industrial sector in any detail because of a perceived lack of relevance. Simultaneously, industry analysts have typically not followed developments in utility planning for similar reasons. Recent events in the industrial marketplace and philosophical changes in utility planning have created tremendous opportunities for the utility planner and the industry analyst to get to know each other better. This paper tries to serve as an information bridge between these two communities.

UTILITY PLANNING

Let us briefly review the experience curve of the electric utility industry. Figure 2-1 shows the trend in the costs of supplying electricity through central-station power plants. The trend was one of almost continuous decline through 1970 or so, followed by an upturn. This upturn in supply-side costs has generated a strong interest in demand-side options.

What are demand-side options? Well, first of all, in the early seventies, when the price of electricity started to go upwards, it was quite obvious that demand-side planning meant discouraging load growth, i.e., energy conservation and load management.

Before too long, though, it became obvious that utilities were faced with different conditions, some requiring peak clipping, others requiring valley filling, still others requiring a load shifting. A new term, demand-side management, came into being: see Figure 2-2. This included the classic load management strategies shown on the left side of the figure, as well as the items shown on the right side: strategic conservation, which, for many utilities, is quite appropriate; yet for others, depending on their fuel costs and capacity margin situation, strategic load growth may be more appropriate.

And finally, the last concept, which is flexible load shape, refers to activities like demand subscription service and other methods of pricing electricity based on variations in reliability levels. Utilities are discovering that different customers have different energy needs, and are realizing it may not be appropriate to charge the same price for different qualities of electric service.

These six load shape objectives form the gist of the demand-side activities that utilities began to look at as they evolved initially from a conservation standpoint and a

FIGURE 2-1
Electric Utility Industry Costs

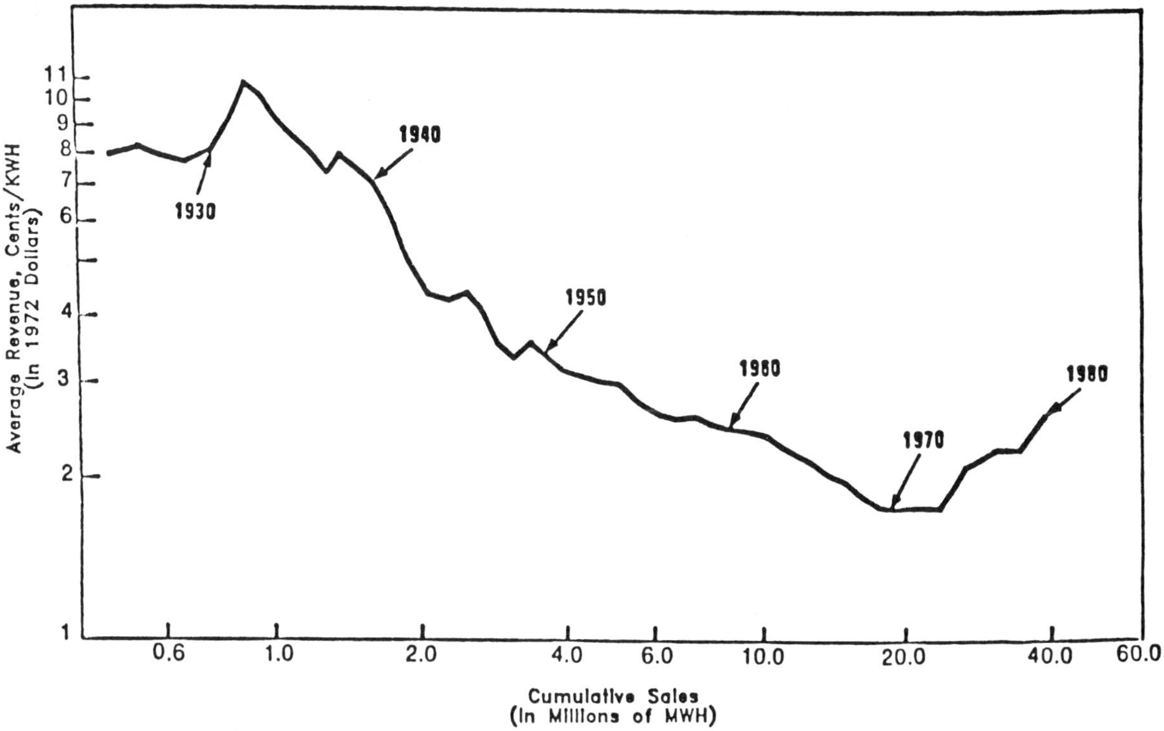

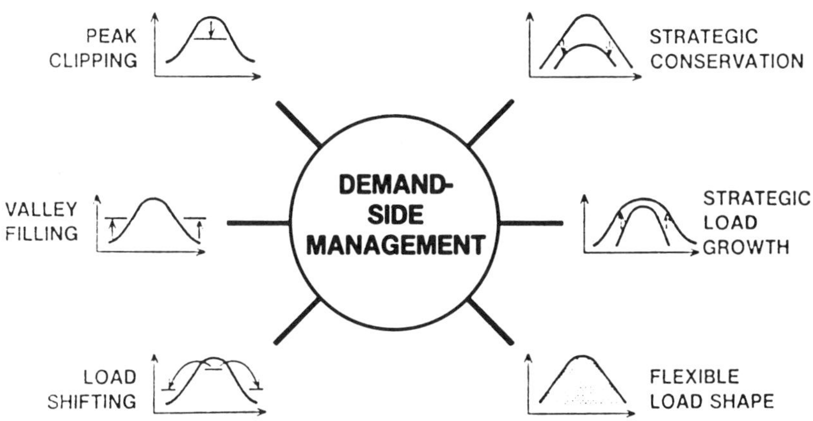

FIGURE 2-2
Demand-Side Management Embraces Several Load Shape Objectives

Source: Reference 1

straight load management standpoint (1). What did this do to utility planning? As Figure 2-3 shows, it placed a new set of alternatives or resources at the planner's disposal. The traditional practice had been to start out with a load forecast and to then run through generation planning, production costing, and financing analyses; to develop rates; and then to look at the available supply-side options and pick a subset of those options for actual implementation. As demand-side planning came along, the portfolio of resources expanded to include not only generation options but also demand-side options. An additional dimension of complexity was added to utility planning.

Now utilities were not only forecasting the load in its natural evolution, i.e., with utilities intervening in the marketplace; they were also required to forecast the load in its strategic evolution, i.e., with utility intervention in the marketplace. Such intervention may take the form of new rate structures, in the form of rebates for efficient appliance purchases, in the form of information on new technology options that might be available to customers to better meet their energy service needs, and so on.

As a result, utility planning on the demand side has evolved into a fairly complex undertaking, shown in Figure 2-4 in a stylized form as a nine-step process. The process begins with utilities specifying their load shape objective(s), any one of the six or combinations of the six shown in Figure 2-2. Next, they prepare an inventory of the demand-side alternatives, such as selected industrial electrotechnologies and economic development rates.

FIGURE 2-3
Demand-Side Management Adds a Major Dimension to Utility Planning

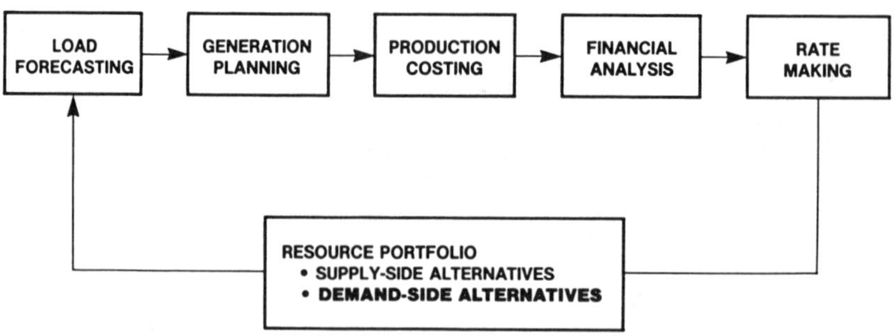

Source: Reference 1

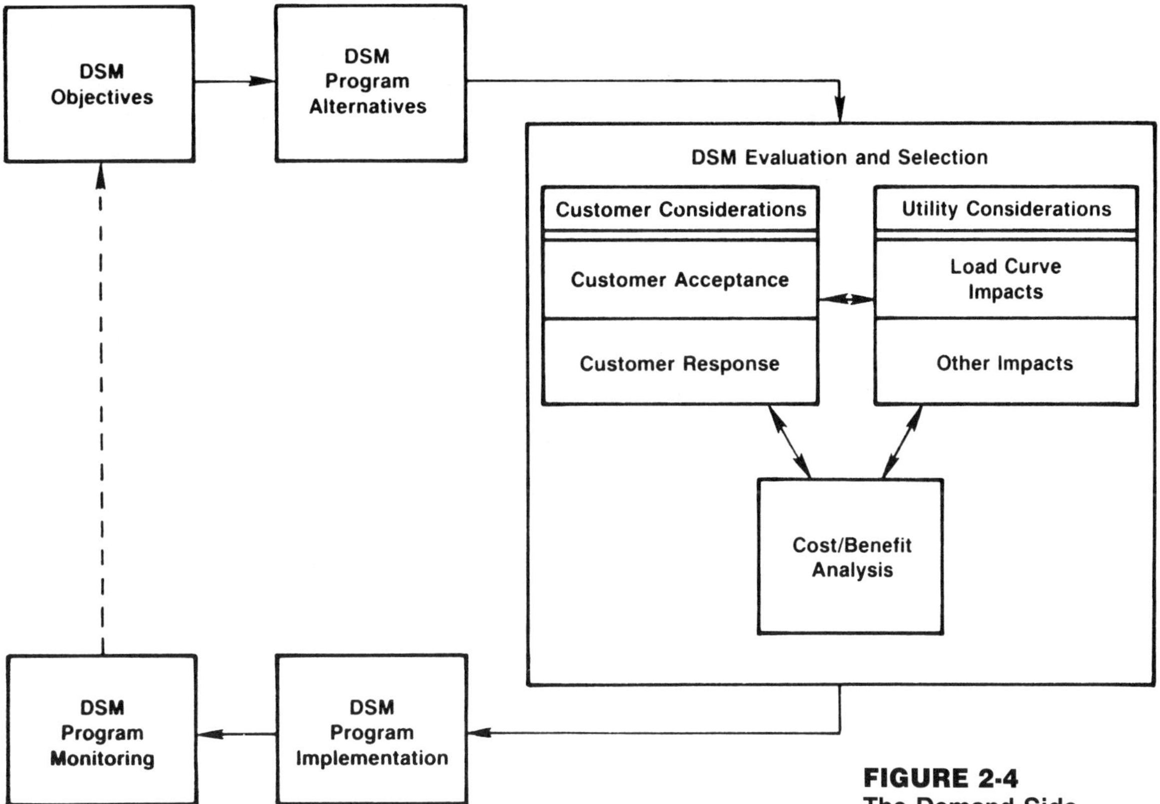

Source: Reference 1

FIGURE 2-4
The Demand-Side Management (DSM) Process

Once these two activities have been completed, there is the real challenge of looking in detail at customer-side impacts, particularly acceptance and response to utility programs, and tracing the implications of that for the utility system. All of that information is then processed together, and out of that big box comes a demand-side plan of action.

The final two steps in the nine-step cycle involve the actual implementation of the plan, and the monitoring of performance on a year-to-year basis. The last mentioned step includes the transmission of feedback to step one, so that suitable revisions may be made. Contemporary utility demand-side planning is a very dynamic process even in this very simplified representation. The figure also indicates the range of information requirements that utilities face as they begin to incorporate demand-side options in their planning framework.

Why are utilities so eagerly interested in demand-side options? Figure 2-5 provides one answer, based on the experience of a large utility in Southern California in the late 1970s (**2**). The various bars correspond to different demand-side programs, such as time-of-use rates, direct load controls, and interruptible rates. The figure compares the levelized cost of demand-side options with the cost of supply-side options. For the system being studied, the figure shows that several demand-cost options are less costly than demand-side options.

Another illustration of the economics of demand-side management is shown in Figure 2-6 (**3**). The particular simulations shown in the figure correspond to the south-eastern EPRI regional system. There are six EPRI regional

FIGURE 2-5
Comparison of Demand-Side and Supply-Side Costs

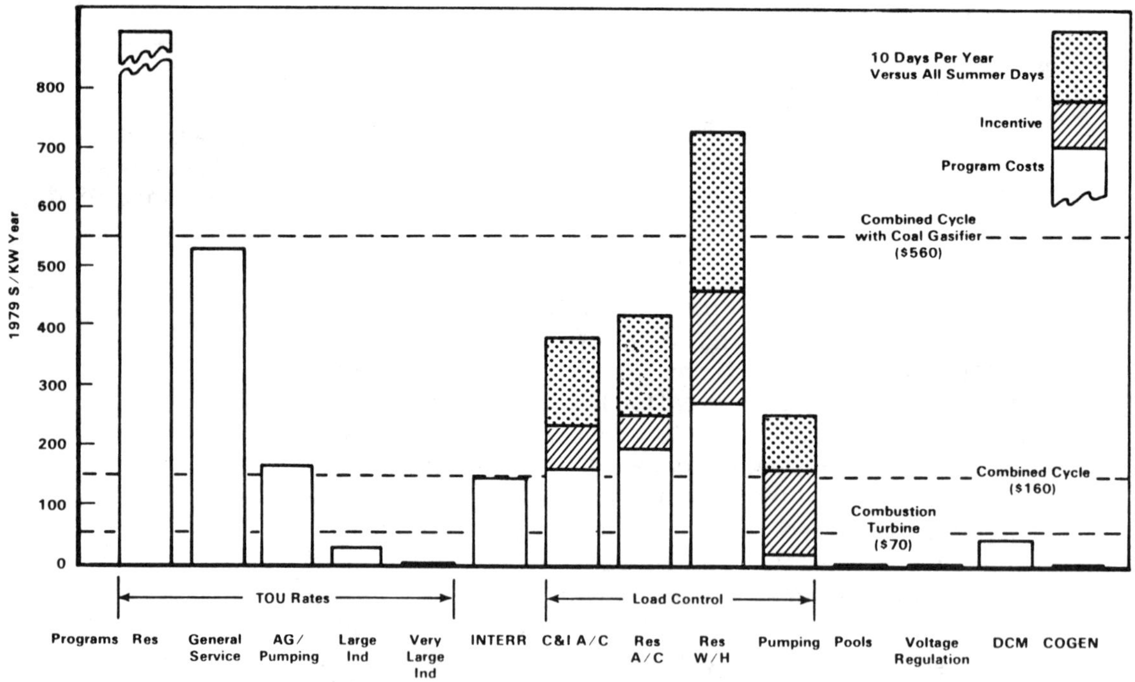

Source: Reference 2

systems altogether, and collectively they represent an aggregation of the nine NERC regions. For the southeastern region, the implications of modifying the regional load shapes for the average cost of electricity in the region were studied.

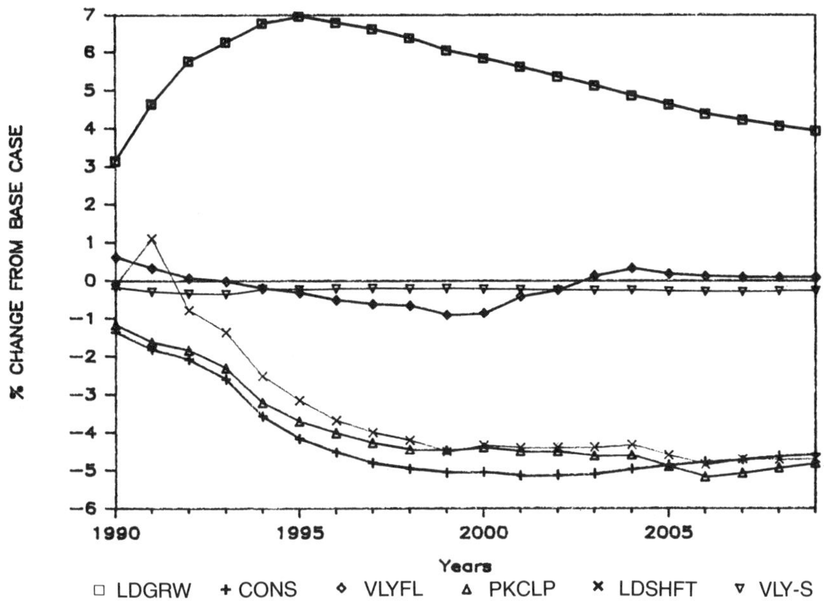

FIGURE 2-6
Average Electricity Cost

Source: Reference 3

The basic message coming from Figure 2-6 is that at least three of the load shape objectives—namely, the ones shown at the bottom: load shifting, strategic conservation, and peak clipping—appear to offer substantial economies as far as the deviations from the base case of average electricity price are concerned.

The chart does not measure the net benefits of demand-side management, only the gross benefits; that is, if you change the load shape, what reductions can you expect in the average cost? What is left out of the aggregation is: What is the cost to utilities of actually implementing the load shape changes? Clearly, in a full cost-benefit analysis, one would look at both sides of the picture.

Even with this clarification, the figure still gets the point across that there are substantial potential benefits that could be tapped through more active participation by the utilities than is visualized in the base case.

As a result of the perceived benefits of demand-side management, many utilities have charged ahead with various types of demand-side programs. Table 2-1 provides some evidence on this point (**4**). In 1983, several million customers were under various types of programs directed at achieving these objectives. For example, approximately three million customers were on some type of peak clipping program (load shedding, direct load control, time-of-use rates, etc.). These programs had an estimated impact on national peak demand of 6,000 megawatts. While the impact estimates are based on several assumptions, the number of customers is based on fairly hard data coming from a couple of national surveys conducted by the Washington-based Investor Responsibility Research Center and by EPRI.

Based on the activity levels that utilities now have in place vis-à-vis demand-side options, and the expansion in those efforts that they expect to occur between now and 1992, Figure 2-7 shows the top 25 utilities—in terms of the estimated reduction in 1992 peak demand. This is an estimate that was made two years ago, so it is not intended as a definitive statement of what the current state of projections is, but as illustrative of the rather broad geographic coverage of demand-side programs.

As we conclude this section, we can state that demand-side management activities appeal to a very wide cross-section of utilities: privately as well as publicly owned, large as well as small, and utilities in virtually all regions of the country. Demand-side management has become a pervasive factor in utility planning, and represents a major new development in utility planning. Utility interest in developments affecting their customers has increased considerably. In the case of the industrial customer, this by definition has translated into an interest in their customer's customers.

TABLE 2-1
Estimated Coverage of Demand-Side Management Programs

Objective	Alternatives	Estimated No. of Customers in 1983	Estimated MW On-Peak Impact in 1983
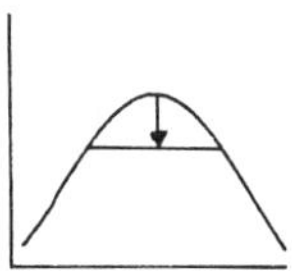	Load Shedding	450,000	
	Direct Load Control	1,280,000	
	Time-of-Day Rates	678,000	
	Interruptible Rates	350,000	
	TOTAL	2,758,000	6,000
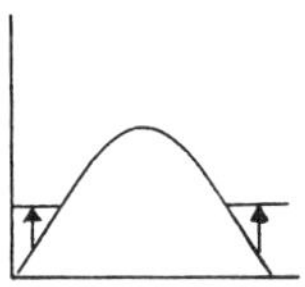	Thermal Energy Storage	9,000	
	Seasonal Rates	25,534,000	
	Off-Peak Rates	175,000	
	TOTAL	25,718,000	300
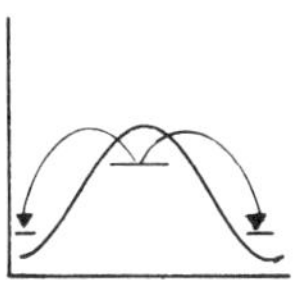	Thermal Energy Storage	1,000	
	Time-of-Day Rates	671,000	
	Appliance Control/Cycling	1,267,000	
	TOTAL	1,939,000	2,000
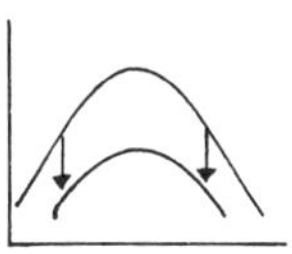	Audits	3,087,000	
	Loans	779,000	
	End-Use Solar	85,000	2,000
	Efficient Energy Use	2,859,000	
	Marginal & Conservation Rates	3,365,000	
	Cogeneration	3,000	3,000
	TOTAL	10,178,000	5,000
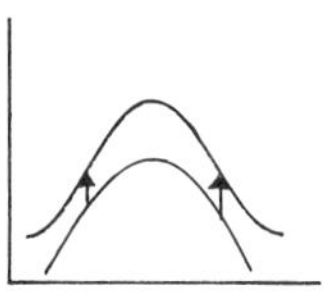	Heat Pumps	66,000	
	Dual Fuel Heating	21,000	
	Promotional Rates	187,000	
	TOTAL	274,000	N/A
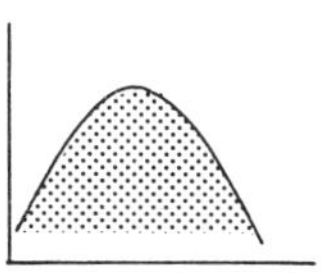	Demand Subscription Service	N/A	
	Variable Reliability Pricing	N/A	--

Source: Reference 4

FIGURE 2-7
Twenty-five Utilities with Significant DSM Programs Estimated 1992 Peak Demand Reductions

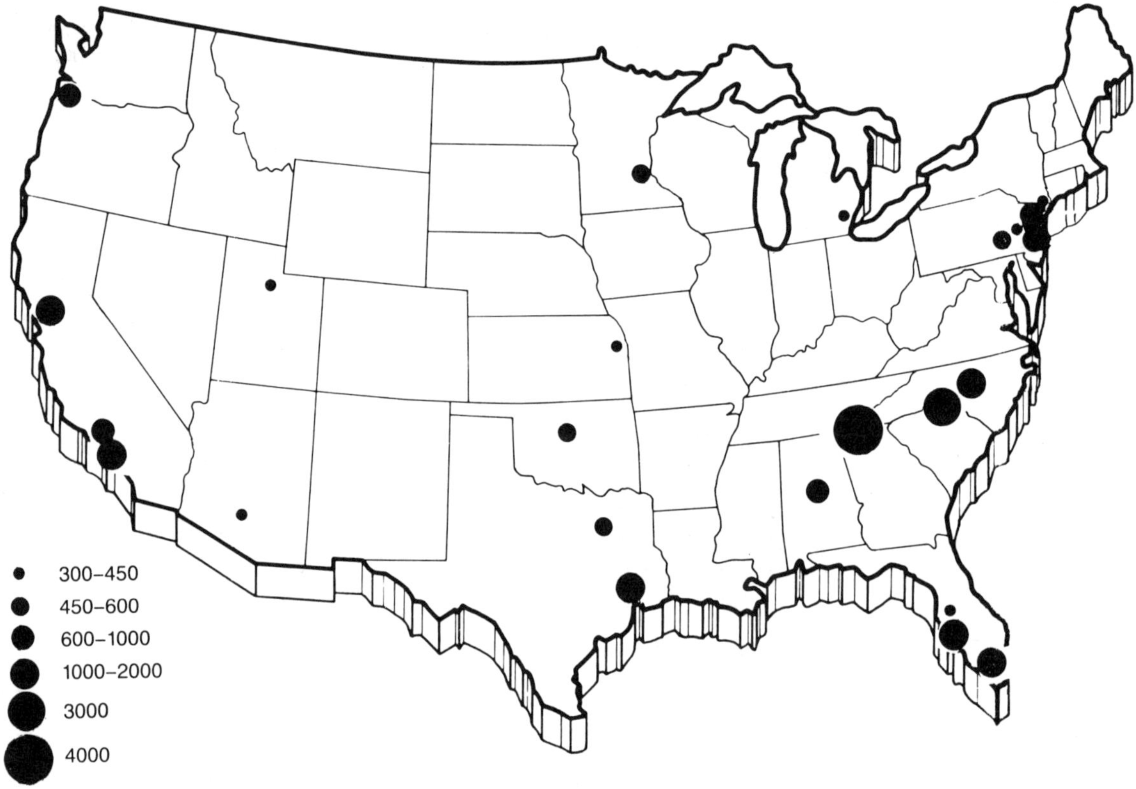

Source: Electric World; Investor Responsibility Research Center

WHAT IS INDUSTRIAL STRUCTURAL CHANGE?

Given the magnitude of interdependencies in any modern economy, there is a temptation to answer the question asked above by saying, "Everything depends on everything else." This will not take us very far in a practical sense, unfortunately.

Let us, instead, examine some of the major changes in the U.S. economy that have occurred over the past several

years. Let us look at the changing composition of the labor force. In Figure 2-8, we go back to the 1860s and trace the evolution of the share of each sector in overall national employment (5). The sectors that are covered include agriculture, services, industry, and information.

The basic message from 1950 onward is that as far as jobs go, America is slowly deindustrializing. This chart has become instrumental in representing the wisdom of one school of thought on structural change, which is that structural change means we are moving away from the industrial activities in the U.S. economy. Some people have expressed a concern that unless this trend is checked, we will become a "nation of hamburger stands."

Still other evidence on deindustrialization is provided in Figure 2-9. This is based on a study done by Mark Ross, a physicist at the University of Michigan, and shows the

FIGURE 2-8
The Changing Composition of the U.S. Labor Force, 1960-1980

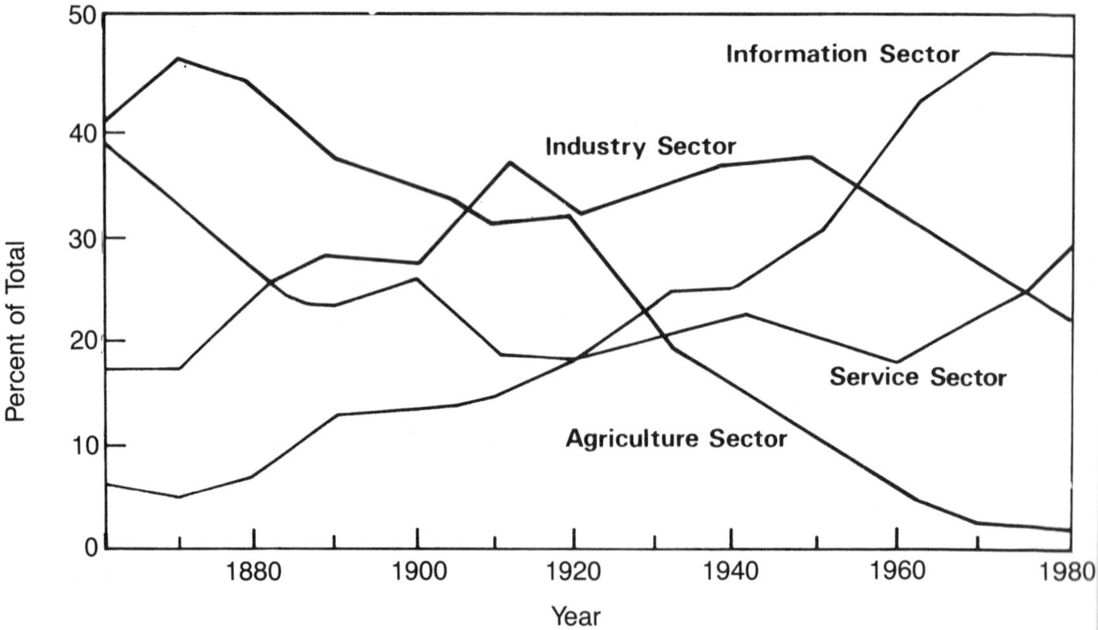

Source: Reference 5

FIGURE 2-9
Materials Consumption and GNP

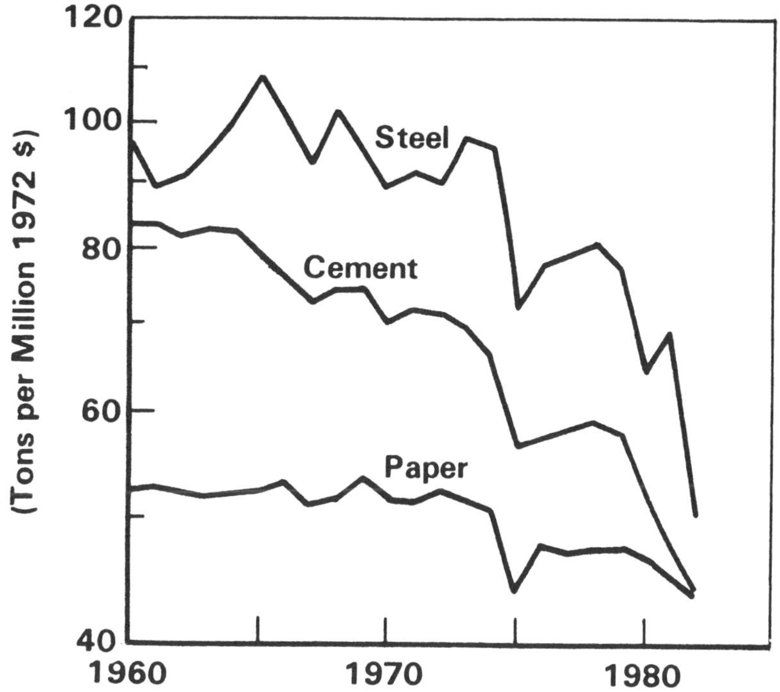

Source: Marc Ross, University of Michigan

relationship between the Gross National Product (GNP) and consumption of certain basic materials.

The figure shows not production but consumption, so it includes the imported values as well as the domestic production. The downward trend in the figure shows that the economy has reached a certain level in its evolution where basic materials, like steel and cement, are contributing less to GNP than other goods and services.

To balance off the discussion, we should point out that there is another school of thought which believes nothing alarming is happening. Normal economic evolution is occurring within the manufacturing sector. This group points out that the GNP share of manufacturing output is not declining. How do we reconcile these diametrically opposed views? In Figure 2-10, we have provided a simplified illustration which draws out some of the most significant

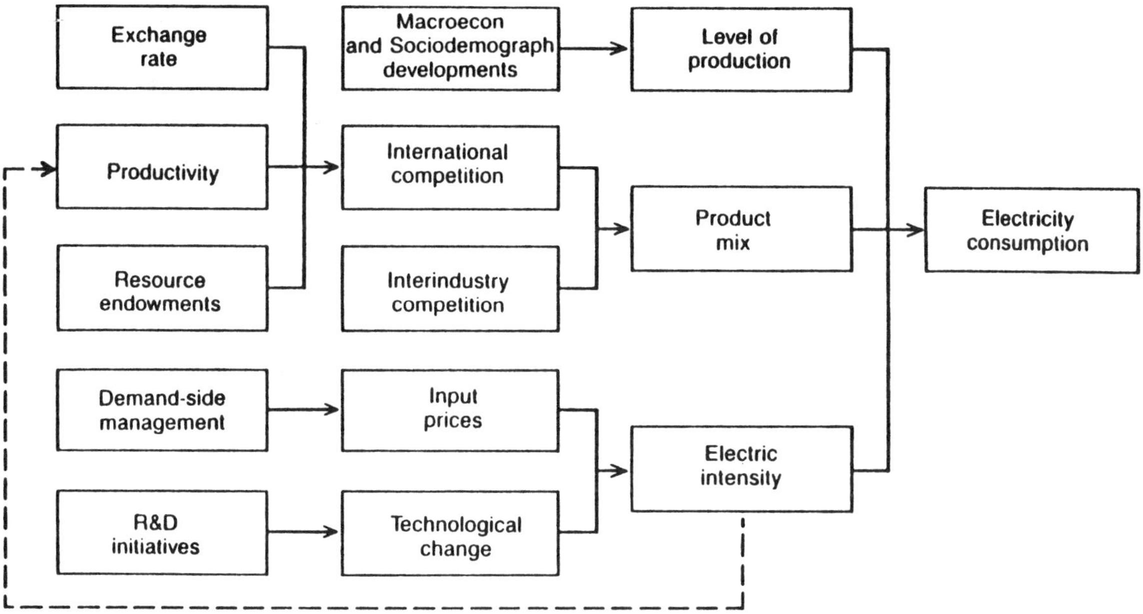

Source: Reference 6

FIGURE 2-10 Structure of Industrial Electricity Demand

structural connections between electricity consumption and its underlying determinants (**6**).

At the very first level, we have found it convenient to break up the structure of electricity consumption into three causal variables: (1) the level of aggregate industrial production, (2) the mix of that industrial production or product mix, and (3) the intensity of electric use (kilowatt hours per ton or kilowatt hours per dollar or product). Next, in a stylized form, we specify what determines each of the first-level elements, i.e., what are the second-level causal elements, and so on.

There are a whole lot of levels that one can specify here. We have just narrowed it down to a few of the more important ones. Conceptually, structural change can refer to a change in any of the key relationships displayed in Figure 2-10. However, we have opted for a narrower definition consistent with widespread usage, i.e., a change in the product mix. This then is the answer to the question raised in the title of this section.

It should now be evident that structural change is only one of the three first-level forces determining industrial energy use. What should also be clear is that once the sources of structural change have been identified, as in the figure, structural change can itself be analyzed and predicted within a certain level of accuracy.

WHAT ARE THE IMPLICATIONS OF INDUSTRIAL STRUCTURAL CHANGE FOR UTILITY PLANNING?

Related questions include: Why is it relevant to utilities? Is it a threat or is it an opportunity? We would argue that structural change is a great opportunity for utilities. Let us look at Figure 2-11, where the share of various cost elements in industrial production costs for certain selected two-digit SIC code industries is shown.

The message that one gets by looking at these data is that, even though representatives of several energy-intensive industries—like steel, like paper, and certainly aluminum—continue to voice their concerns about the impact of rising costs of electricity on their profitability, by and large, for many industries, these costs of electric energy are a very small share of their total costs. If we assume that all industries generally are more concerned with improving their overall profitability, i.e., ultimately improving their productivity, then they will focus more on the major elements in their cost structure—labor and raw materials—than they will on electricity.

Admittedly, this point can be argued in terms of degree, but, certainly, as an order-of-magnitude type of statement, it would appear to be valid. Since raw materials and labor dominate the cost structure of most industries, therefore as part of any productivity improvement exercise, those would be the cost elements that the industrial customer can be made to emphasize through utility intervention in the marketplace, i.e., strategic marketing by utilities. There is a large potential here for capitalizing on the unique attributes of electricity in improving productivity (**7**).

One side comment needs to be made about the use of electrotechnologies to improve productivity. We are not advocating the use of automation to create a "reserve army of the unemployed," to use the Marxian phrase. All we are saying is that in particular parts of the industrial sector

FIGURE 2-11
Structure of Industrial Production Costs, 1967 and 1980

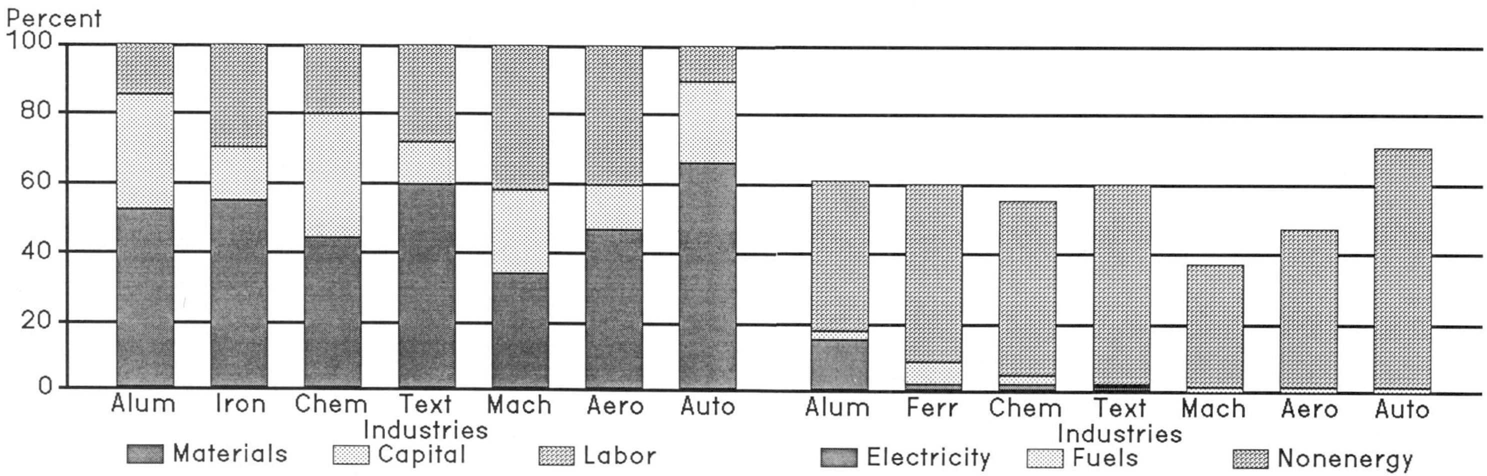

certain production activities might be more efficiently performed through automation. Of course, in other parts of the industrial sector, as well as in several other sectors, such as the information and services sectors, this may not be true.

Process heating is another end-use market where significant penetration opportunities exist for electrotechnologies, in this case through induction heating. This technology is being heavily researched by EPRI and other organizations in terms of the potential applications for electricity.

At EPRI, we have developed models to simulate the likely share of electric energy in overall production activities under a variety of scenarios. As an example, we have found that, between 1980 and the year 2000, the share of electrotechnologies in the pulp and paper industry is expected to increase significantly from 25 percent all the way to 60 percent (**8**).

In the case of iron and steel, there is an opportunity, through electric arc furnaces, for the share of steel made from electric-based processes to increase from 28 percent to 50 percent. These results are, of course, based on several assumptions and are not to be used as forecasts of the future. However, they are useful in getting estimates of the opportunities that can be tapped by utilities through strategic pricing, information, and research and development programs.

For example, in Figure 2-12, alternative simulations of the percent of purchased electricity in total energy are depicted for the iron and steel industry. The electric share is shown as rising monotonically in the base case. In a scenario where electricity is priced at zero cost (the free electricity case), the share does not reach 100 percent, because of physical limitations. Another case was run showing the implications of expensive electricity, where the rate of growth of electric energy prices was made to be twice as high as in the base case. Of course, that resulted in a lower penetration, but still the overall slope remained positive.

So, there are significant opportunities indicated by these marketing studies. Whether these are actual opportunities that can be realized will, of course, require utility-specific

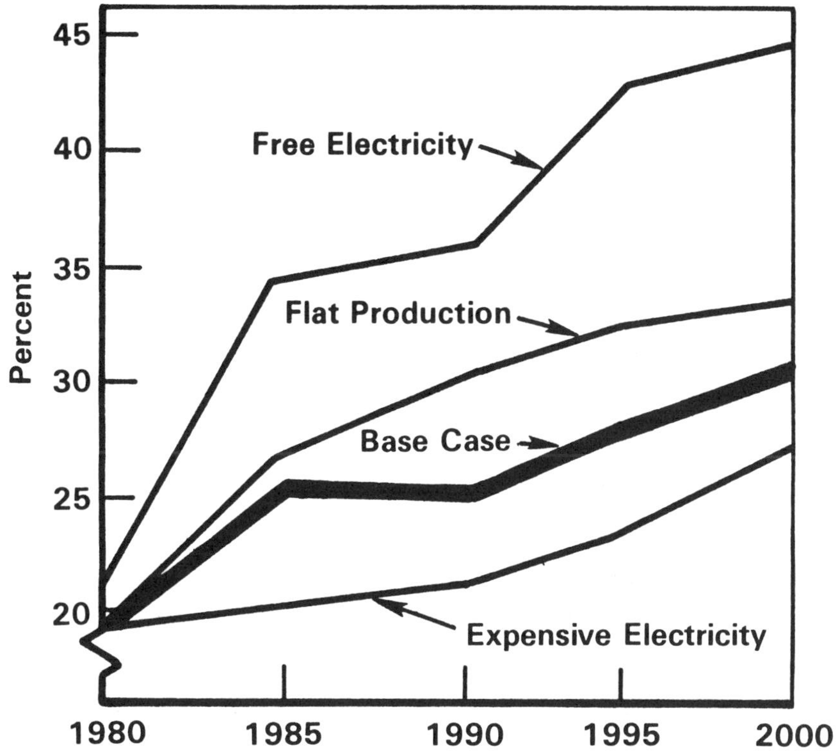

FIGURE 2-12
Share of Purchased Electricity in Purchased Energy, Iron and Steel Industry

Source: Reference 8

analysis and much more detailed data than we had access to when these studies were done.

Taking a broad view of the industrial sector as a whole, another internal EPRI study has identified the likely market for a variety of industrial electrotechnologies (**9**). These technologies represent a cross-section of existing as well as new industrial end-use technologies that are intensive users of electricity and also offer a substantial opportunity to the industrial customers in improving productivity and to utilities in adding desirable types of loads. As shown in Figure 2-13, induction heating shows up as the one with the most significant potential, followed by induction melting and flexible manufacturing.

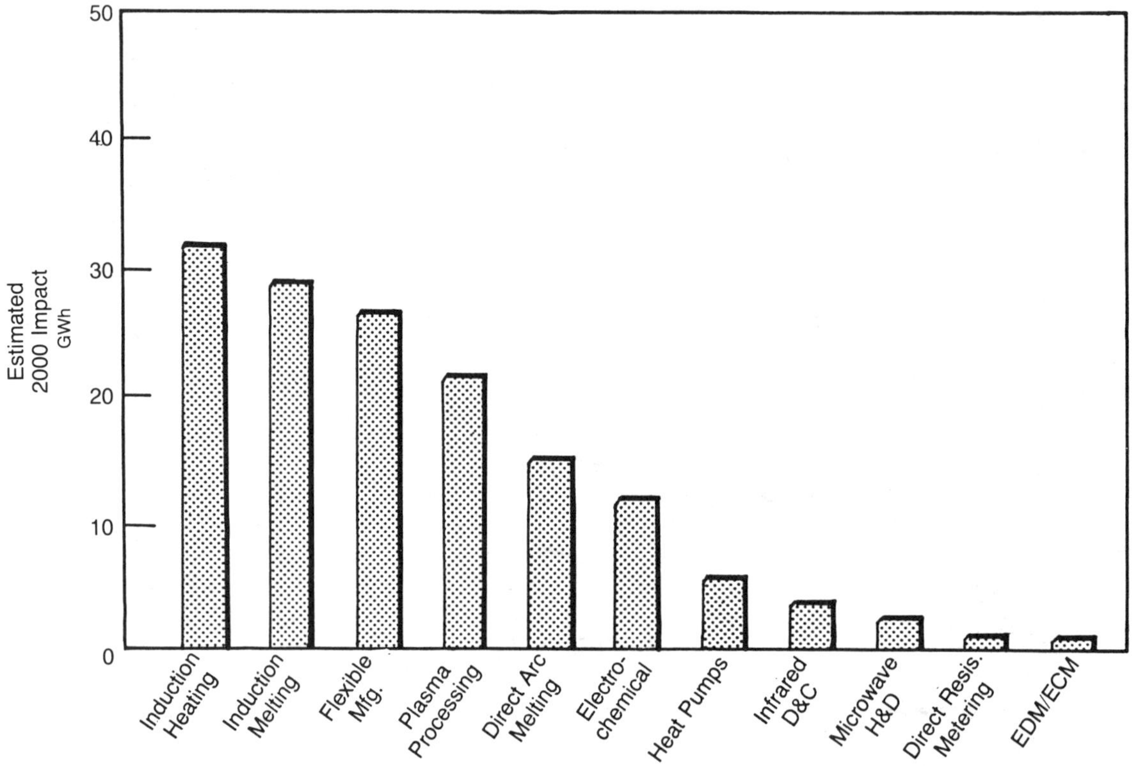

Source: Internal Planning Study, EPRI Industrial Program

FIGURE 2-13
Industrial Electrotechnologies with Significant Energy Impacts

Of all the factors influencing industrial electric demand, the one of most relevance from the standpoint of strategic marketing is the indirect effect of energy intensity on electricity consumption. This effect has rarely been quantified. It involves the favorable impact of new electrotechnologies on productivity, thereby changing the international competitiveness of U.S. industry and in turn changing the product mix. It is that feedback route which, perhaps more than anything else, offers a marketing opportunity for the utility.

As we reflect on how to tap the marketing opportunities afforded by industrial electrotechnologies to electric utilities, let us examine a conversation that took place in

Hanoi in 1975 between two army colonels, one American and one Vietnamese (**10**). The American officer said, "You know, you never defeated us on the battlefield." And the Vietnamese replied, "That may be so, but it is also irrelevant."

American tactics were successful; American strategy was not! So, in terms of the utility's involvement with demand-side activities, the message is that utilities need to adopt industrial marketing as a strategic concept—not just as a collection of piecemeal programs, but as a concept that can lead them to strategic victory in a competitive environment.

Within the context of determining industrial marketing strategy, utilities need to "seize, retain, and exploit the initiative" (**11**). In cases where electricity is currently a small share of the aggregate market, this means that flanking attacks are the most efficient way to achieve the marketing goals.

Electricity has several unique attributes. Some are direct attributes and some are indirect. The indirect attributes focus on productivity implications, and if utilities can focus on that particular line of attack, it may offer to them the most leverage in terms of favorably utilizing the opportunities posed by structural change.

One version of that particular statement is shown as a simple planning matrix in Figure 2-14, where we plot the growth of output of industrial sectors on one axis and their electric intensity on the other axis. The dashed lines show how utilities might be able to favorably shape industrial structural change through selected marketing policies.

For example, the steel industry is a high electric-intensity industry. It is forecast, in most scenarios, to be a low output-growth industry. However, certain segments of the steel industry could be substantially accelerated along that path through suitable incentives provided by utilities. Perhaps that is an avenue that could be pursued.

Of course, an open question is: As far as certain industries are concerned, is it a lost cause? How much would it take to get the steel or textile industries to increase their rate of growth? Would it be better to target high-growth industries for electrification? What is the leverage the

FIGURE 2-14
Planning Matrix

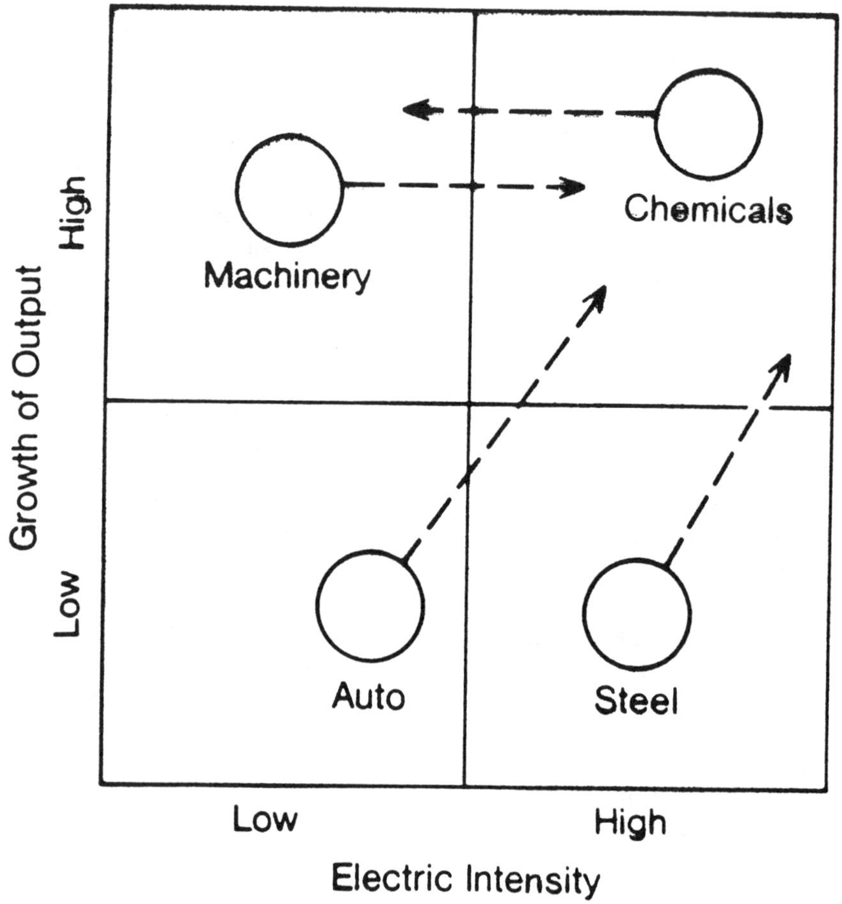

Source: Reference 6

utilities have vis-à-vis the natural evolution of the industrial sector? Those are questions that ultimately will influence the evolution of the utility industry and, to the extent that utilities can influence industrial structural change, will provide an opportunity for the market planners in the utility industry to look at and examine with some seriousness.

EPRI is developing a tool to help utilities address those types of questions (**12**). Called the industrial end-use planning methodology (INDEPTH), this tool consists of three modules—an econometric, a process, and an equipment module—which can be used to answer questions about industrial electrification at several different levels of detail.

CONCLUSION

In conclusion, the following can be stated. First, utilities now need to be forecasting electricity sales with and without participation on the demand side, i.e., demand-side management.

Second, significant industrial structural change is occurring in the United States. Several of the other papers in this volume provide ample evidence on that.

Third, microfactors such as productivity effects, as well as macrofactors such as the exchange rate, are both important in analyzing industrial structural change. Much of the recent discussion in leading newspapers and magazines presents a simplified macropicture of structural change which implies that if only the dollar would come back to a normal level, business-as-usual conditions would be restored to American industry. That is only half, or maybe even less than half, of the picture, if we take a longer-term view. There are productivity changes dealing with cost competitiveness and other factors that could still persist even if the dollar were to return to a more normal level. So, our contention is that both effects have to be analyzed.

Fourth, electrotechnologies can help both utilities and industrial customers cope with industrial structural change.

And finally, industrial structural change offers utilities numerous long-term opportunities, provided, first of all, that we understand what is driving our customers. This means not only our industrial customers but the customer's customer, because in many cases industrial customers are selling to other customers. So, if we take a more detailed look at what is motivating them and their customers, by looking at their cost structure and whether or not they are competitive with their competitors, including foreign as well as those in the United State but outside of the local utility service area, then we will get a better appreciation of the opportunities. Armed with such understanding, utilities should be able to develop effective demand-side programs involving marketing and other signals that benefit them and also boost customer productivity.

REFERENCES

1. Battelle Columbus Division, and Synergetic Resources Corporation. Demand-Side Management. EA/EM-3597, Vols. I-III. Palo Alto, CA: Electric Power Research Institute, August 1984.

2. M. Douglas Whyte, "Load Management." In Resource Planning Associates (eds.), Proceedings of the Rate Design Study Regional Conferences. EPRI EURDS 92, October 1980.

3. Ahmad Faruqui, Clark W. Gellings, and Rajat Deb, "Value of Demand-Side Management: A Case Study." In D. E. Jones, R. H. Males, and A. Faruqui (eds.), Strategic Planning and Marketing for Demand-Side Management: Selected Seminar Papers. EA-4308. Palo Alto, CA: Electric Power Research Institute, November 1985.

4. Ahmad Faruqui and Clark W. Gellings. "The Impact of Demand-Side Management on Load Growth: Some Tentative Results." In Terry Morlan (ed.), Energy Forecasting. New York: American Society of Civil Engineers, 1985.

5. Robert U. Ayres. The Next Industrial Revolution: Reviving Industry Through Innovation. Ballinger Publishing, 1984.

6. Ahmad Faruqui, Pradeep C. Gupta, and Joseph B. Wharton. "Ten Propositions in Modeling Industrial Electricity Demand." In A. M. Bolet (ed.), Forecasting United States Electricity Demand: Trends and Methodologies. Boulder, Col.: Westview Press, 1985.

7. Philip S. Schmidt. Electricity and Industrial Productivity: A Technical and Economic Perspective. New York: Pergamon Press, 1984.

8. Ahmad Faruqui and Barbara Pierce. "A Process-Based Simulation of Electricity Use in Three Energy-Intensive Industries." In A. Faruqui and K. Yamaji (eds.), American and Japanese Perspectives on Energy Analysis Research. EPRI EA-4067. June 1985.

9. Synergic Resources Corporation. "Markets for Electric End-Use Technologies." Memorandum for EPRI, 1984.

10. Col. Harry G. Summers. On Strategy. Novato, CA: Presidio Press, 1982.

11. Ahmad Faruqui. "Competitive Marketing for Electric Utilities: An Application of the Classic Principles of Warfare," Long Range Planning, 1987, forthcoming.

12. Battelle Columbus Division. Industrial End-Use Planning Methodology: Recommended Design. EA-4019. Palo Alto, CA: Electric Power Research Institute, May 1985.

3

ALTERNATIVE ECONOMIC GROWTH SCENARIOS AND IMPLICATIONS FOR UTILITY STRATEGIC PLANNING

Bruce G. Humphrey
Edison Electric Institute

INTRODUCTION

The turbulence of the energy and economic environment over the course of the past 15 years has led many utilities to incorporate alternative futures into their planning process. This paper briefly reviews scenario development techniques, discusses examples related to energy, presents some specific results of scenario analysis, and speculates about the implications for utility planning.

Scenario development requires an imaginative blend of

logic, observed facts, and intuition. Ideally there will be a seamless connection between the events and relationships of the past and the possibilities for the future. That is not to say that there will not be surprises and discontinuities in trends. Rather, the point is that a surprise or a trend cannot be recognized as such without some sort of connection or comparison with events of the past and present.

Scenarios, whether driven by historical data or sheer speculation, are a popular method for exploring the future. The Conference Board (1) occasionally publishes information about how scenarios are developed and used by specific companies. The topic is often addressed in books on corporate or strategic planning. Within the electric utility industry alternative futures are widely applied in resource planning. A recent survey by the Edison Electric Institute (EEI) found that more than 80 percent of the responding companies make alternative forecasts for planning purposes. Half of these apply probabilities to the outcomes. These alternative forecasts are not limited to the kilowatts or kilowatthours which may have to be served. Nearly two-thirds of the utilities explicitly consider the uncertainty in the future costs and performance of demand-side and supply-side options from which a resource plan will be selected. The survey did not reveal how extensive the considerations are of factors either internal or external to the industry. Yet, it is clear that alternative futures, or scenarios, are central to utility planning.

One of the works that is sometimes given much of the credit for taking the study of the future out of the exclusive domain of the science-fiction writers and making it an important component of government and business planning is the book by Herman Kahn and Anthony J. Weiner (2). The popularity of that book made scenarios a standard technique for speculating about the future. The book does propose a number of forecasts about political and economic trends. One irony is that Kahn and Weiner mention energy only in passing. It does not warrant an entry in the index. Yet, obviously, energy has become one of the central political and economic problems in the era between 1967, when the book was published, and 2000. Nevertheless, since 1967 scenario building has become a systematized technique of business planning.

SCENARIO DESIGN

There are a number of conceptual frameworks and schematic diagrams that can be used as aids in scenario development. A few of them are presented here. But, the reader should note the references cited and the bibliographies in those references for a more thorough view of the topic. One framework that is often cited in the literature [George A. Steiner (3)] is that used by the General Electric Company. The conceptual flow is described in the following steps.

1. **Prepare Background:** The first step is to review all of the factors which are relevant to the question at hand. This includes social change, demographics, economic conditions, governmental policies and regulations, and changing technology. This step includes not only the collection of information, but also an assessment of how each factor individually and in concert with others can affect the outcome of a business decision.
2. **Select Critical Indicators:** The background material must be brought into sharper focus. Key variables are identified. Speculation begins about possible future events related to these key variables. During this phase of the process at General Electric, a Delphi panel is created.
3. **Establish Past Behavior for Each Indicator:** This is a phase for modeling and analysis. Historical relationships and trends are established and the reasons for these relationships and trends are analyzed.
4. **Verify Potential Future Events:** Based upon the material gathered and the relationships analyzed, potential future events are identified. Forecasts are prepared and the probabilities of future events are assessed. Models can play a central role here for documenting assumptions and specifying future relationships.
5. **Forecast Each Indicator:** With input from the Delphi panel, the range of future values for the key variables is established.
6. **Write Scenarios:** After all of the preceding work is completed, the scenarios are written.

There are various schematic diagrams to assist the scenario writing process. An elementary presentation is contained in Forecasting by Joel Kurtzman (**4**). Possible outcomes can be arrayed in a matrix format (Figure 3-1). These possibilities can be presented informally or probabilities can be assigned to each box. The most likely path across the diagram is called the "grand scenario."

FIGURE 3-1
Matrix Format of Grand Scenario Diagram

	PLANT CONSTRUCTION	OPERATION	FUNDING
BEST	ON TIME OR EARLY	OPERATION ON SCHEDULE	FULL COST RECOVERY
MOST LIKELY	0 - 2 YEARS LATE	OPERATION CHRONICALLY DELAYED	LONG-TERM PHASE-IN
WORST	GREATER THAN 2 YEARS	CANCELLED	COSTS DISALLOWED

In reality, the probability of any individual outcome is conditional upon the outcomes in the other boxes and upon external factors. Another tool which is widely used is a branching point diagram (Figure 3-2). This diagram maps the flow of events through time and can have probabilities assigned to the branches. It is commonly applied in decision analysis.

EEI has had some recent experience in scenario development and analysis. Several years ago Choice over Chance by William Thompson, Jerome Karaganis, and Kenneth Wilson (**5**) explored economic and energy options for the future. This book provided an imaginative example of how the dichotomy between policy choices and environmental uncertainties can be the fulcrum for scenario design. Choice over Chance presents a matrix of outcomes for each important variable (Figure 3-3). The matrix illustrates the interaction of the policy choices that are available to

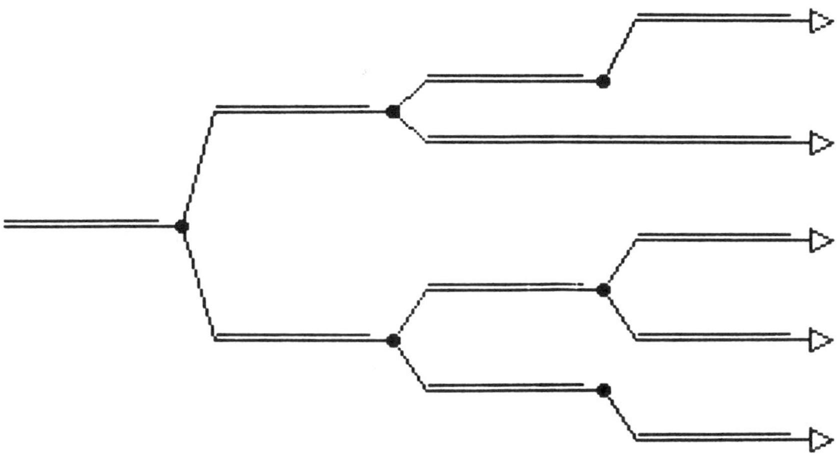

FIGURE 3-2
Branching Point Diagram

consumers, business, and government, and the element of chance which affects all outcomes. In analyzing economic growth, a set of possible values for productivity growth, labor force participation, oil price changes, and other variables can be derived by explicitly considering choice and chance forces.

		CHANCE	
		High	Low
CHOICE	High	HiHi	HiLo
	Low	LoHi	LoLo

FIGURE 3-3
Choice Over Chance Scenario Design

Machiavelli argued that half of people's actions are ruled by chance and the other half by the choices of people themselves (fortuna versus virtu). Scenario development in general and the explicit consideration of the trade-off between choice and chance factors offer the potential for redressing that balance. In fact, one of the principal conclusions of Choice over Chance is that wise policy choices can largely overcome the uncertainties of the future.

Each of these devices provides assistance when developing scenarios. They provide frameworks for both facts and intuition. Joel Kurtzman also suggests a list of questions that a scenario should answer.

1. What is the purpose of the scenario?
2. What are the relevant facts?
3. Which are the most important factors and variables?
4. What is the main theme of the scenario?
5. How do the main facts interact?
6. With all this in mind, what is the present situation?
7. What is the most probable scenario of events?
8. How does the scenario stand up when basic facts are altered?
9. Is the scenario consistent and clear?
10. What are some alternate scenarios?

In summary, there is some rationale to scenario design. Good scenarios often rely heavily on the lessons of the past, not only in terms of outcomes but also in terms of relationships between variables. Good scenarios are also designed with a purpose in mind—they are not random collections of possible events. Finally, scenarios are not ends in themselves. They are a means to planning effectively for the contingencies of the future.

TWO ECONOMIC SCENARIOS: TREND AND CYCLE

The Edison Electric Institute has had a long-running interest in the relationship between electricity and the economy. Recently EEI worked with Data Resources, Inc. (DRI) to develop alternative economic growth scenarios. These scenarios relied heavily on standard DRI forecasts but were enriched with sectoral and industrial detail by DRI's Interindustry Service. The motivation for this work was to understand structural change in the economy, which many observers claimed was occurring at an accelerated pace. Certainly some industries and some regions were experiencing changes that were and are having a profound effect on some electric utilities.

The possibilities for scenarios relating to structural change in the economy are endless. However, the pattern of electricity sales in recent years suggested two scenarios which could be the foundation for more. Figure 3-4 shows what has happened to the ratio of the growth of national electricity sales (kWh) and the growth of GNP. Over the course of 35 years the growth of electricity sales has declined from being a multiple of economic growth to being less than economic growth. In fact, electricity growth has not exceeded economic growth since 1976. As electricity growth came closer and closer to economic growth, implications arose for the pattern of electricity sales over the course of the business cycle. The high growth that insulated electricity from economic change had disappeared by the 1970s.

In 1982, for the first time in nearly 40 years, the electric utility industry generated less electricity than in the previous year. Even during past economic downturns, sales had actually increased (except for a very small decline during the recession in 1974 resulting from the oil embargo).

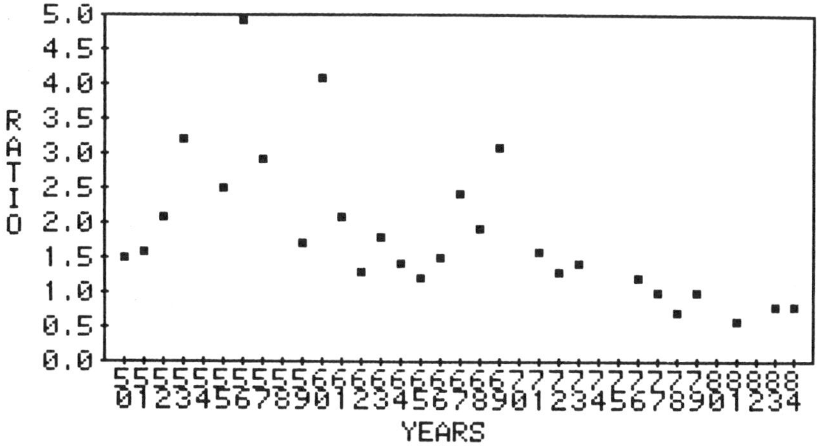

FIGURE 3-4
Growth Rate Ratio of kWh and GNP (1950–1984)

Table 3-1 presents the historical record. The index of coincident indicators published by the Department of Commerce is used as the measure of the state of the economy. The index is based upon total nonagricultural employment, industrial production, personal income, and sales in manufacturing and trade. It is a widely used indicator of movements into and out of recessions. The table shows that even when recessions have been severe, electricity sales have generally continued to grow.

TABLE 3-1
Business Cycle Indicators
(percent change from previous year)

Year	Index of Coincident Indicators	Total Electricity Sales
1954	-6.7	+6.9
1958	-7.6	+2.0
1961	-1.0	+5.5
1970	-2.0	+6.4
1974	-1.4	-0.1
1975	-8.1	+1.9
1980	-3.6	+2.0
1982	-6.6	-2.4

The results shown for 1982 reflect the interaction of long-term trends and unusual short-term circumstances. Long-term trends indicate a slowing rate of electricity growth and a shift in the mix of consumption. Short-term circumstances in 1982 relate to the business cycle and the weather. From an economic structure perspective, the industrial sector was experiencing a severe recession. The recession was particularly severe for the major electricity users. Electricity consumption is concentrated in a few sectors. Industries which account for less than 20 percent of total industrial production account for more than 50 percent of the share of electricity sales to industrial customers (primary metals; chemicals, paper and allied products; stone, clay, and glass; and, petroleum refining). In sum, because the long-run trend growth of electricity has declined and because a sizable share of sales is concentrated in cyclical industries, the electricity industry may continue to be greatly affected by business cycles in the future.

The reality of the business cycle is not always reflected in economic forecasts. Figure 3-5 shows the plot of a typical forecast. The saw-toothed path of the past becomes a smooth line when projected into the future. This example is taken from a DRI (**6**) trend forecast of GNP. But, forecasts of most other variables by DRI and others exhibit the same behavior. The uneven past somehow becomes a smooth future. At the structural level we know that some industries are more cyclical than others. Consequently, in designing scenarios for the purpose of analyzing structural change, the development of a cyclical version of the future may yield some insight—especially for the electricity industry as it becomes more sensitive to cyclical forces.

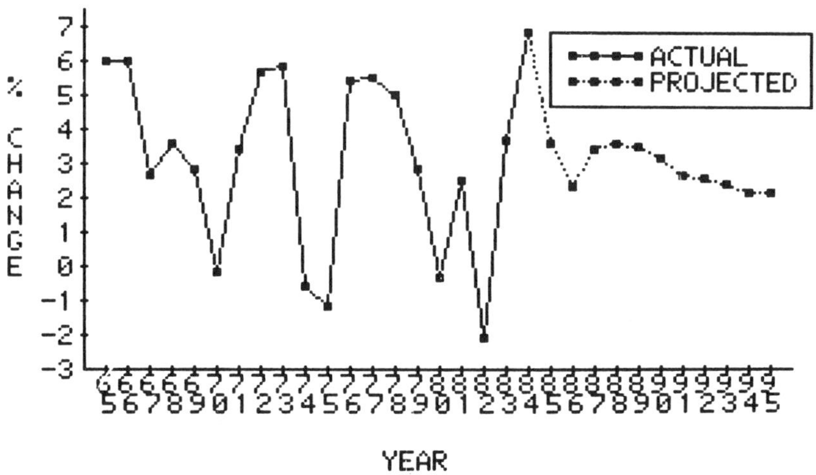

FIGURE 3-5
Gross National Product Annual Rate of Change

Two scenarios were prepared—a conventional trend forecast and a cycle forecast. The trend forecast assumes no major economic disruptions and is characterized by smooth growth. Developing a cycle forecast required making exogenous assumptions which would stimulate cyclical behavior. Traditional business cycle forces were combined with an exogenous shock reminiscent of the last decade. Monetary policy was such that nonborrowed reserves fluctuated. Interest rates, investment, and consumption expenditures exhibited traditional cycles. And, energy prices were increased.

The results of the alternative scenarios are presented in Table 3-2. Both cases indicated that economic performance during the balance of the 1980s and the first half of the 1990s will be better than it was during the past decade but not as good as the 1960s. Of the two scenarios, however, the trend results are more favorable. GNP growth is higher, inflation is lower, unemployment is lower, and productivity improves faster. Overall, a cyclical world is just not as comfortable as a trend world. This is partially due to the assumptions necessary to induce the cycles. But, it is also true that cycles diminish the growth of the capital stock.

TABLE 3-2
Historical and Projected Comparison of the Economy's Performance (annual averages in percent)

	1960-71	1972-83	Trend 1984-95	Cycle 1984-95
Real GNP Growth	3.9	2.6	3.1	2.8
Inflation (GNP Deflator)	3.1	7.0	5.7	6.8
Unemployment	4.9	7.1	7.3	7.8
Labor Force Growth	1.8	2.4	1.4	1.4
Labor Productivity Growth	2.4	1.2	1.9	1.6

Capital investment lost or not made in a recession is not fully recovered in the economic upswing. Consequently, the capital stock which is the foundation of labor productivity growth becomes lower in the cycle case than in the trend case. The net effect is to reduce potential output.

The nature of economic growth also affects the structural composition of manufacturing activity. Table 3-3 shows the growth performance of a selection of manufacturing industries. In the high-growth category some of the industries are the strong performers from the past. Others show particularly strong recoveries from the doldrums of 1979 to 1983. Some are strongly affected by economic conditions in the business cycle scenario—rubber and plastics, transportation equipment, and fabricated metal products. Among the industries expected to experience medium growth, primary metals fares poorly in the cycle case relative to the trend. The low-growth industries are those which are victims of slow demand growth and import competition. These industries are generally already recognized as having problems in this regard and these problems are exacerbated by business cycles—apparel and products, textile mill products, and leather and products.

Clearly cyclical conditions have a much stronger effect on some industries than on others. Table 3-4 presents a much finer disaggregation identifying the fastest growing sectors. (The previous table came from a 70-sector breakdown of the economy. This finer breakdown divides the economy into 400 sectors.) This greater disaggregation brings out the dynamism of structural change. First, of the 20 fastest growing sectors in the trend scenario, only five were in the top 20 in the 1963 to 1978 period. These long-term strong performers are semiconductors, computing equipment, optical instruments, medical instruments, and plastic materials. Second, greater disaggregation shows how business cycles affect structural change. Four of the top ten in the cycle scenario fall out of their top ten rankings in a trended economic future. Again, some industries are simply more sensitivie, positively or negatively, to business cycles. And, this sensitivity is reflected in the long-run structure and performance of the economy.

The composition of industrial growth has a measurable effect on energy growth as shown in Table 3-5. For electric utilities there is a 25 percent decline in the rate of output

TABLE 3-3
Average Annual Rates of Growth in Manufacturing Sectors (ranks based on trend projection)

	1959-69	1969-79	1979-83	Trend 1983-95	Cycle 1983-95
HIGH					
Nonelectrical Machinery	6.8	4.1	-2.1	6.3	5.9
Rubber and Plastics Products	9.7	7.7	1.8	6.1	5.4
Electrical Machinery	8.9	4.6	1.4	5.4	5.2
Instruments	7.6	4.3	-2.4	5.4	4.9
Chemicals and Products	8.1	6.0	0.3	5.1	4.7
Transportation Equipment	5.6	2.2	-3.3	4.7	3.9
Fabricated Metal Products	4.3	3.2	-5.2	4.4	3.6
MEDIUM					
Primary Metals	4.8	0.6	-8.4	4.2	3.3
Ordnance	8.5	-3.9	6.1	4.0	3.9
Stone, Clay, and Glass	3.0	4.1	-3.4	4.0	3.6
Furniture and Fixtures	4.6	3.5	1.4	3.1	2.7
Printing and Publishing	4.6	2.5	2.7	3.0	2.7
Miscellaneous Manufactures	5.1	3.2	-1.3	3.0	2.6
Paper and Products	5.7	2.7	2.1	3.0	3.0
LOW					
Apparel and Products	2.9	2.3	-3.2	2.8	2.1
Petroleum Products	3.9	2.9	-4.4	2.6	2.3
Textile Mill Products	4.8	2.6	-0.8	2.6	1.8
Food and Products	3.3	3.3	1.5	2.5	2.4
Lumber and Wood Products	3.1	2.4	0.1	2.4	2.4
Tobacco Products	1.0	1.9	-1.2	1.6	1.3
Leather and Products	0.2	-2.9	-3.6	-0.2	-1.3
TOTAL	5.7	3.3	-0.9	4.4	4.0

TABLE 3-4
Fastest Growing Sectors (1986–1995)

Name	Rank in Cycle Scenario	Rank in Trend Scenario
Semiconductors	1	1
Computing Equipment	2	2
Electronic Components	3	3
Elec. Measuring Instr.	4	4
Communications	5	7
Tanks & Components	6	20
Nonferrous Rolling	7	8
Medical Instruments	8	15
X-Ray Apparatus & Tubes	9	13
Plastic Materials	10	11

TABLE 3-5
Energy Output Growth
(average annual rates in percentage points)

Sector	1963-70	1970-78	1978-86	Trend 1986-95	Cycle 1986-95
Pet. Refining	4.6	3.6	-0.3	0.4	0.0
Coal Mining	3.7	1.0	2.9	3.8	3.3
Electric Utilities	6.9	5.4	2.2	2.5	1.9
Gas Utilities	6.3	-0.6	-1.2	-0.5	-1.0

growth. This compares with a 10 percent difference in the trend versus cycle GNP growth rates. This is further evidence of how sensitive the electric utility industry has become to business cycles and will continue to be in the future. Note that the trend case of 2.5 percent electricity growth is almost identical to the projection of 2.4 percent growth prepared by the North American Electric Reliability

Council (**7**). Even this modest growth, however, is threatened by future cyclical economic behavior.

CONCLUSION

These specific alternative economic growth scenarios and alternative scenarios in general have implications for utility strategic planning. The specific trend and cycle scenarios highlight the importance of recognizing the real world economy of fluctuations and short-term instability. The fluctuations create strategic challenges and operational opportunities. Certainly, some customers are more cyclical than others. In addition, a cogeneration partner/customer may be more cyclical than the customer base as a whole. In the area of rate design, there may be some opportunity for countercyclical capacity and energy charges. Interest rates have traditionally been highly cyclical. As a major player in financial markets, electric utilities can either profit or suffer from the timing of financial activities. Fuel inventory management may also have a cyclical dimension. At a more general level, but still tied to these specific scenarios, cyclicality favors variable cost investments which allow greater cost control in fluctuating conditions over fixed cost investments which impose a fixed cost structure over time. This supports the case for small increments of new capacity with a high variable cost component.

The implications are certainly more numerous than those listed here. To generate additional implications, consider how many economic variables and assumptions are made in a corporate planning model. Then make cyclical rather than smooth trend assumptions. The number of possible outcomes and strategic choices obviously becomes very large.

Given the very large number of outcomes and the inaccuracy of forecasts as a first principle, there is one overarching implication. This final implication is more important than choosing a specific set of planning options to fit any given scenario. Given recent economic history and the potential for short- and long-term structural change, a utility planner must focus on a mix of capabilities in order to cope with a wide range of possible futures. Scenario development becomes one of the inputs for positioning a company to deal with unanticipated surprises—surprises which even imaginative scenario design does not encompass. Strategic planning, to which scenarios contribute, then becomes an exercise in assessing whether the set of possible utility capabilities will cover the future environment rather than predicting what the environment will be.

REFERENCES

1. Rochelle O'Connor. Facing Strategic Issues: New Planning Guides and Practices. New York: The Conference Board, 1985.
2. Herman Kahn and Anthony J. Weiner. The Year 2000: A Framework for Speculation on the Next Thirty-Three Years. New York: Macmillan, 1967.
3. George A. Steiner. Strategic Planning. New York: The Free Press, 1979.
4. Joel Kurtzman. Futurecasting. Palm Springs: ETC Publications, 1984.
5. William F. Thompson, Jerome J. Karaganis, and Kenneth D. Wilson. Choice over Change: Economic and Energy Options for the Future. New York: Praeger, 1981.
6. Stanley J. Feldman and David McClain. Structural Change in the United States: Perspectives on the Future. Lexington, KY: Data Resources, Inc., November 1984. (Prepared for and available from the Edison Electric Institute.)
7. Electric Power Supply & Demand for 1985-1994. Princeton, NJ: North American Electric Reliability Council, 1985.

QUESTION-AND-ANSWER SESSION FOLLOWING PRESENTATION

MR. SAM SUGIYAMA, Bonneville Power Administration: Given your last statement—and let's put ourselves back in the late 1960s when all of the nuclear plants were being planned or they were in progress—how do you convince top management to consider some of the alternatives, because, apparently at least in the Pacific Northwest, there were forecasts that, I think, were low, and would have tracked the current period reasonably well. How do you convince them to even consider that?

MR. HUMPHREY: My impression is that they are already convinced. Again, let me refer back to the survey that we have recently completed on utility planning practices.

Virtually every respondent to this survey of investor-owned utilities—and we got coverage of more than 96 percent of the kilowatthours provided by investor-owned utilities, so we virtually got the industry in this survey—provided us information about the range of supply-side and demand-side options that are currently in or are being put into their resource plan: demand-side management, all sorts of conservation programs, and all kinds of supply-side options as well. Large coal and nuclear plants are certainly there, but smaller coal plants and alternative generation technologies as well.

MR. SUGIYAMA: I didn't phrase the question very well. I was thinking of going back and reliving 1967. One of the possibilities, looking from now, is that demand growth could be extremely high. Something could happen that would give us a demand growth that, say, was the same as historical trends looking from the early '70s—Bob Marlay and your chart—if we examined the trends. That is a possibility but it seems extremely unlikely.

Back in the late '60s, what seemed extremely unlikely was a 1 to 2 percent growth in load, in the Pacific Northwest, anyway.

I guess part of the answer might be that you just try to make sure you include that in the range, but management is really not too impressed with a forecast that seems to be totally outside of many expectations.

Is my question or my thought clear? Going back to '67, convincing or trying to—

MR. HUMPHREY: And the answer is political. Obviously the regulatory, political, and economic incentives in place today argue for not building anything at all, regardless of what the forecast is. This is a topic that is obviously being addressed at the political level.

I don't want to characterize the attitude of management toward building or not building, or a high-growth case or a low-growth case. I guess I would put on my advocacy hat of

EEI and simply say that member companies are required to be prudent, and political and economic and financial incentives in place today drive the resource plans that are being put in place today.

But let me also say that, again, based on the results that we're getting back from this survey, the degree of flexibility in those plans is much greater than, say, in the early '70s when a 1200 MW nuclear plant was contemplated.

MR. SUGIYAMA: I will try not to occupy very much more time. The reason I asked is that, in our last long-term forecast, we presented to management a draft that had a range that was unacceptable, and so we had to narrow the range, and that was the reason for asking. Do you have any suggestions for convincing them we should maintain that range?

MR. HUMPHREY: I will treat that comment as a piece of data.

4

INTRODUCTION TO DELPHI-BASED SCENARIO GENERATION

William R. Huss
Battelle Columbus Division

ABSTRACT

This paper represents a transcript of a series of four workshop sessions designed to produce a set of alternative scenarios of the future of industrial electricity consumption in the United States until 2000. The workshops relied on the expertise of those in attendance to provide a list of influencing factors and their respective interrelationships or cross-impacts. A computer software package called BASICS was then used to process the data and produce several alternative scenarios. The predominant scenario was very optimistic for the electric utility industry and showed large increases in electricity use and the market penetration of electrotechnologies.

OBJECTIVES AND AGENDA

The four workshop sessions which are scheduled over the next 2-1/2 days are designed to accomplish several objectives. First, the intent is to generate a conference scenario or a picture of how this group views the future of electricity consumption in the industrial sector. Although the results of these workshops will be preliminary in nature, they should provide some useful perspectives. The second purpose is to identify what the group feels are the key influencing factors affecting future industrial electricity consumption. Finally, these workshops are intended to be a vehicle for encouraging group participation and interaction. Each of the key influencing factors will be discussed in small group sessions and their relationship to each other will be analyzed.

Let me briefly describe the agenda for the workshop titled "What Will Be the Electricity Consumption (GWh) in the Industrial Sector in the United States to the Year 2000?" In Session I, I will introduce you to the process of scenario analysis, and explain how it is performed at Battelle as well as at other organizations. I will also present a set of influencing factors which we have prepared at Battelle for your discussion and revision. At the end of the first session, you will vote on the most important influencing factors.

After lunch tomorrow, we well begin Session II. After I present the results of the voting, we will break up into small groups to discuss and, to a limited extent, research the influencing factors, their current and past trends, and interactions which may exist between them. Each group will then present the results of their analysis in Session III later in the afternoon. That evening Battelle will enter the data from the small groups into the computer and our scenario generation software package called BASICS (Battelle Scenario Inputs to Corporate Strategies) will develop sets of alternative scenarios.

At the fourth and final session, Thursday morning, I will present the computer printouts, discuss the results, describe the algorithm, and answer any questions you may have about the process.

SCENARIO ANALYSIS PROCESS—SESSION I

A scenario can be defined as "a description of a consistent set of factors which provides a framework for planning." The key word is "description" because typically the results tend to be presented in essay form supported by considerable numerical documentation. Therefore the process is much more qualitative than standard econometric models typically found in utility planning organizations. The qualitative nature of scenario analysis is both an advantage and a disadvantage. It is not based on a solid statistical foundation like some econometric models; yet, scenario analysis can be useful for much longer time frames (20 to 40 years) and can incorporate factors which are difficult to measure, such as the cohesion of the Organization of Petroleum Exporting Countries (OPEC), consumer attitudes toward nuclear generation, and behavioral changes such as how residential thermostats are set.

A second key to scenario analysis is that it attempts to define a **consistent** set of influencing factors. It is inconsistent, for example, to generate a scenario which includes low energy prices and strong industrial growth with a high cohesion of OPEC. By analyzing the interaction between influencing factors, scenarios will not include these types of inconsistencies.

Another method of using scenario analysis is as a framework for planning. What are the alternative futures facing the electric utility industry? Which futures are most probable and which will have the highest impact on the industry? Utility managers need to address these issues and develop strategies which take advantage of future conditions. Results of these scenario analyses can even be used to drive corporate financial, generation expansion, and load forecasting models to analyze how alternative business environments will affect the profitability of the industry and its ability to meet demand in a reliable and efficient manner.

STEPS TO CONDUCT SCENARIO ANALYSIS

There are seven steps which typically comprise scenario analysis as performed at Battelle. The first step is to define the topic, which for us is "What Will Be the Electricity Consumption (GWh) in the Industrial Sector in the United

States to the Year 2000?" Topics should include what is being forecast, a unit of measure, a forecast horizon, and geographic bounds.

Step 2 is to develop a comprehensive list of the major factors which affect the topic, such as GNP, technological advances, consumer attitudes, etc. Often this step is completed using various brainstorming techniques and interviews with experts in the field. Once this list of influencing factors has been generated, a screening process occurs whereby the list of influencing factors is reduced in number and those that remain are strictly defined.

In the third step, these factors (which we often call "descriptors") are further researched and essays are prepared describing various methods of measuring the factors, documenting previous trends, and discussing alternative future projections. At this point, two to five states or alternative outcomes are defined for each descriptor as well as an initial estimate of the probability of occurrence for each state. For example, a descriptor called "GNP" might have three states, less than 2 percent real growth, 2 to 4 percent, and greater than 4 percent. The initial or a priori probabilities of occurrence might be one-third, one-third, one-third—or any combination of values which add up to one. Also as a part of Step 3, a cross-impact guide is prepared which summarizes the direction and magnitude of the relationship between descriptors.

In Step 4, a full cross-impact matrix is prepared relating every descriptor **state** to every other one. In other words, one might ask a series of questions such as:

- What is the effect of high GNP on high electricity sales?
- What is the effect of medium GNP on high electricity sales?
- What is the effect of low GNP on high electricity sales?
- What is the effect of high GNP on medium electricity sales?
- What is the effect of medium GNP on medium electricity sales?

Clearly, the list of questions is long. If GNP and electricity sales have three states each, a total of nine questions must be answered to determine the effect of GNP on electricity sales. An additional nine questions must be answered to define the relationship in the other direction, the effect of electricity sales on GNP. In our workshops here, we will not have time to complete the entire cross-impact matrix. The computer program, however, does have an automatic procedure which can create the matrix using the cross-impact guide prepared as part of Step 3.

In Step 5, the data are entered into the computer and printouts of alternative scenarios are generated. Typically essays are then written to translate the numerical results into a "story" or description of alternative outcomes. In Step 6, the effect of low-probability, high-impact events, such as an oil embargo, are tested. Finally, in Step 7 analysis is performed to relate the scenario outcomes to implications for the client organization (electric utilities in this case) and eventually to the development of planning strategies.

Obviously, in the time we have, only a small portion of this total procedure can be completed. We will, however, work through the first three steps in enough detail that the computer algorithm can be exercised and some results presented.

DEVELOPING AREAS OF INFLUENCE—SESSION II

Since the topic for this workshop has already been defined, we can proceed directly to Step 2, developing areas of influence. Prior to this workshop, Battelle industry and technical experts developed a preliminary list of areas of influence which appear in Figure 4-1. These areas of influence have been divided into four categories: technology, economic, political/regulatory, and energy. In addition, three alternative states have been defined for each area of influence. Now I would like to ask those of you in the audience to add other factors which you think we may have missed. It is not necessary to comment on the merits of the factors already identified because your opinions will be reflected in the results of the voting exercise.

MR. PAUL WERBOS, Department of Energy: "We might want to add what Bob Marlay has called the Marlay Index, the energy-weighted or electricity-weighted industrial

FIGURE 4-1
Preliminary List of Areas of Influence

Key Descriptors That May Influence U.S. Industrial Electricity Consumption Over the Next 15 Years

Technology	Economic	Political/Regulatory	Energy
Renewable Energy	GNP Growth Rate	Sulfur Emissions	Price Nat Gas
Robot/Flexible Mfg	Prime Int Rate	State Air/Water Regs	Price Coal
Central Station	Capital Invest	Public Regs Attitude	Price Crude Oil
Safe Nucl Disposal	Industrial Prod	PUC's Oversight	Price Indus Elec
New Battery/Storage	Unemployment Rate	Depreciation Schedule	Indus Elec Volume
Coal Desulfurized	Consumer Spending	Local Tax Abatements	Indus Conservation
Reduce Trans Loss	Inflation Rate	Import Restrictions	North Amer Reserves
Electric Auto	Federal Deficits	Cohesion of OPEC	Resid/Comm Elec Use
Elec Motors Effic	Dow Jones Average	Energy Price Regs	World Energy Demand
Electro-Furnaces	Industrial Wages	Regs for Nucl Disposal	Elec/Energy Ratio
	Marley Index	Fed Appli Effic Stds	Cogeneration
	$ Exchange Rate	Loss of Nat Mono Status	
	Energy Weight Output		

production index. The states would perhaps be less than 2 percent growth, 2 to 3 percent, 3 to 4 percent, and greater than 4 percent."

MR. FRANK LIN, New England Power Service: "What about the possible extension of federal and state appliance efficiency standards? The states would be more stringent, less stringent, and about the same."

If there are no more additions, we can move on to the voting exercise. Each of you should have a blue sheet (Figure 4-1) listing the areas of influence which we have discussed in this session. Make sure you append the list with the two additions which were just presented. We would like each person to circle the eight most important factors. You can vote on eight from one category or spread your votes across all categories. The results will be tabulated this evening and presented to you at the beginning of the second session tomorrow.

SMALL GROUP DISCUSSIONS

Figure 4-2 shows the results of yesterday's voting exercise. As you can see, there was fairly strong consensus for about six or seven factors—in other words those receiving more than 20 votes. I have circled what the group feels are the 13 most important factors plus the item we are trying to forecast, industrial electricity consumption. Therefore, we have 14 factors which we want to investigate further. Each of you was assigned to a group as you entered and one person from each group was asked to be the group spokesperson. Figure 4-3 shows an example of the data sheet which each group should complete.

There are three tasks which you should accomplish during the small group discussions. First, each group should discuss the states or alternative outcomes for your factor. Are the states that have been presented the most reasonable ones in your mind or should they be modified? These states should be written in the blanks in the upper left-hand corner of your data form (Figure 4-3). Next, the group needs to define a priori or initial probabilities for each state. This

FIGURE 4-2
Preliminary List of Areas of Influence and Voting Results

Key Descriptors That May Influence U.S. Industrial Electricity Consumption Over the Next 15 Years and Votes Received

Technology	Votes	Economic	Votes	Political/Regulatory	Votes	Energy	Votes
Renewable Energy	7	GNP Growth Rate	28	Sulfur Emissions	10	Price Nat Gas	26
Robot/Flexible Mfg	26	Prime Int Rate	7	State Air/Water Regs	7	Price Coal	7
Central Station	8	Capital Invest	26	Public Regs Attitude	9	Price Crude Oil	15
Safe Nucl Disposal	7	Industrial Prod	29	PUC's Oversight	8	Price Indus Elec	20
New Battery/Storage	5	Unemployment Rate	2	Depreciation Schedule	5	Indus Elec Volume	38
Coal Desulfurized	4	Consumer Spending	5	Local Tax Abatements	1	Indus Conservation	10
Reduce Trans Loss	1	Inflation Rate	4	Import Restrictions	12	North Amer Reserves	1
Electric Auto	2	Federal Deficits	14	Cohesion of OPEC	5	Resid/Comm Elec Use	1
Elec Motors Effic	18	Dow Jones Average	0	Energy Price Regs	9	World Energy Demand	5
Electro-Furnaces	18	Industrial Wages	6	Regs for Nucl Disposal	2	Elec/Energy Ratio	5
		Marley Index	27	Fed Appliance Effic Stds	24	Cogeneration	25
		$ Exchange Rate	2	Loss of Nat Mono Status	1		
		Energy Weight Output	1				
Total Votes by Category	96		151		93		153

FIGURE 4-3
Example of Data Sheet

Specify A Priori Probabilities for Descriptor ____________

State	*A Priori* Probability
____________	______
____________	______
____________	______
____________	______

Cross Impact Guide

Descriptor ____________
Impact when state (high) ____________ occurs

What is the relationship on each of the following descriptor states when state ____________ occurs, more likely or less likely, and what is the strength (strong, moderate, weak) of that relationship?

	Descriptor	State (high)	Relationship (More/Less Likely)	Strength	Explanation
1.			______	______	
2.			______	______	
3.			______	______	
4.			______	______	
5.			______	______	
6.			______	______	
7.			______	______	
8.			______	______	
9.			______	______	
10.			______	______	
11.	Industrial Elec Volume	much higher	______	______	

exercise also allows you to test whether you have defined three reasonable states. If one of the probabilities is close to zero or to one, you may wish to redefine the states so that a more equal spread of probabilities is obtained. The third and more rigorous part of the exercise is to complete the cross-impact guide.

Let us try an example. For the descriptor GNP, the highest state is "greater than 4 percent real growth." What we want to determine is the relationship between the occurrence of this state and the occurrence of the highest state for all of the other descriptors. To do this, we assume that the one thing we know for sure about the future is that GNP will increase at greater than 4 percent. Given this additional piece of information the question is: How would we revise the initial probabilities estimated earlier? We can revise the initial probabilities either up or down and either significantly, moderately, or slightly. Specifically, let us discuss the relationship which the occurrence of high GNP has on high capital investment. ("High" is one state of the capital investment descriptor.) In a world where you know for sure that GNP will be growing faster than 4 percent, one would tend to increase the initial probability that there would be high capital investment. This relationship is probably strong. When performing this type of exercise, I also recommend that you write down a few words describing the logic which you used to estimate the relationship. This will help significantly later on when you review and modify the cross-impact matrix and when you need to explain your logic to others.

Let me remind you that for the third session, scheduled for later this afternoon, each group should prepare a 2- or 3-minute presentation summarizing the discussion and results of your session. I have given each group one blank viewgraph and a marking pen to assist in this presentation.

SUMMARY OF SMALL GROUP DISCUSSIONS —SESSION III

For this next session, I would like each group leader to spend roughly 2 or 3 minutes discussing the results of the session we had earlier this afternoon. Please mention what your descriptor is, the probabilities of occurrence for each state, and what you think are the key cross-impacts and key relationships between the descriptors.

MR. TERRY MORLAN, Northwest Power Planning Council: "Our descriptor is **natural gas prices.** The probabilities that we assigned to the various scenarios for natural gas prices included a fairly low 10 percent probability that natural gas prices would escalate greater than 3 percent. For a 1 to 3 percent escalation in natural gas prices, we assigned a 40 percent probability; and for a less than 1 percent increase in natural gas prices, we assigned a 50 percent probability.

In terms of the relationship of this descriptor to other descriptors, there were four areas—robotics, efficient motors, federal budget deficits, and appliance standards—where we did not see any obvious relationship between high natural gas prices and the corresponding high categories in those descriptors. The four areas where we saw relationships were generally economic factors—GNP growth, capital investment, industrial production index, and crude oil prices. Our assumption was that, in the long run, strong economic activity would tend to be associated with higher oil prices, and natural gas prices are likely to be correlated with oil prices. We saw a positive relationship between economic activity and high gas prices because high economic activity would lead to high prices, not the other way around. In some cases we saw a direct relationship—for example, the higher the demand for natural gas, the more likely you would see electric furnaces. Of course there was an expected correlation between higher electricity prices and higher gas prices because with higher gas prices, you tend to have higher electricity demand and more capacity expansion and higher electricity rates."

MS. KATHRYN PRICE, Pennsylvania Power and Light Company: "For **robots and flexible manufacturing** we felt that there was only a 10 percent probability that the price would be the same 15 years from now as it is today. Although the difference between "greatly increases" and "increases" is judgmental, we assigned a 30 percent probability to "increases" and a 60 percent probability to "greatly increases." In terms of relationships, the one we felt most strongly about was capital investment, because it is going to take dollars to build robots and make changes in flexible manufacturing. Other descriptors we placed in the same category included electrofurnace market share, electric motor efficiencies, Industrial Production Index increases, and GNP growth. For two other descriptors—industrial electric volume and the Marlay Index—we felt there was also going to be a relationship, though not as strong as with the previous descriptors. For industrial electric volume, we were not sure about potential offsetting problems—the robot might be using more electricity, but the workplace would require less lighting and heating.

We felt that there was little relationship between robots and flexible manufacturing, and the price of fuel. With robots the process could be more flexible and efficient; the processes could be scheduled for off-peak periods so that electrical and fuel prices would not increase as much. We did not find any correlation with appliance efficiencies and the federal deficit."

MR. EDWARD FISCHLER, Georgia Power Company: "Our descriptor was **motor efficiencies.** The first thing we did was to change the descriptor slightly. Rather than consider just the changes in motor efficiency, we considered efficiency improvements of the motor-driven processes across the range of loads on the motor. We assigned a probability of 40 percent to achieving efficiency levels over 90 percent. We gave the more moderate efficiency levels of 70 to 90 percent a 50 percent probability. The probability of achieving a level of only 70 percent or less efficiency was felt by the group to be slight, so it was assigned a 10 percent probability.

For the cross-impacts, there were several that we thought were likely to occur. We strongly thought that the improvement of the efficiency of motor-driven processes would cause an increase in GNP growth, capital investment, and industrial production mix. We also thought, though not as strongly, that appliance efficiencies would improve. As far as industrial electricity volume, because EPRI is conducting research into motor process efficiencies with the objective of increasing sales in the industrial sector, we believe there is a strong relationship. As for the price of electricity—we felt fairly strongly that improving the industrial process and the motor process would make an increase in the real price of electricity less likely."

MR. LARRY B. BUTLER, Southern California Edison Company: "We split our descriptor, **GNP growth**, into three categories. The first category was productivity measures: robotics, motor efficiency, and electric furnaces. We felt that GNP would unquestionably have a positive effect, because there should be a correlation between measures that improve productivity and something that is related to productivity. The effect would be much stronger for robotics; however, while motors might increase efficiency between 90 and 100 percent, there is no guarantee that anybody will purchase them, so we assumed a smaller effect.

The second category was economic variables. We felt that the Industrial Production Index and the Marlay Index were close to equaling GNP. The federal deficit would have a crowding-out effect, and by ultimately lowering capital intensity, would reduce GNP. Capital investment and GNP have a strong relationship. Obviously GNP affects capital investments—high GNP leads to high capital investment and vice-versa. For the price descriptors, we felt that high GNP would imply higher prices—especially for electricity prices.

As for the operating probabilities, we felt that only 10 percent would be greater than 4 percent; 80 percent would be in the 2 to 4 percent range; and 10 percent would be in the below 2 percent range."

MR. RICHARD TATE, Georgia Power Company: "We made the assumption that our descriptor—**Industrial Production Index**—was similar to GNP in terms of behavior. We changed the states slightly by adding a quantitative measure: the high state was greater than 4 percent annual escalation; the medium state was 2 to 4 percent; and the low state was less than 2 percent annual escalation. The probabilities that we assigned were 20 percent probability for the high state, 50 percent probability for the medium state, and 30 percent probability for the low state.

In terms of relationships, we felt that there was a strong relationship between our descriptor and robotics, motor efficiency, GNP growth, capital investment, and the Marlay Index. In the category of weak relationships we included natural gas, crude oil, electricity prices, and cogeneration—

because of the external forces on these prices and, with the nuclear plants coming on-line, the effect of other generation problems on price external to production.

We saw little relationship between our descriptor and appliance efficiency standards, electricity production, and the federal deficit."

MR. E. L. HILLSMAN, Oak Ridge National Laboratory: "We also changed our descriptor—the **federal deficit**—by raising all of the intervals by $50 billion. No one believed that a deficit of under $100 billion was very likely between now and 1995. It is fairly easy to identify a number of descriptors that would be strongly impacted by high deficits—GNP growth, capital investment, Industrial Production Index, the Marlay Index, and the volume of industrial electricity. With regard to robotics and electric furnaces, high deficits would also tend to reduce the likelihood of reaching the high states of these descriptors. For the price of natural gas, the price of electricity, and cogeneration, we felt that high deficits would depress supply and demand—we were not sure which would have the greater effect. We did not see any relationship between the federal deficit and motor efficiency, which we viewed as a problem in technology development."

MR. FRED NORRELL, Alabama Power Company: "For the **Marlay Index**, we took the middle-of-the-road case, which indicated no significant further shift in the Marlay Index from its present level, and we assigned it a 40 percent probability. We assigned a 30 percent probability to the other two cases: (1) a further shift away from electrical users, indicating less electric intensity; and (2) a reversal of the recent shift in the Marlay Index indicating greater electric intensity. Very simply, we found two conditions that would bring about a move away from electric intensity—a large federal deficit and the price of electricity increasing at a very rapid rate. All of the other conditions we found consistent with a reversal in the Marlay Index, that is, a shift toward electric intensity."

MR. PAUL WERBOS, U.S. Department of Energy: "We were all in agreement that the chances of lower **appliance efficiency standards** were about 10 percent. After thinking about it, we decided that the odds were that there would be appliance standards between now and the year 2000. We did not see very large impacts. The one moderate impact we saw was on motor efficiency, where increased appliance standards would tend to apply more motor efficiency to standards. We saw a very slight effect against robotics, because we had an image of companies like G.E. spending more money on appliance efficiency and not having as much money left over for R&D. Everybody could see appliance standards reducing energy demand a little and therefore having a weak effect on reducing energy prices. In terms of capital investment and GNP growth, we saw a weak relationship—but we felt there were both positive and negative factors."

MS. SHANNON M. LARSON, Boston Edison Company: "We kept the states for the **price of crude oil** the same—that the price of crude oil was going to grow at greater than 3 percent per year. We assigned that a probability of 35 percent. For the middle state—zero to 3 percent per year—we assigned a probability of 45 percent. And for the lower state, that is a decline in real crude oil prices, we assigned a 20 percent probability. The high real price of crude oil will increase the relationship between motor efficiency, electrofurnace technology, federal deficit, and the Marlay Index. We feel that appliance efficiencies will become more likely with high crude oil prices. The price of electricity will grow at a faster rate because of high crude oil prices, and greater increases in cogeneration will be more likely. Less likely impacts would be robotics (because we felt there would not be as much capital available for investment), GNP growth, capital investment, industrial production index, and industrial electric sales."

MR. JOHN GROCKI, Public Service Electric and Gas: "For our descriptor, **electricity price for the industrial sector**, we left the states as they were yesterday: 15 percent probability for less than zero; 50 percent probability for greater than 2 percent; and 35 percent probability for greater than 2 percent. The economic variables—GNP growth, capital investment, and the Industrial Production Index—had significant relationships, but the price of feedstocks, natural gas, and crude oil had even stronger relationships with the descriptor. Cogeneration and the Marlay Index were also significant for the greater than 2 percent state. We saw no relationship between electrical price for the industrial sector and the federal deficit and appliance efficiencies."

MR. THOMAS MOORE, Tampa Electric Company: "We are 90 percent sure that **cogeneration** will increase in the future. We noticed several strong relationships. One is investment: if cogeneration increases, so does investment. Another is the price of electricity. Relationships that were less strong included the price of natural gas and the price of crude oil. All of the other relationships we thought were moderate."

MR. WAYNE LUCAS, Kentucky Utilities: "We left the states for our descriptor—**industrial electricity consumption**—as they were yesterday: 5 percent probability for greater than 7 percent; 15 percent probability for 3 to 7; 50 percent probability for zero to 3; and 30 percent for less than zero. The group agreed that the correct descriptors were in the model, with the possible exception of crude oil."

RESULTS OF THE SCENARIO ANALYSIS—SESSION IV

I would like to welcome you to the fourth and final session of our scenario generation workshop. Last night, Mr. Edward Honton of Battelle and I entered the data generated in yesterday's small group meetings into the computer. Some minor modifications were made to assure consistency between groups. There was also a tendency to identify an unusually high number of very strong relationships, so in some cases we reduced the magnitude of the relationship. Revisions such as these are typically a part of the scenario analysis process, just as the initial specification of an econometric model is revised based on an analysis of the diagnostics.

Each of you should have a number of computer printouts in front of you which display the results of the scenario analysis. The first sheet (Figure 4-4) shows a list of the descriptors, states, and initial probabilities. Figure 4-5 shows the full cross-impact matrix which was developed using the cross-impact guides completed in the small group meetings.

Using Figure 4-5, I would like to briefly describe the algorithm used in the analysis. Please feel free to ask questions or contact me at Battelle if you need further explanation. In the left-hand column, you should see listed the initial probabilities. For example, robotics—greatly increasing has an initial probability of 0.60; robotics—increasing has an initial probability of 0.30; and robotics—decreasing has an initial probability of 0.10. The algorithm, in effect adjusts these probabilities up and down through time until all are equal to either zero or one. Those equal to one are assumed to occur and those equal to zero are assumed not to occur. Thus a single scenario is defined.

To begin, however, the program needs a starting point. A starting point is defined as setting one of the descriptor states either to occur or not to occur. Thus, if you have 43 descriptor states (13 descriptors with 3 states each and 1 descriptor with 4 states), there are 86 possible starting points because each state could be set either to occur or not to occur. In order to obtain a full range of results, we typically try all possible starting points which for our analysis is 86 simulations.

In order to explain the algorithm, let me select a single starting point, say robotics—greatly increasing occurring. When this state is set to occur, robotics—increasing and robotics—decreasing are automatically set not to occur (or zero). Next the computer reads down the column of the cross-impact matrix associated with robotics—greatly increasing. For example, let us proceed down the first column until we reach real GNP growth greater than 4 percent. The entry in that cell is a "1". This means that given the occurrence of robotics—greatly increasing, the probability of GNP greater than 4 percent should be increased slightly. We use a -3 to +3 scale to indicate the direction and magnitude of the relationship. The computer then modifies the initial probability of GNP greater than 4 percent from 0.10 to a slightly higher value using a formula which I do not

FIGURE 4-4
Computer Printout of List of Descriptors, States, and Initial Probabilities

09/18/85 20:28:06

U.S. Electricity Consumption till 2000 9–18–85

	Descriptor	States	A Priori Probabilities
1.	Robot/Flexbl Mfg	gr incr	0.60
		incr	0.30
		same	0.10
2.	Eff Motor Drv Pr	90–100%	0.40
		70–90%	0.50
		< 70%	0.10
3.	Electro-Furnaces	> 25%	0.50
		15–25%	0.30
		< 15%	0.20
4.	Real GNP Growth	> 4.0%	0.10
		2.0–4.0%	0.80
		< 2.0%	0.10
5.	Capital Invest	High	0.30
		Medium	0.40
		Low	0.30
6.	Indus Prod Index	> 4%	0.20
		2.0–4.0%	0.50
		< 2.0%	0.30
7.	Federal Deficits	> $250 B	0.35
		$150–250	0.45
		< $150 B	0.20
8.	Marley Index Shf	reversal	0.30
		no chang	0.40
		further	0.30
9.	Appl Effic Stnds	More	0.50
		Same	0.40
		Less	0.10
10.	Price Nat Gas	>3.0%/yr	0.10
		1.0–3.0%	0.40
		1.0%	0.50
11.	Pr Crude Oil	>3.0%/yr	0.35
		0.0–3.0%	0.45
		< 0.0%	0.20
12.	Pr Indus Elec	>2.0%/yr	0.15
		0.0–2.0%	0.50
		<0.0%/yr	0.35
13.	Lvl Cogeneration	Grt Incr	0.40
		Slt Incr	0.50
		Same/Dwn	0.10

FIGURE 4-5
Computer Printout of Full Cross-Impact Matrix Results

09/18/85 20:28:47 U.S. Electricity Consumption till 2000 9-18-85

OCCURRENCE MATRIX

	1.	2.	3.	4.	5.	6.	7.	8.	9.	10.	11.	12.	13.	14.	SUM VALS	NON ZERO
1. Robot/Flexbl Mfg																
0.60 gr incr	0 0 0	0 0 0	0 0 0	2 0 -2	2 0 -2	1 0 -1	-2 0 2	1 0 -1	-1 0 1	0 0 0	-1 0 1	-2 0 2	1 0 -1	0 0 0 0	0	18
0.30 incr	0 0 0	0 0 0	0 0 0	0 1 0	0 1 0	0 1 0	0 0 0	0 1 0	0 0 0	0 0 0	0 0 0	0 0 0	0 1 0	0 0 0 0	5	5
0.10 same	0 0 0	0 0 0	0 0 0	-2 0 2	-2 0 2	-1 0 1	2 0 -2	-1 0 1	1 0 -1	0 0 0	1 0 -1	2 0 -2	-1 0 1	0 0 0 0	0	18
2. Eff Motor Drv Pr																
0.40 90-100%	2 0 -2	0 0 0	0 0 0	1 0 -1	0 0 0	1 0 -1	0 0 0	1 0 -1	1 0 -1	0 0 0	2 0 -2	1 0 -1	1 0 -1	2 1 -1 -2	0	20
0.50 70-90%	0 1 0	0 0 0	0 0 0	0 1 0	0 0 0	0 1 0	0 0 0	0 1 0	0 1 0	0 0 0	0 1 0	0 1 0	0 1 0	0 1 1 0	10	10
0.10 < 70%	-2 0 2	0 0 0	0 0 0	-1 0 1	0 0 0	-1 0 1	0 0 0	-1 0 1	-1 0 1	0 0 0	-2 0 2	-1 0 1	-1 0 1	-2 -1 1 2	0	20
3. Electro-Furnaces																
0.50 > 25%	2 0 -2	0 0 0	0 0 0	1 0 -1	0 0 0	0 0 0	-2 0 2	1 0 -1	0 0 0	2 0 -2	2 0 -2	-2 0 2	2 0 -2	-1 0 0 1	0	18
0.30 15-25%	0 1 0	0 0 0	0 0 0	0 1 0	0 0 0	0 0 0	0 0 0	0 1 0	0 0 0	0 1 0	0 1 0	0 -1 0	0 1 0	0 -1 -1 0	3	9
0.20 < 15%	-2 0 2	0 0 0	0 0 0	-1 0 1	0 0 0	0 0 0	2 0 -2	-1 0 1	0 0 0	-2 0 2	-2 0 2	2 0 -2	-2 0 2	1 0 0 -1	0	18
4. Real GNP Growth																
0.10 > 4.0%	1 0 -1	1 -1 -1	0 0 0	0 0 0	2 1 -2	1 0 -1	-3 0 3	1 0 -1	0 0 0	-1 0 1	-2 -1 2	-1 0 1	0 0 0	0 0 0 0	-1	21
0.80 2.0-4.0%	0 1 0	0 1 0	0 0 0	0 0 0	0 1 0	0 0 0	0 0 0	0 1 0	0 0 0	0 -1 0	1 0 1	0 -1 0	0 0 0	0 0 0 0	4	8
0.10 < 2.0%	-1 0 1	-1 -1 1	0 0 0	0 0 0	-2 -1 2	-1 0 1	3 0 -3	-1 0 1	0 0 0	1 0 -1	2 -1 -2	1 0 -1	0 0 0	0 0 0 0	-3	21
5. Capital Invest																
0.30 High	3 0 -3	3 0 -3	1 0 -1	3 0 -3	0 0 0	2 0 -2	-3 0 3	1 0 -1	1 0 -1	1 0 -1	-3 0 3	1 0 -1	2 0 -2	2 1 -1 -2	0	28
0.40 Medium	0 1 0	0 1 0	0 1 0	0 1 0	0 0 0	0 0 0	0 1 0	0 1 0	0 1 0	0 1 0	0 0 0	0 1 0	0 1 0	0 0 0 0	10	10
0.30 Low	-3 0 3	-3 0 3	-1 0 1	-3 0 3	0 0 0	-2 0 2	3 0 -3	-1 0 1	-1 0 1	-1 0 1	3 0 -3	-1 0 1	-2 0 2	-2 -1 1 2	0	28
6. Indus Prod Index																
0.20 > 4%	1 0 -1	1 0 -1	1 0 -1	3 0 -3	2 0 -2	0 0 0	-2 0 2	3 1 -3	0 0 0	0 0 0	-1 0 1	-1 0 1	0 0 0	0 0 0 0	1	19
0.50 2.0-4.0%	0 1 0	0 1 0	0 1 0	0 1 0	0 1 0	0 0 0	0 -1 0	1 2 1	0 0 0	0 0 0	0 -1 0	0 -1 0	0 0 0	0 0 0 0	6	11
0.30 < 2.0%	-1 0 1	-1 0 1	-1 0 1	-3 0 3	-2 0 2	0 0 0	2 0 -2	-3 1 3	0 0 0	0 0 0	1 0 -1	1 0 -1	0 0 0	0 0 0 0	1	19

FIGURE 4-5
Computer Printout of Full Cross-Impact Matrix Results (continued)

```
 7.
Federal Deficits
  0.35  > $250 B  0  0  0   0  0  0   0  0  0  -2  0  2  -2  0  2  -1  0  1   0  0  0  -1  0  1   0  0  0   0  0  0   2  0 -2   0  0  0  -2  0  2   0  0  0  0   0 12
  0.45  $150-250  0  0  0   0  0  0   0  0  0   0  1  0   0  1  0   0 -1  0   0  0  0   0 -1  0   0  0  0   0  0  0   0  1  0   0  0  0   0  1  0   0  0  0  0   2  6
  0.20  < $150 B  0  0  0   0  0  0   0  0  0   2  0 -2   2  0 -2   1  0 -1   0  0  0   1  0 -1   0  0  0   0  0  0  -2  0  2   0  0  0   2  0 -2   0  0  0  0   0 12

 8.
Marley Index Shf
  0.30  reversal  1  0 -1   1  0 -1   1  0 -1   2  0 -2   2  0 -2   3  0 -3  -2  0  2   0  0  0   0  0  0  -1  0  1  -1  0  1  -1  0  1   1  0 -1   0  0  0  0   0 22
  0.40  no chang  0  1  0   0  1  0   0  1  0   0  1  0   0  1  0   0  1  0   0 -1  0   0  0  0   0  0  0   0 -1  0   0 -1  0   0 -1  0   0  1  0   0  0  0  0   3 11
  0.30  further  -1  0  1  -1  0  1  -1  0  1  -2  0  2  -2  0  2  -3  0  3   2  0 -2   0  0  0   0  0  0   1  0 -1   1  0 -1   1  0 -1  -1  0  1   0  0  0  0   0 22

 9.
Appl Effic Stnds
  0.50  More      0  0  0   3  0 -3   0  0  0   1  0 -1   0  0  0  -1  0  1   0  0  0   0  0  0   0  0  0   0  0  0   3  0 -3   0  0  0  -1  0  1   0  0  0  0   0 10
  0.40  Same      0  0  0   0  1  0   0  0  0   0  1  0   0  0  0   0  0  0   0  0  0   0  0  0   0  0  0   0  0  0   0  1  0   0  0  0   0  0  0   0  0  0  0   3  3
  0.10  Less      0  0  0  -3  0  3   0  0  0  -1  0  1   0  0  0   1  0 -1   0  0  0   0  0  0   0  0  0   0  0  0  -3  0  3   0  0  0   1  0 -1   0  0  0  0   0 10

10.
Price Nat Gas
  0.10  >3.0%/yr -1  0  1   0  0  0   2  0 -2   1  0 -1   0  0  0   1  0 -1   0  0  0   1  0 -1  -1  0  1   0  0  0   3  0 -3   1  0 -1  -2  0  2  -2 -1  1  2   0 22
  0.40  1.0-3.0%  0  0  0   0  0  0   0  1  0   0 -1  0   0  0  0   0  1  0   0  0  0   0  1  0   0  0  0   0  0  0   0  0  0   0  0  0   0 -1  0   0  0  0  0   1  5
  0.50  < 1.0%    1  0 -1   0  0  0  -2  0  2  -1  0  1   0  0  0  -1  0  1   0  0  0  -1  0  1   1  0 -1   0  0  0  -3  0  3  -1  0  1   2  0 -2   2  1 -1 -2   0 22

11.
Pr Crude Oil
  0.35  >3.0%/yr -1  0  1   0  0  0   2  0 -2   1  0 -1   0  0  0   1  0 -1   0  0  0   1  0 -1  -1  0  1   3  0 -3   0  0  0   1  0 -1  -2  0  2  -1 -1  1  1   0 22
  0.45  0.0-3.0%  0  0  0   0  0  0   0 -1  0   0 -1  0   0  0  0   0  1  0   0  0  0   0  1  0   0  0  0   0  0  0   0  0  0   0  0  0   0 -1  0   0  0  0  0  -1  5
  0.20  < 0.0%    1  0 -1   0  0  0  -2  0  2  -1  0  1   0  0  0  -1  0  1   0  0  0  -1  0  1   1  0 -1  -3  0  3   0  0  0  -1  0  1   2  0 -2   1  1 -1 -1   0 22

12.
Pr Indus Elec
  0.15  >2.0%/yr -1  0  1  -3  0  3  -2  0  2   1  0 -1   1  0 -1   1  0 -1   0  0  0   1  0 -1  -1  0  1   1  0 -1   3  0 -3   0  0  0   0  0  0   3  2  1 -2   4 24
  0.50  0.0-2.0%  0  0  0   0  0  0   0  0  0   0 -1  0   0  1  0   0  1  0   0  0  0   0  1  0   0  0  0   0  0  0   0  0  0   0  0  0   0  0  0   1  1  0 -1   3  7
  0.35  <0.0%/yr  1  0 -1   3  0 -3   2  0 -2  -1  0  1  -1  0  1  -1  0  1   0  0  0  -1  0  1   1  0 -1  -1  0  1  -3  0  3   0  0  0   0  0  0  -3 -2 -1  2  -4 24
```

FIGURE 4-5
Computer Printout of Full Cross-Impact Matrix Results (continued)

```
13.
Lvl Cogeneration
  0.40  Grt Incr -1  0  1   0  0  0   1  0 -1   1  0 -1   1  0 -1   1  0 -1  -1  0  1   2  0 -2   0  0  0  -2  0  2   2  0 -2   3  0 -3   0  0  0   2  1 -1 -2   0 24
  0.50  Slt Incr  0  0  0   0  0  0   0  1  0   0  1  0   0  1  0   0  1  0   0  0  0   0  1  0   0  0  0  -1 -1  1   0  1  0   0  1  0   0  0  0   0  1  1  0   8 12
  0.10  Same/Dwn  1  0 -1   0  0  0  -1  0  1  -1  0  1  -1  0  1  -1  0  1   1  0 -1  -2  0  2   0  0  0   2  0 -2  -2  0  2  -3  0  3   0  0  0  -2 -1  1  2   0 24

14.
Volume Electrici
  0.05  >7.0%/yr  1  0 -1   3  0 -3   3  0 -3   3  0 -3   3  0 -3   2  0 -2  -3  0  3   3  0 -3  -1  0  1   3  0 -3  -2  0  2  -2  0  2  -2  0  2   0  0  0  0   0 26
  0.15  3.0-7.0%  1  0  0   2 -1 -1   1  0 -1   1  1 -1   1  1 -1   1  1 -1  -1  0  1   1  1 -1  -1  0  0   1  0 -1  -1  0  1  -1  0  1  -1  0  1   0  0  0  0   4 29
  0.50  0.0-3.0%  0  0  0  -1 -1  1  -1  0  1  -1  1  1  -1  1  1   0  1  0   1  0 -1  -1  1  1   0  0  0  -1  0  1   0  0 -1   0  0 -1   0  0 -1   0  0  0  0   0 22
  0.30  Some Dec -1  0  1  -3  0  3  -3  0  3  -3  0  3  -3  0  3  -2  0  2   3  0 -3  -3  0  3   1  0 -1  -3  0  3   2  0 -2   2  0 -2   2  0 -2   0  0  0  0   0 26


SUM OF VALUES     1  6  0   1  1  0   0  4  0   0  8  0   0  9  0   1  8 -1   0 -1  0   1 14  1  -1  2  0  -1 -1  1   0  1  1  -1 -1  0  -1  4  0   1  2  1 -1

NON ZERO ENTRIES 23  6 22  16  9 16  18  6 18  28 14 28  18 11 18  25 10 25  18  3 18  27 15 27  15  2 14  19  5 19  28  9 29  23  7 24  21  8 22  15 16 15 15
```

have time to present here but is available upon request. In a similar manner all of the other initial probabilities are adjusted up or down based on the occurrence of robotics—greatly increasing. The computer then selects the probability that is closest to either zero or one and sets the associated descriptor state either to occur (one) or not occur (zero). The cross-impact matrix is again accessed, the probabilities adjusted, and the process repeated until all probabilities have been set to either zero or one. This completes one simulation and the computer moves on to try other starting points.

Figure 4-6 displays the outputs. Across the top, you will see, in sequence, the numbers 1 through 25. These represent 25 different outcomes or scenario types defined by the column of zeros and ones below. Remember that a one means that the descriptor state occurred and a zero means that the state did not occur. The second row shows the frequency or number of times that each scenario type occurred. If you add all of the frequencies together, you should get the total number of starting points, which in this case is 86. Many of the scenario types are very similar and perhaps upon further study could be combined. It is important to note that the analyst should pay particular attention not only to the most likely scenarios but also to the low-probability, high-impact ones as well.

At the right of the page is displayed the total number of times that each descriptor state occurred out of the 86 simulations. Robotics—greatly increasing occurred 31 times (36 percent); robotics—increasing occurred 52 times (60 percent); and, robotics—decreasing occurred 3 times (3 percent). This represents some change in the group's initial estimates, which were 60, 30, and 10 percent, respectively.

The main-line scenario for the group seems to be a very optimistic one if you believe the posterior probabilities displayed in the second to last column. I guess that is somewhat to be expected since this is an electric utility group. The main-line scenario seems to predict increasing robotics, motor efficiency above 70 percent, high penetration of electrofurnaces, moderate to high GNP growth, medium to high capital investment, medium to high productivity, smaller federal deficits, a reversal in the Marlay Index, no change in efficiency standards, moderate increases

FIGURE 4-6
Computer Printout of Output Resulting from Scenario Analysis

09/18/85 20:30:56

U.S. Electricity Consumption till 2000 9-18-85

Outcome Type	1	2	3	4	5	6	7	8	9	10	11	12	13	14	15	16	17	18	19	20	21	22	23	24	25	A Priori Prob	Total Occurs	Posterior Prob	
Frequency	8	7	6	5	3	3	2	2	2	2	1	1	1	1	1	1	1	1	1	1	1	1	1	1	1				
1. Robot/Flexbl Mfg																													
gr incr	0	1	0	1	0	0	1	0	0	0	1	1	0	1	1	1	0	1	0	0	1	1	1	0	0	0.60	31	0.36	gr incr
incr	1	0	1	0	1	1	0	1	1	1	0	0	0	0	0	0	1	0	1	1	0	0	0	1	1	0.30	52	0.60	incr
same	0	0	0	0	0	0	0	0	0	0	0	0	1	0	0	0	0	0	0	0	0	0	0	0	0	0.10	3	0.03	same
2. Eff Motor Drv Pr																													
90–100%	1	1	0	1	0	1	1	0	1	0	1	0	0	1	1	1	0	1	0	1	0	0	1	0	0	0.40	43	0.50	90–100%
70–90%	0	0	1	0	1	0	0	1	0	1	0	1	1	0	0	0	0	0	1	0	1	1	0	1	1	0.50	41	0.48	70–90%
< 70%	0	0	0	0	0	0	0	0	0	0	0	0	0	0	0	0	1	0	0	0	0	0	0	0	0	0.10	2	0.02	< 70%
3. Electro-Furnaces																													
> 25%	1	1	0	1	1	1	1	0	1	1	0	0	0	1	1	1	0	1	0	1	1	1	1	0	0	0.50	56	0.65	> 25%
15–25%	0	0	0	0	0	0	0	0	0	0	1	0	0	0	0	0	0	0	0	0	0	0	0	1	0	0.30	11	0.13	15–25%
< 15%	0	0	1	0	0	0	0	1	0	0	0	1	1	0	0	0	1	0	1	0	0	0	0	0	1	0.20	19	0.22	< 15%
4. Real GNP Growth																													
> 4.0%	0	1	0	1	0	0	1	0	0	0	1	0	0	1	0	1	0	1	0	0	1	1	1	0	0	0.10	29	0.34	> 4.0%
2.0–4.0%	1	0	1	0	1	1	0	1	1	1	0	1	1	0	1	0	0	0	1	1	0	0	0	1	1	0.80	54	0.63	2.0–4.0%
< 2.0%	0	0	0	0	0	0	0	0	0	0	0	0	0	0	0	0	1	0	0	0	0	0	0	0	0	0.10	3	0.03	< 2.0%
5. Capital Invest																													
High	1	1	0	1	1	1	1	0	1	0	1	0	0	1	1	1	0	0	0	1	1	1	1	0	0	0.30	48	0.56	High
Medium	0	0	1	0	0	0	0	1	0	1	0	1	1	0	0	0	1	1	0	0	0	0	0	1	1	0.40	34	0.40	Medium
Low	0	0	0	0	0	0	0	0	0	0	0	0	0	0	0	0	0	0	1	0	0	0	0	0	0	0.30	4	0.05	Low
6. Indus Prod Index																													
> 4%	1	1	0	1	0	1	1	0	1	0	1	0	0	1	1	1	0	1	0	1	0	0	1	0	0	0.20	46	0.53	> 4%
2.0–4.0%	0	0	1	0	1	0	0	1	0	1	0	1	1	0	0	0	0	0	1	0	1	1	0	0	1	0.50	34	0.40	2.0–4.0%
< 2.0%	0	0	0	0	0	0	0	0	0	0	0	0	0	0	0	0	1	0	0	0	0	0	0	1	0	0.30	6	0.07	< 2.0%
7. Federal Deficits																													
> $250 B	0	0	0	0	0	0	0	0	0	0	0	0	0	0	0	0	0	0	0	0	0	0	0	0	1	0.35	7	0.08	> $250 B
$150–250	0	0	1	0	0	0	0	1	1	1	0	1	1	0	0	0	1	0	1	1	0	0	0	1	0	0.45	32	0.37	$150–250
< $150 B	1	1	0	1	1	1	1	0	0	0	1	0	0	1	1	1	0	1	0	0	1	1	1	0	0	0.20	47	0.55	< $150 B

FIGURE 4-6
Computer Printout of Output Resulting from Scenario Analysis (continued)

8.																													
Marley Index Shf																													
reversal	1	1	0	1	1	1	1	0	1	0	1	0	0	1	1	1	0	1	0	1	1	1	1	0	0	0.30	55	0.64	reversal
no chang	0	0	1	0	0	0	0	1	0	1	0	1	1	0	0	0	0	0	1	0	0	0	0	1	1	0.40	27	0.31	no chang
further	0	0	0	0	0	0	0	0	0	0	0	0	0	0	0	0	1	0	0	0	0	0	0	0	0	0.30	4	0.05	further
9.																													
Appl Effic Stnds																													
More	0	0	0	0	0	0	1	0	0	0	0	0	0	1	0	0	0	0	0	1	0	0	0	0	0	0.50	9	0.10	More
Same	1	1	1	1	1	1	0	1	1	1	1	1	1	0	1	1	1	1	1	0	1	1	1	1	1	0.40	76	0.88	Same
Less	0	0	0	0	0	0	0	0	0	0	0	0	0	0	0	0	0	0	0	0	0	0	0	0	0	0.10	1	0.01	Less
10.																													
Price Nat Gas																													
>3.0%/yr	1	0	0	0	1	1	0	0	1	1	0	0	0	0	0	0	0	0	0	1	0	0	0	0	0	0.10	31	0.36	>3.0%/yr
1.0-3.0%	0	1	0	1	0	0	1	0	0	0	0	0	0	1	0	1	0	1	0	0	0	0	1	0	0	0.40	27	0.31	1.0-3.0%
< 1.0%	0	0	1	0	0	0	0	1	0	0	1	1	1	0	1	0	1	0	1	0	1	1	0	1	1	0.50	28	0.33	< 1.0%
11.																													
Pr Crude Oil																													
>3.0%/yr	1	0	0	0	1	1	0	0	1	1	0	0	0	1	0	0	0	1	0	1	0	0	1	0	0	0.35	38	0.44	>3.0%/yr
0.0-3.0%	0	1	0	1	0	0	1	0	0	0	0	0	0	0	0	1	0	0	0	0	0	0	0	0	0	0.45	22	0.26	0.0-3.0%
< 0.0%	0	0	1	0	0	0	0	1	0	0	1	1	1	0	1	0	1	0	1	0	1	1	0	1	1	0.20	26	0.30	< 0.0%
12.																													
Pr Indus Elec																													
>2.0%/yr	1	0	0	0	1	1	0	0	1	0	0	0	0	0	0	0	0	0	0	1	0	0	0	0	0	0.15	28	0.33	>2.0%/yr
0.0-2.0%	0	1	1	1	0	0	1	1	0	1	1	1	1	1	1	1	1	1	1	0	1	1	1	0	1	0.50	55	0.64	0.0-2.0%
<0.0%/yr	0	0	0	0	0	0	0	0	0	0	0	0	0	0	0	0	0	0	0	0	0	0	0	1	0	0.35	3	0.03	<0.0%/yr
13.																													
Lvl Cogeneration																													
Grt Incr	1	1	0	1	1	1	1	0	1	0	1	0	0	1	1	0	0	1	0	1	0	1	1	0	0	0.40	52	0.60	Grt Incr
Slt Incr	0	0	1	0	0	0	0	1	0	1	0	1	1	0	0	1	1	0	1	0	1	0	0	1	1	0.50	31	0.36	Slt Incr
Same/Dwn	0	0	0	0	0	0	0	0	0	0	0	0	0	0	0	0	0	0	0	0	0	0	0	0	0	0.10	3	0.03	Same/Dwn
14.																													
Volume Electrici																													
>7.0%/yr	1	1	0	0	0	0	0	0	0	0	0	0	0	1	0	0	0	0	0	1	0	0	1	0	0	0.05	21	0.24	>7.0%/yr
3.0-7.0%	0	0	0	1	1	1	1	0	1	1	1	0	0	0	1	1	0	1	0	0	1	1	0	0	0	0.15	36	0.42	3.0-7.0%
0.0-3.0%	0	0	1	0	0	0	0	0	0	0	0	1	0	0	0	0	0	0	1	0	0	0	0	1	0	0.50	18	0.21	0.0-3.0%
Some Dec	0	0	0	0	0	0	0	1	0	0	0	0	1	0	0	0	1	0	0	0	0	0	0	0	1	0.30	11	0.13	Some Dec

in energy prices with increases in electricity prices being slightly lower than those for gas or oil, increased cogeneration, and an increase in the volume of electricity consumed by the industrial sector (including cogeneration) at an annual rate of 3 to 7 percent.

5

INDUSTRIAL ELECTRICITY CONSUMPTION AND CHANGING ECONOMIC CONDITIONS

Robert C. Marlay
U.S. Department of Energy

INTRODUCTION

Increasingly, a better understanding of the U.S. economy and its energy-intensive industrial sectors is becoming a prerequisite for improved forecasting of energy demand (**1–3**). While recent experiences may lend some confidence to our ability to anticipate the energy needs of the residential (**4,5**), commercial (**6**), and transportation (**7**) sectors, there is a growing consensus that energy demand in the industrial sector is undergoing fundamental change (**8–10**).

Current patterns of industrial energy use are dramatically different from past trends (**11**). It is no longer possible, for example, to assume a general rate of growth for a convenient macroeconomic aggregate, such as gross national product or the Federal Reserve Board's index of industrial production, apply correction factors to account for historical relationships and improving technology, and derive a reasonable forecast. Industry's requirements for energy use are now seen as volatile and uncertain (**12**).

Underlying these developments, the industrial sector has begun to behave economically like the heterogeneous composite that it really is. The prognosis for each element is shaped by an array of forces whose scope, in certain circumstances, may include governmental policies and competitive prices in some of the world's least significant economies (**13–20**). Moreover, the principal mechanisms by which changes in industrial energy demand are being brought about have proven to be remarkably facile, changing easily and quickly over time, and are reversible.

As a manifestation of this trend toward, for lack of a better word, increasing dynamism of the U.S. economy and its relationship with the rest of the world, the 15 years between 1970 and 1985 witnessed a number of profound changes. In particular, certain segments of heavy industry, identified with intensive use of energy, labor, and raw materials, suffered an accelerating decline in their relative economic importance. This decline was also reflected in industry's aggregate demand for heat and power.

In 1984, for example, a year of robust economic activity, utility sales of electricity to industry totaled 850 billion kilowatt hours. This amount was nearly one-third less than what historical trends dating from the 1960s and early 1970s would have otherwise predicted. More significantly, it was considerably less than what most forecasters and energy planners had expected.

The resulting shortfall, combined with staggering economic losses for certain investors who had relied too heavily upon such expectations, led to a heightened awareness of forecasting uncertainty. It also challenged our basic understanding of energy demand. One consequence of increased uncertainty is that risk-averse individuals will likely invest less in future energy supply facilities than what society at large would prefer. This, in turn, presents risks to society of the kind of which we are all aware.

A better understanding of energy demand in general, and

of demand in the least understood industrial sector in particular, would help to reduce both forecasting uncertainty and its associated risks. If successful, the information upon which long-term energy planning decisions are based would improve, and investor confidence in the legitimate need for certain new energy supply facilities might be restored.

In this paper, I examine the changing nature of energy demand in U.S. industry, with a special focus on electricity. I portray a number of mechanisms of change, characterize their respective roles in reducing growth in industrial energy demand, and suggest a number of possible links between these mechanisms and more fundamental and underlying causes. Finally, I present evidence on the changing nature of the industrial sector of the U.S. economy, and suggest that the extent of these changes can only be explained within a broad context of causal factors that extends beyond the parameter of energy and its price.

HISTORICAL PERSPECTIVE

Between the end of World War II and 1973, industrial use of all forms of energy grew steadily at rates averaging between 2.7 and 3.6 percent per year, slightly less than the growth rate for industrial economic output in general (**21**). For the purposes of context and perspective, this growth and certain well-documented projections of future energy demand made in the 1970s are illustrated in Figure 5-1 (**22,23**).

In direct contrast to this historical pattern of growth, industry's use of energy after 1973 declined. By 1982, it was at its lowest level since 1967. It rebounded somewhat with an expanding economy in 1984, but was still less than that of 15 years earlier. Yet, throughout this post-1973 period, industrial economic activity continued to expand at an average rate of about 2.5 percent per year (**21**).

Exhibiting a similar pattern, historical trends in industrial use of electricity, including both purchased and self-generated power, increased annually through 1973 at rates far outpacing industrial economic growth. In the 1950s, electricity growth exceeded 10 percent per year, and in the 1960s well over 5 percent per year (**21**). Figure 5-2 presents data for the period 1947 through 1984, separately for

FIGURE 5-1
Historical Trends in Industrial Energy

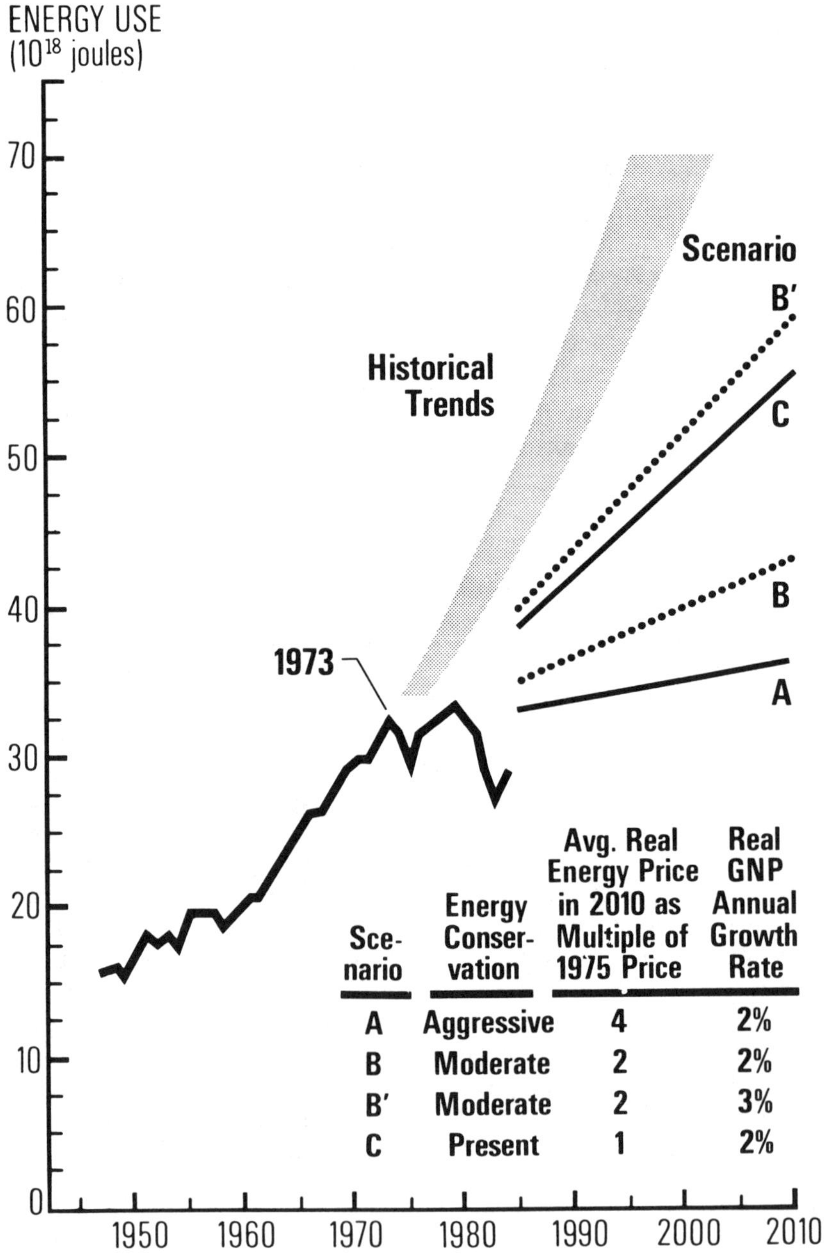

Scenario	Energy Conservation	Avg. Real Energy Price in 2010 as Multiple of 1975 Price	Real GNP Annual Growth Rate
A	Aggressive	4	2%
B	Moderate	2	2%
B′	Moderate	2	3%
C	Present	1	2%

FIGURE 5-2
Industrial Use of Purchased and Self-Generated Power, 1947-1984

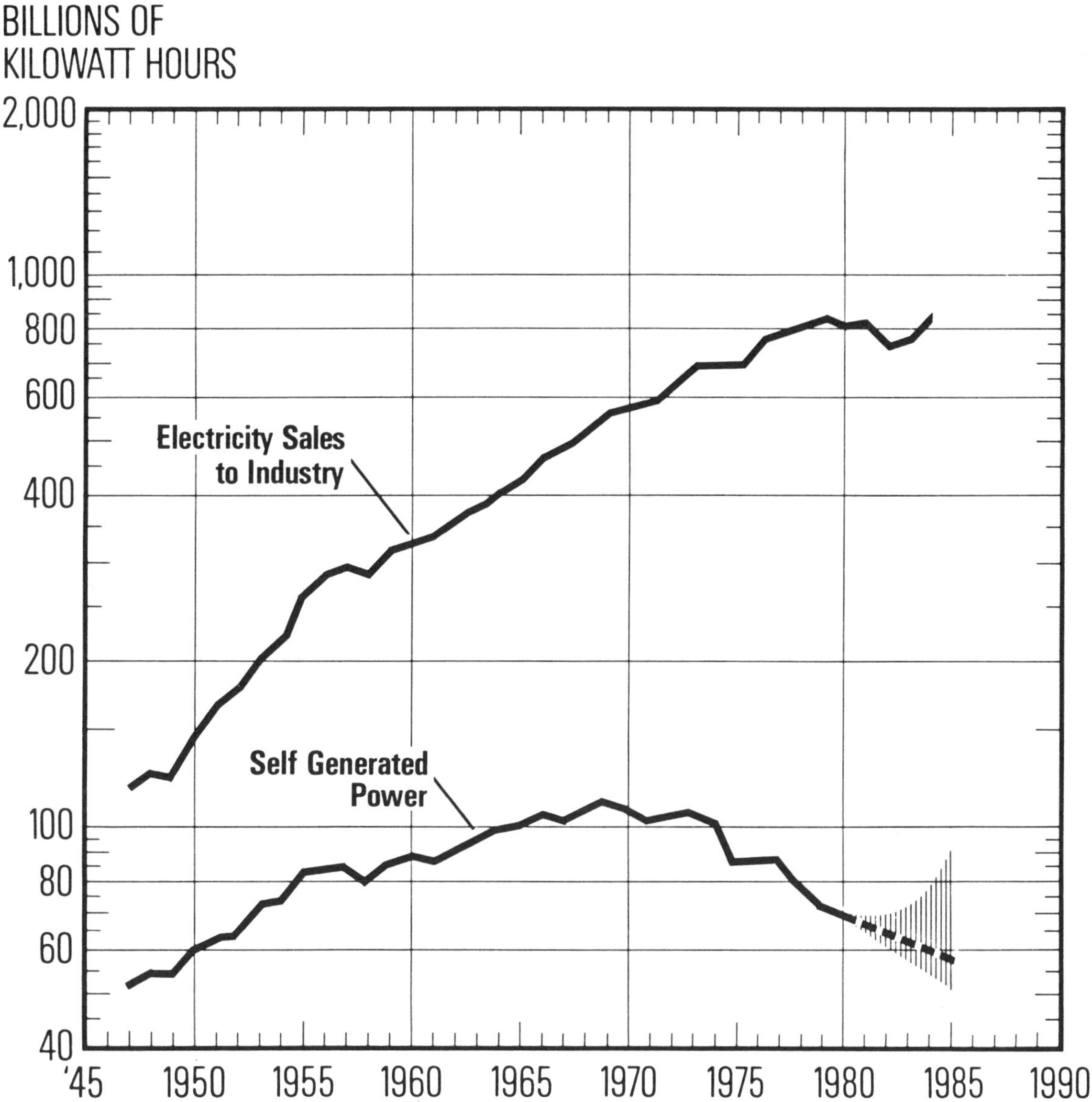

purchased and self-generated power, and Figure 5-3 presents data over the same period for the total of the two combined. From Figure 5-3 it can be seen that industrial use of electricity peaked in 1979 at 915 billion kilowatt hours, and declined thereafter, and that industrial demand for electricity actually began to exhibit a fundamentally different pattern of growth as far back as 1973.

The extent of this change and its cumulative impact on the electric power industry may be visualized from Figure 5-4, where pre-1973 trends, assumed to be 5 percent per year, are extended through 1984 and contrasted to actual use. The difference between actual use and this extended trend line, called the base case, equates roughly to the forgoing consumption in 1984 of power that otherwise would have been required from about eighty 1000-megawatt power plants operating at a typical capacity factor of 60 percent. At $3,000 per kilowatt of capacity, this change may have affected as much as $240 billion in plant investments, as valued in current terms.

Clearly, the period from 1973 to 1984 witnessed significant changes in the historical patterns of growth in industrial energy and electricity demand. But what were the specific mechanisms contributing to these developments? Beyond these mechanisms, what were the more fundamental and underlying causal factors? Are they permanent or only of a passing nature? Important for analysis, what methodologies may be applied quantitatively to measure and distinguish among the relative effects of the contributing parts to the whole?

CONCEPTUAL APPROACH AND RESEARCH METHODOLOGY

In developing a quantitative approach to these questions, the amount of energy consumed by industry is portrayed as having three sources of variation. In the manufacture of any one kind of product, energy consumption varies with both the level of production, or output, and the energy efficiencies of the technologies of production.

In the more general treatment of industry as a whole, however, there is an added complexity. This stems from the fact that there are more than 10,000 different classes of products produced by industry and potentially many

FIGURE 5-3
Industrial Use of Electricity from All Sources, 1947-1984

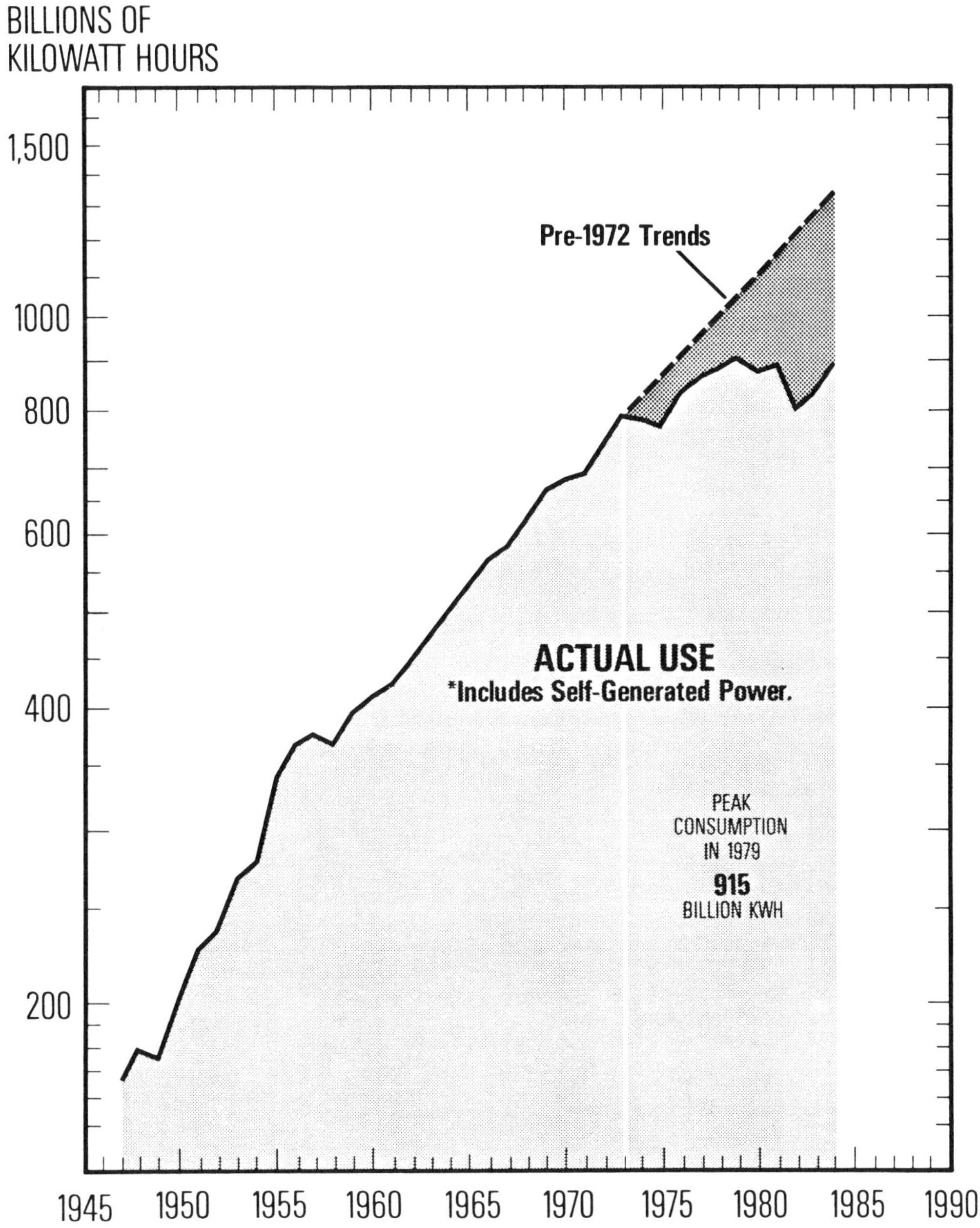

FIGURE 5-4
Comparison of Actual Electric Power Use and Pre-1973 Trend Base Case

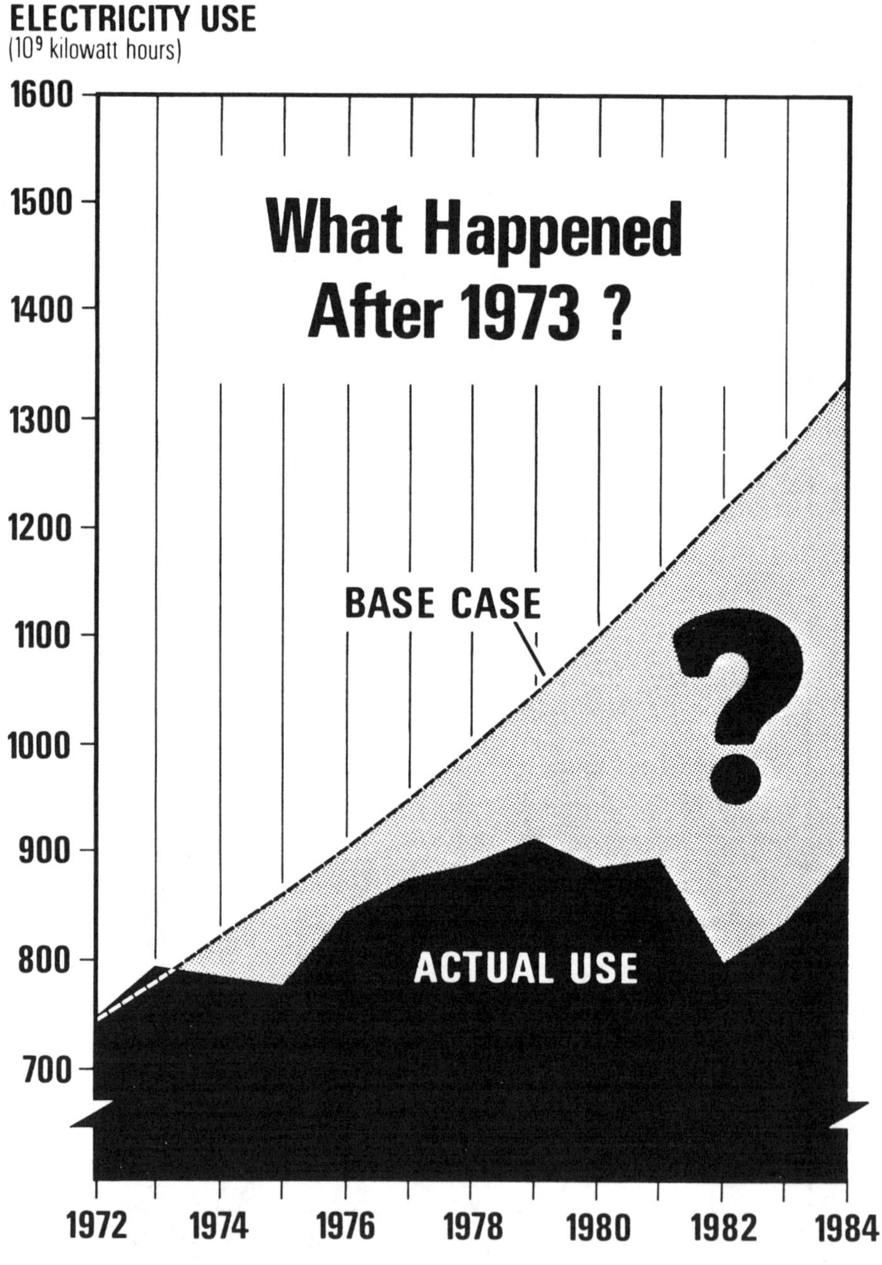

variations of products within each class. Some products differ from others by as much as 100-fold in their intensity of energy use per unit of economic value.

Further, the mix of these products can change quickly. In response to many different factors, new products enter the marketplace as others drop out. Domestic products may be displaced by imports. Hence, when industry is viewed more generally as an economic aggregate, as is often the treatment in energy and economic models, the amount of energy consumed by industry depends additionally on the precise nature and composition of that which is produced.

In the research discussed below (**21**), the changing patterns of industrial production, energy use, and energy productivity were examined for 472 industries in the mining and manufacturing sectors of the U.S. economy over a 38-year historical period from 1947 through 1984. Data on industrial production (economic output) were developed for each industry from the quinquennial indexes of industrial production, published by the Bureau of the Census (**24**), and from monthly, quarterly, and annual indexes of industrial production, published by the Federal Reserve Board (**25**). Data were also developed for each industry on the costs and quantities of 21 forms of energy use and on a variety of economic statistics, including value of shipments, value added, value of year-end inventories, and the costs of capital, labor, and materials. Each industry was thus characterized as to its production history and its use of energy and economic resources, namely capital, labor, energy, and materials.

From the underlying data on output, aggregate measures, or so-called indexes of industrial production, were constructed using selected energy and economic parameters as weights.[1] Each measure was constructed to reveal a different aspect of industrial production, either intensive (relative measure) or consumptive (absolute measure) in its use of particular resources.

An electricity-weighted aggregate measure of industrial production, for example, gives proportionally more weight

---

[1]A factor-weighted, aggregate measure of industrial production is given by:

to those elements of industrial production that consume large amounts of electricity. A value-added weighted measure, by contrast, weights each industry by its relative economic importance. Such a measure is similar to that used by the Federal Reserve Board to monitor industrial economic activity, and was constructed for use as a standard reference in making comparisons.

Once the output measures were constructed, weighted by the various forms of energy, it was a straightforward matter to construct indexes of industrial energy productivity.[2] This was accomplished by dividing the fuel-specific, energy-

---

fn. 1, cont'd.

$$w_j^I(t) = \frac{\sum_{i=1}^{I} \frac{y_i(t)}{y_i(t_o)} \cdot x_{ij}(t_o)}{\sum_{i=1}^{I} x_{ij}(t_o)} \qquad (1)$$

where I specifies the set of industries to be included in the aggregation, j defines the energy or economic parameter, or factor, to be used as the weight, $y_i(t)$ measures production (output) of industry i, belonging to the set of industries I, in time period t; $x_{ij}(t)$ is a quantity of energy or economic value of factor j, for industry i, in time period t; and t_o is a reference time period.

[2]Industrial energy productivity, with respect to the use of a specific form of energy j, in time period t, for a given set of industries I, is a dimensionless ratio of j energy-weighted output divided by j energy input, and is given by:

$$p_j^I(t) = \frac{\sum_{i=1}^{I} \frac{y_i(t)}{y_i(t_o)} \cdot x_{ij}(t_o)}{\sum_{i=1}^{I} x_{ij}(t)} \qquad (2)$$

weighted aggregate measures of industrial production by their complementary measures of energy input. By defining energy productivity in this way, the effects on industrial energy demand of improvements in the efficiencies of process technologies could be measured independently of the effects, if any, of the changing level and composition of industrial output.

This distinction is important because it gives rise to different implications as to the underlying causes of changing energy demand. For example, the aluminum industry, at full production, accounts for roughly 10 percent of total industrial electricity use. In terms of its relative economic importance, however, it contributes less than one-half of 1 percent to total industrial value added.

Suppose, in the extreme, that domestic output of the aluminum industry fell from full production to zero. As a result, the amount of electricity used by the aluminum industry would also fall to zero. Total industrial demand for electricity, because of aluminum's large share, would fall by 10 percent. Industrial economic activity, however, as measured by value added, would fall by something on the order of only one-half of 1 percent.

Certain aggregate measures of industrial energy efficiency, such as energy use per constant dollar of industrial value added, would "improve" as a result, in this case by

---

fn. 2, cont'd.

The parameters here are the same as those of Equation 1 in footnote 1, with the exception that the denominator is variable with respect to time. As applied, a substitution is made in the denominator, using aggregate time series data on industrial energy consumption,

$x_j^{I'}(t)$, where

$$x_j^{I'}(t) = \sum_{i=1}^{I} x_{ij}(t) \qquad (3)$$

and I' is of a known specification of I industries.

about 10 percent. Such improvements could easily be interpreted as promising signs of industry's improving energy efficiency. While this may be true in one sense, no fundamental improvements in the technologies of production were made at all. The apparent gains in efficiency observed at the aggregate level would be fully explained in this case by a single change in the composition of industrial production. A large electricity-consumptive element dropped out, taking with it proportionally 20 times more energy than economic value.

Confusion concerning how much of reduced energy demand is due to improvements in process efficiency and how much is due to underlying changes in the composition of output may be avoided by using a different specified measure of energy productivity. For example, an electricity-weighted aggregate measure of industrial production, by virtue of the aluminum industry's large weighting factor, would drop by the same amount as electricity demand, that is, by 10 percent. In this construction, a ratio of electricity-weighted output over electricity input would show that overall industrial process efficiency had remained unchanged, and that the drop in electricity demand was attributable solely to compositional changes in the aggregate nature of industrial production.

ENERGY EFFECTS OF COMPOSITIONAL CHANGE

Applying these measurement concepts and methodologies to the assembled data, the energy effects of compositional change can be quantified, as follows. Figure 5-5 presents the "standard" value-added weighted index of industrial production for the years 1947 through 1984. Its construction uses what the Bureau of the Census calls quantity-based unit relatives, or annual indexes of industrial production, for each of 472 industries in mining and manufacturing, weighted in their aggregation to the whole by each industry's relative contribution to total industrial value added.

The standard index of Figure 5-5, a good measure of overall industrial economic activity, shows industrial production since 1973 growing at smoothed rates of about 2.5 percent per year. The chart inset at the bottom of

FIGURE 5-5
Standard Index of Industrial Production*

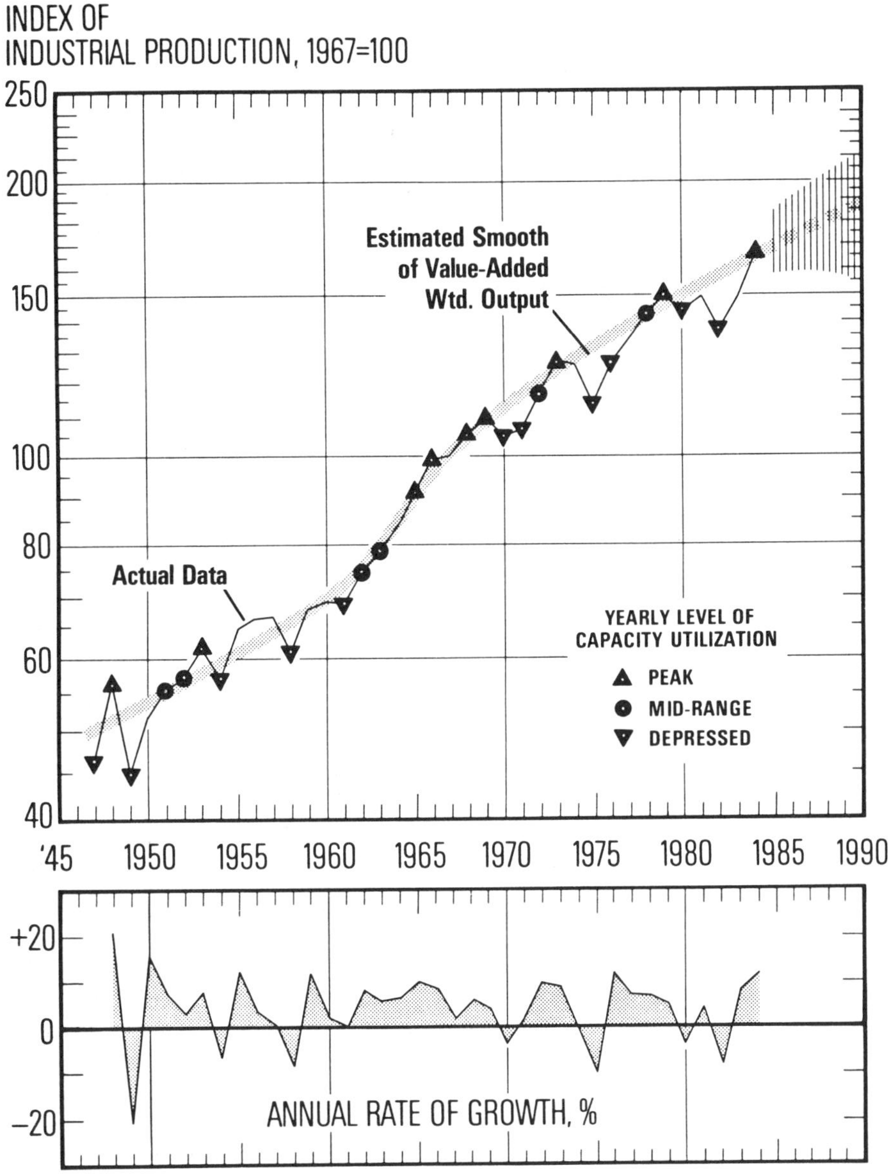

* **Value-Added Weighted Measure of Output for Mining and Manufacturing, 472 Industries.**

Figure 5-5 graphically displays the annual rate of growth for each year. Growth in industrial production in 1984, for example, is shown to exceed 10 percent.

Figure 5-6 presents what others have named the "Marlay Index" of industrial production, wherein the underlying data on industrial output, identical to those used above for the standard index, are aggregated using energy weights. In this discussion these are restricted to electricity. By comparing the standard index to the Marlay Index, precise information may be revealed, year to year, about the positive and negative energy effects of compositional change, or so-called shifts in output mix.

The industries that dominate the electricity-weighted index are those that consume large amounts of electricity. Some of these industries are energy intensive, such as aluminum, steel, chemicals, cement, and others. Some are not energy intensive, but are energy consumptive, by virtue of their large size. That is, they may have relatively low-cost shares (of total production costs) for energy, such as textiles, automobile manufacturing, and food processing, but in the aggregate consume large amounts of electricity because they are large industries.

In comparing the two indexes of Figures 5-5 and 5-6, one can observe distinct and dramatic differences, particularly in the post-1973 period. Focusing first on the electricity-weighted measure, its growth path is strongly positive during the 1960s, but slowed considerably over the 1973-1984 period. What is most significant, however, is its gradual and steady divergence from the standard, value-added weighted index. A quantitative measure of this divergence, called the "shift factor," is displayed graphically in the chart shown at the bottom of Figure 5-6. It is calculated as an arithmetic difference in the annual growth rates of the two indexes. The shift factor indicates for each year whether the composition of industrial output moved toward (+) or away from (-) electricity consumptive production, and by how much.

The plot of the shift factors shown at the bottom of Figure 5-6 reveals that the divergence in any given year between industrial economic activity in general, and that of electricity consumptive production in particular, can exceed ±5 percent. Over the 30 years shown on Figure 5-6, such divergences occurred with regularity. In years past, it

FIGURE 5-6
Marlay Index of Indutrial Production*

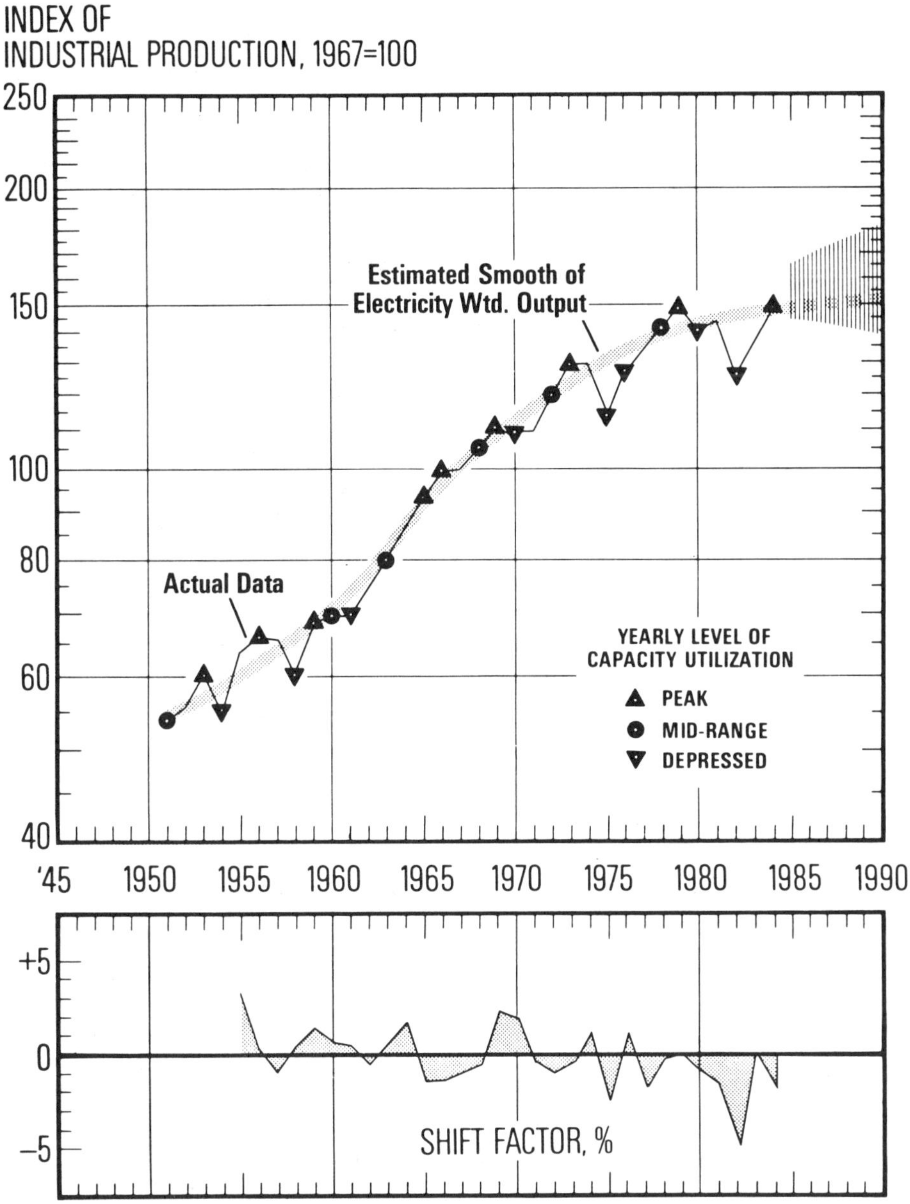

* Electricity Weighted Measure of Output for Mining and Manufacturing, 472 Industries

appears, the economy first moved away from electricity-consumptive production, and then back again, in periods ranging from three to six years.

In the period prior to 1971, electricity-consumptive production grew on the average slightly faster than production in general, by about +0.3 percent per year. This would imply, other things being equal, that industrial electricity demand would have tended to grow slightly faster than industrial production in general, due to compositional effects alone.

After 1971, however, a different pattern emerged. On the average, electricity-consumptive production fell behind industrial production in general by an unprecedented -1.1 percent per year. Such a large average divergence, combined with the fact that it was apparently sustained and increasing for more than ten years, suggests that fundamentally different patterns of industrial production began to manifest themselves in the early 1970s. Clearly, electricity-consumptive production, as represented by the Marlay Index, declined by a substantial amount in relative economic importance compared to that of the rest of mining and manufacturing, as represented by the standard index.

The effect of this decline on the growth rate in industrial electricity demand was measurable and significant. The change in compositional effect from +0.3 percent per year prior to 1971 to -1.1 percent per year after 1971 resulted in a net slowing of industrial electricity demand attributable to this effect by -1.4 percent per year. Observing the size and direction of the shift factors shown in Figure 5-6, this trend appeared to be accelerating in the negative direction in the early 1980s.

A better understanding of this trend may be gained by examining, in detail, specific data for the underlying industries. Table 5-1 lists several industries which are large electricity users. These industries collectively account for 60 percent of total industrial electricity use. These same industries account for about one-fifth of industrial value added. Their changing patterns of growth can be seen from the production statistics shown on Table 5-1.

In the analysis of these statistics on Table 5-1, the year of the Arab oil embargo (1973) is chosen somewhat arbitrarily to separate two distinct periods of time. The comparison of the pre- and post-1973 trends shows that production

TABLE 5-1
Industrial Electricity Consumption and Changing Growth Patterns (1954–1984)

SIC Code	Industry	Cumulative % of Total: Industrial Elec. Consumption	Cumulative % of Total: Value Added	Average Annual Growth Rate: Jan 54 Oct 73 %	Average Annual Growth Rate: Oct 73 Dec 84 %	Average Annual Growth Rate: Apparent Change %
--	All Mining and Manufacturing	100	100	--	--	--
3334	Primary Aluminum	10	--	5.9	-1.1	-7.0
2819	Industrial Inorganic Chemicals	19	1	4.3	-0.3	-4.6
3312	Blast Furnaces and Steel Mills	26	4	1.8	-5.1	-6.9
2818	Industrial Organic Chemicals	31	5	9.7	2.9	-6.8
2621	Papermills, Ex. Building Paper	35	6	3.8	2.3	-1.5
2911	Petroleum Refining	39	8	3.6	0.1	-3.5
2631	Paperboard Mills	42	8	5.6	0.8	-4.8
2812	Alkalies and Chlorine	44	9	3.9	-1.3	-5.2
3313	Electrometallurgical Products	47	9	2.5	-2.3	-4.8
1311	Crude Petroleum & Natural Gas	48	12	2.6	-1.4	-4.0
3241	Cement, Hydraulic	50	13	2.4	-0.7	-3.1
2221	Weaving Mills, Synthetics	51	13	8.0	3.7	-4.3
2813	Industrial Gases	52	14	10.3	4.2	-6.1
3714	Motor Vehicle Parts & Access.	54	16	3.5	0.3	-3.2
3711	Motor Vehicles	55	18	4.4	-0.7	-5.1
1211	Bituminous Coal	56	19	1.6	3.3	1.7
1011	Iron Ores	57	19	3.8	-4.4	-8.2
3079	Miscellaneous Plastic Products	58	20	13.7	3.7	-10.0
2821	Plastics Materials & Resins	58	21	11.2	6.9	-4.3
3352	Aluminum Rolling & Drawing	59	21	7.3	1.9	-5.4
2421	Sawmills and Planing Mills	60	21	0.4	-2.2	-2.6
2824	Organic Fibers, Noncellulosic	60	22	16.6	3.5	-13.1

slowed significantly for almost all the industries shown. In the pre-1973 period, virtually all industries had strong positive growth rates. In the following ten-year period, however, only one-third had positive growth rates. All but one, bituminous coal, slowed in their growth rates from those of prior trends, as is apparent from the list of negative numbers in the far right column of Table 5-1.

This analysis shows that shifts in the composition of industrial output away from electricity-consumptive elements were not only measurable and significant, but broadly based. Hence, in addition to the general post-1973 slowing of economic growth, which was experienced by industry at large, there were unprecedented compositional changes in both energy-intensive and energy-consumptive industries, which resulted in the marked negative effect on the growth rate in industrial electricity demand.

EFFECTS OF IMPROVED ENERGY PRODUCTIVITY

Intuitively, it would seem that the higher energy prices of the post-1973 period would motivate significant improvements in the energy efficiencies of the technologies of production, or more precisely, improved energy productivity. Hence, in attempting to explain the reduced growth rate in electricity demand after 1973, it is prudent to ask also how much was due to this price-oriented market behavior.

The term "energy productivity," as it is used in this paper, is a formal construct, described earlier and shown mathematically in Figure 5-7. Essentially, it is a ratio of electricity-weighted output over electricity input, where electricity input is defined to include both utility sales of electricity to industry and industrially generated power. It measures the **net** extent to which other inputs to production may have been substituted for electricity, and vice versa. Other inputs are typically characterized as capital, labor, nonelectric forms of energy, and materials. By the nature of its construction, the measure is independent of the effects, if any, of the changing level and composition of industrial production, down to the level of sector disaggregation represented by 472 industries.

$$p_j^I(t) = \frac{\sum_{i=1}^{I} \frac{y_i(t)}{y_i(t_o)} \cdot x_{ij}(t_o)}{\sum_{i=1}^{I} x_{ij}(t)}$$

An Energy (Elec.) Weighted, Aggregate Measure of Industrial Production or OUTPUT

DIVIDED BY:

A Comparable Measure of ENERGY INPUT

YIELDS:

A Measure of Industrial PROCESS EFFICIENCY

ABSENT:

The Energy Effects of SHIFTS IN OUTPUT MIX

FIGURE 5-7 Mathematical Formulation of Industrial Energy Productivity

Using the assembled data, numeric values were developed and the results for the period 1949 through 1984 are presented graphically in Figure 5-8. The dominant feature of Figure 5-8 is its sharp negative slope prior to 1960, followed by a gradual flattening out over the next 20 years. A negative slope means that increasing amounts of electricity were used per unit of output, reflecting industry's strong preference for this form of energy as an input to production. A negative slope does not necessarily mean that no technological efficiency gains were made. It does indicate, however, that industry preferred to substitute increasing amounts of electricity for other inputs to production, and that these increases more than offset, in the aggregate, any decreases that might have occurred from reverse substitution and other forms of technical change.

While there is considerable irregularity in the productivity data shown on Figure 5-8 from 1972 through 1984, the trend over this period appears to be almost flat, particularly if one ignores the deep recession years of 1975 and 1982. A flat productivity curve implies a constant value over time. If this condition were to prevail in the future, growth in industrial electricity demand would not be affected by net changes in technology and factor substitution, but would instead take the identical form of that of the index of

FIGURE 5-8
Industrial Electric Energy Productivity (1949-1984)*

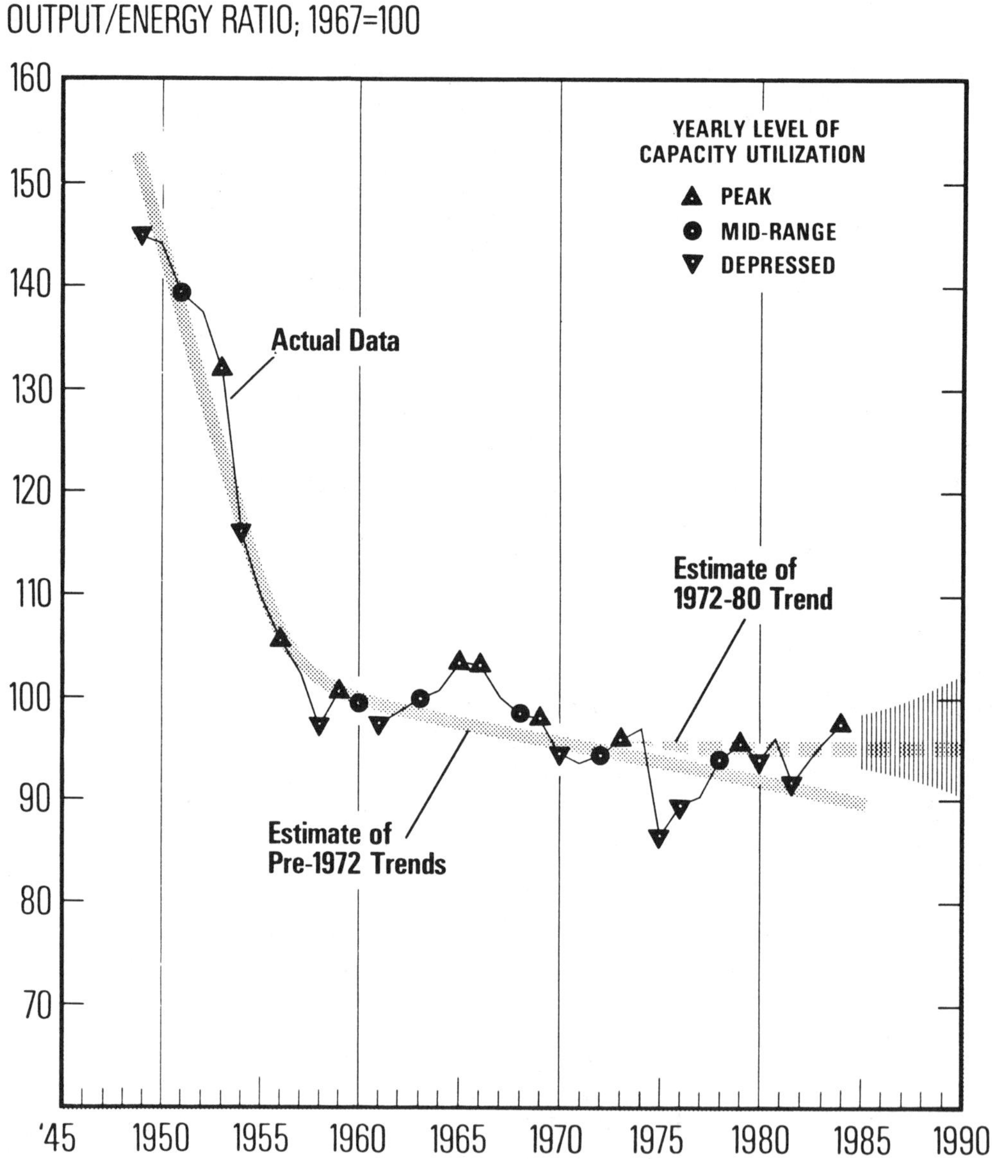

* Electricity Weighted Measure of Output Divided by Electricity Input, 472 Industries in Mining and Manufacturing.

electricity-consumptive production (i.e., the Marlay Index). In 1984, the Marlay Index, shown in Figure 5-6, was exhibiting a smoothed and slowing growth rate of 0.3 percent per year.

Finally, the slightly positive divergence of this curve from the negative slope of prior trends in the years 1972 through 1984 means that improvements in energy productivity over this period, to the extent that they occurred, were greater than those that would have been expected had past trends continued. Hence, compared to past trends, there was a small negative effect on the growth rate in electricity demand due to these developments. The net effect is estimated to be on the order of -0.7 percent per year.

SUMMARY OF RESULTS

Returning to the question of what happened to industrial electricity demand after 1973, represented graphically by Figure 5-4, specific year-to-year quantitative information can now be developed within an internally consistent framework that fully explains the growing divergence between the projections of past trends and actual use. This information is presented for electricity in Figure 5-9 and, for the purposes of completeness and comparison, for electricity and combustible fuels in Figure 5-10.

In general, a slowdown in overall economic growth, compared to that experienced in the 1960s and early 1970s, accounted for a large portion of the decline in the growth rate of industrial energy demand in total, and of electricity in particular. Apart from slower growth, compositional changes in the mix of industrial output had additional and significant energy effects—depressing growth rates in demand for both fuels and electricity. Improved energy productivity, by contrast, had a major effect—depressing demand for combustible fuels—but surprisingly only a minor effect on the demand for electricity.

Summarizing these conclusions in quantitative terms for electricity, slower growth in industrial electricity demand after 1973 was found to be the result of three distinct mechanisms:

1. Slower economic growth (which had a depressing effect on demand of about -1.7 percent per year);

2. Changes in the composition of industrial output away from the electricity-consumptive elements of production (-1.4 percent per year); and
3. A modest improvement in net energy productivity relative to historically declining trends (-0.7 percent per year).

The magnitude of the energy effects over time (1973-1984) for each mechanism are illustrated in bar charts shown at the top of Figures 5-9 and 5-10.

INDUSTRIALLY GENERATED POWER

Figure 5-11 illustrates the past and more recent trends in the role played by industrially (or "self-") generated power. In 1970, industry produced about 16 percent of its total requirement for electricity from its own generating plants. Although some of this power was from hydroelectric sources, 97 percent was derived from fossil fuels burned in conventional steam boilers. In this discussion, self-generated power should not be confused with cogenerated power. In industry, the latter is a small but burgeoning component of the former, and involves the simultaneous generation of electricity and use of the waste heat from combustion.

By 1980, only 10 years later, industrial use of self-generated power had fallen by nearly 50 percent. Throughout the 1970s, there was an apparent shutdown of 10 or more gigawatts of industrial generating capacity, roughly equivalent to 10 of today's largest electric generating stations. The power not self-generated by industry was replaced by purchased power from the central electric utilities, boosting utility sales in this sector by about 50 billion kilowatt hours per year, or by about 0.8 percent per year in 1980.

Hence, the declining status of industrially generated power was found to have a significant positive effect on the apparent growth rate in utility sales to industry over the period 1970-1980. The otherwise slowing growth rate in total industrial electricity demand during this period was partially masked in the case of utility sales by a positive 0.8 percent per year increase attributed to industry's sharply reduced use of self-generated power and its simultaneous

FIGURE 5-9
Changing Trends in Industry's Use of Electricity

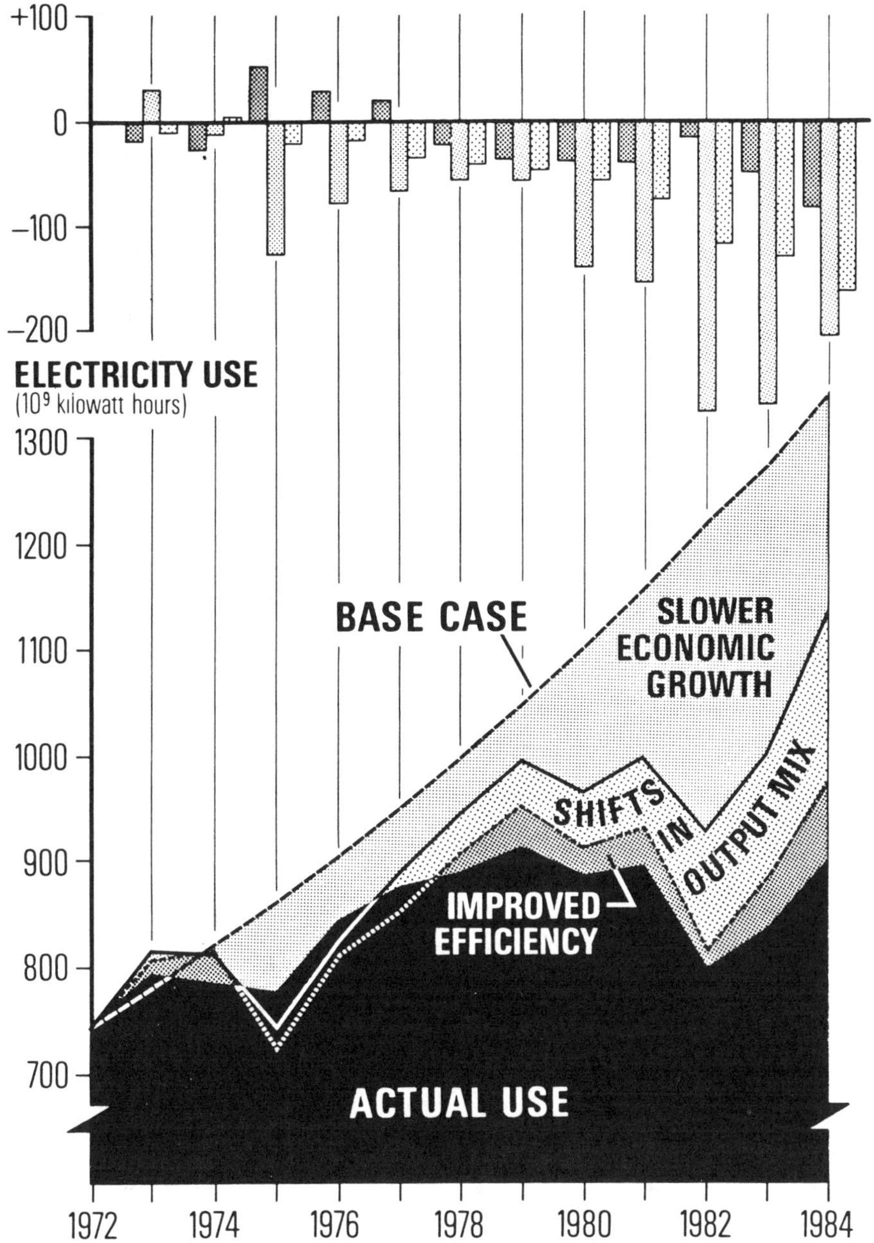

FIGURE 5-10
Changing Trends in Industry's Use of Electric Power and Fuels

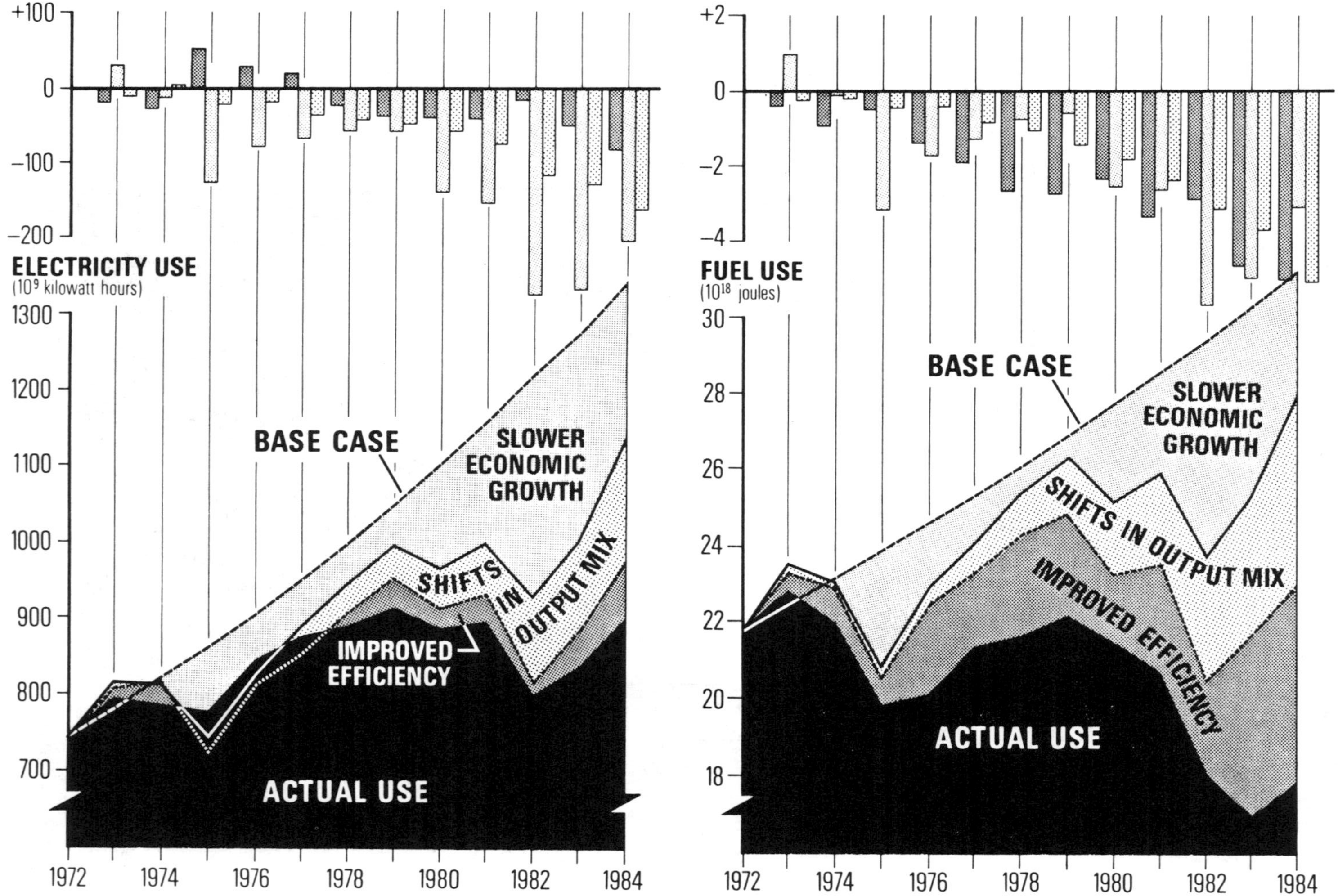

replacement with additional purchased power from the central utilities.

Because of changing conditions, however, an opposite trend may emerge over the next decade. The real price of oil, once a major fuel of self-generators, may decline. Curtailments of natural gas, which had earlier shut down self-generators, ended in the early 1980s, and this fuel may become more available to industrial users. The price of replacement power purchased from the utilities, which had earlier been a bargain at the regulated prices of the 1970s compared to the four-fold increase in the price of oil, may become relatively more costly compared to other fuels. These developments, should they prevail, combined with a continued preferential tax and regulatory treatment of cogeneration, may result in a revival of self-generated power of both the more traditional and cogenerated forms.

If this scenario were to become a reality, it would result in what I call a "double whammy" effect on utility sales to industry. First, the positive effect on utility sales caused by industry's abandonment of self-generated power in the 1970s would vanish. This would manifest itself as an immediate loss of 0.8 percent growth. Second, future growth in industrial demand for electricity, which ordinarily would have been met by utility sales, and is likely to have been included in baseline forecasts, would not materialize. This would manifest itself as an additional loss of growth in industrial sales of about 0.8 percent per year. Hence, the changing status of industrially generated power, found to be a significant factor affecting utility sales to industry in the past, may be even more so in the future, adding an additional element of uncertainty that must be addressed in future planning and predictive work.

UNDERLYING CAUSAL FACTORS

From the above discussions, we have seen that industrial demand for electricity purchased from central utilities was found to be measurably affected by a number of phenomena, two of which are not well studied—the changing composition of industrial production and the changing status of self-generated power. But in fact, these phenomena are simply manifestations of more fundamental and underlying causal factors. Among these is the dramatically rising price of energy, which certainly played an important role, but as I hope to show below, not an exclusive one.

Concerning energy prices, Figure 5-12 presents data on the real prices manufacturers paid for purchased fuels and power from 1974 through 1984. In general, real energy prices declined from 1947 through 1969. Beginning in 1970, however, prices of oil and coal began to rise steeply, followed in 1974 by extraordinary additional increases brought on by the Arab oil embargo. Real prices of regulated forms of energy, electricity, and natural gas lagged beyhind those of oil and coal by a number of years, but generally followed the pattern set by oil. Except for coal, prices of all forms of energy peaked in 1981 or 1982 at levels between two and four times those of 1969, in real terms. Meanwhile, real prices of other inputs to production (not shown), namely capital and nonenergy materials, did not rise nearly so steeply and, in the case of labor, declined in real terms.

In theory, as the price of energy rises faster than those of the other inputs to production, manufacturers will attempt to reduce their energy costs by eliminating waste and substituting where possible cheaper and alternative inputs. The result is improved energy productivity. Further, manufacturers will attempt to recover their increased energy costs by raising the prices of their products. But as the prices of energy-intensive products rise faster than those of other goods and services, business customers and consumers alike will buy less and substitute others in their place. The result in this case is a shift in the composition of product demand, and hence output, away from energy-intensive products. Finally, econometric evidence provided by others (**26**) suggests that increased energy prices were also responsible, at least in some part, for slower economic growth during the 1970s. Thus, the dramatic increases in energy prices of the 1970s may be linked in theory to all the energy effects and mechanisms identified in earlier discussions and on Figures 5-9, 5-10, and 5-11, but in varying degrees.

In the case of accelerated improvements in energy productivity, higher energy prices undoubtedly played a motivating role. Causality is highly probable, response behavior is statistically correlated, and leverage is strong. But other factors do exist. Principal among these are the

FIGURE 5-11
Changing Trends in Industrially Generated Electric Power

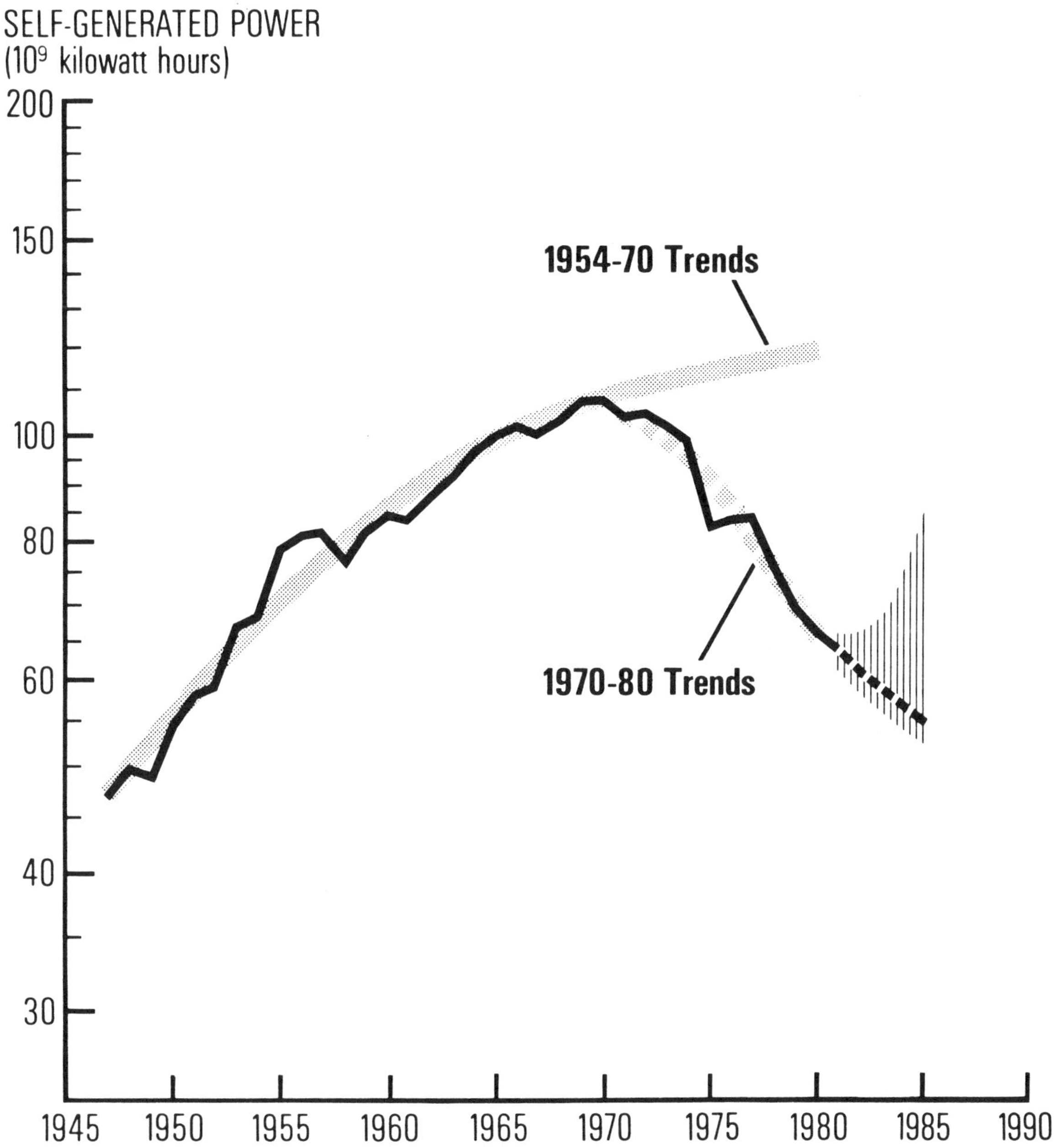

FIGURE 5-12
Real Prices of Purchased Energy

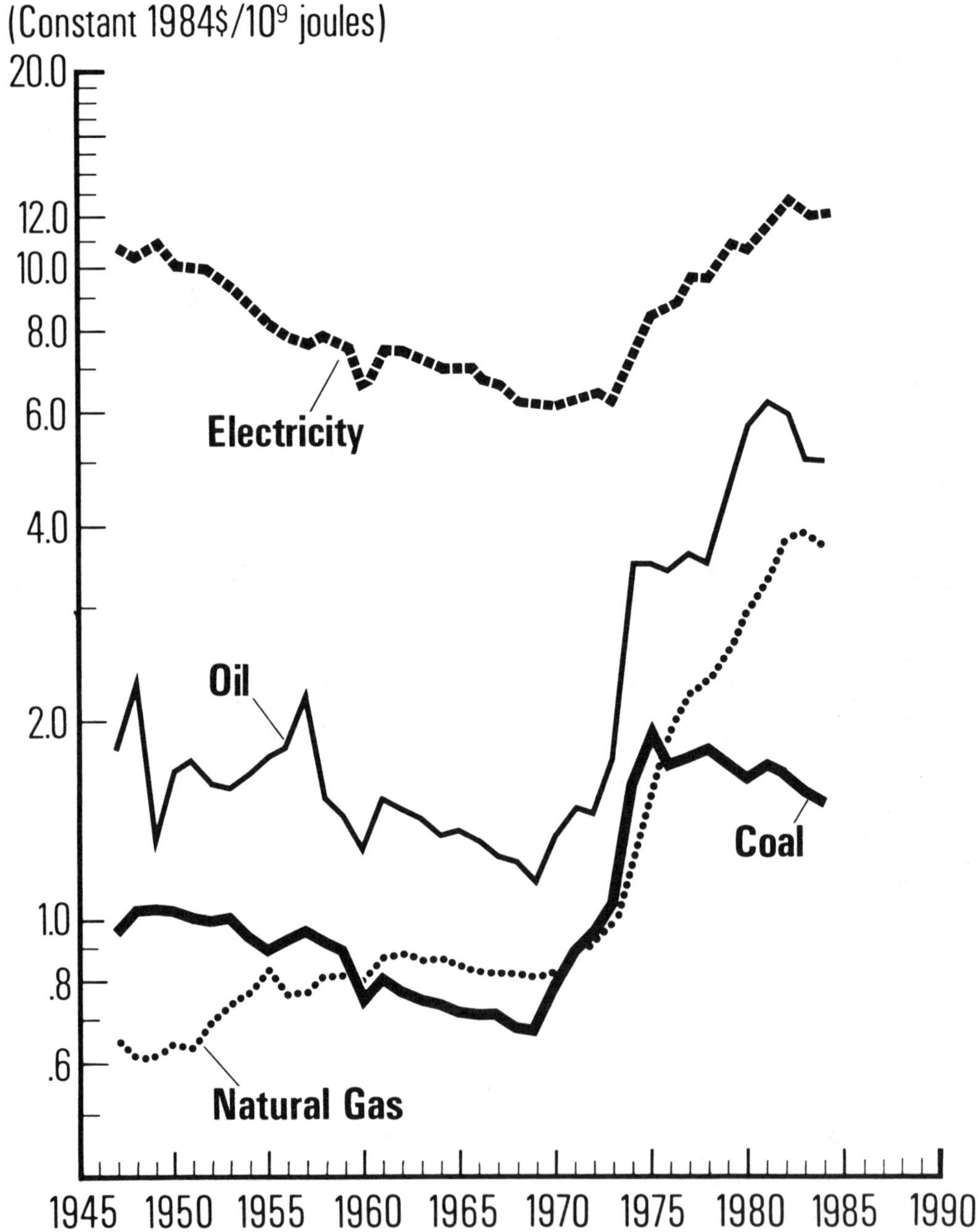

ever-present competitive pressures to improve total factor productivity, and continuing investments in research and development aimed at providing the technological means toward this end. These factors, it is noted, are largely independent of energy prices.

In the case of self-generated power, its decline was not so much an instance of higher energy prices in general, but of higher energy prices in particular, combined with government intervention in the marketplace. For ten years, utilities offered better prices than industry could offer itself, especially since industrial use of natural gas was forbidden and the effective cost of industrial-use coal increased dramatically with the advent of the 1970-era environmental regulations and their enforcement. Speculation about future trends, should prices and other factors change, was discussed earlier.

In the case of compositional changes, higher energy prices likely affected the demand for industrial products in the manner reasoned above. Causality is plausible and behavior is statistically correlated, but leverage in most cases is weak. On average, the costs of the other factors of production (95 percent) far outweigh those of energy (5 percent). Further, the hypothesis that higher energy prices played a major role in restructuring U.S. industry from 1970 to 1984 is not entirely satisfying. It does not explain, for example, the decline observed among energy-consumptive industries that are **not** energy intensive, nor does it explain the many other compositional changes (discussed below) that also took place throughout industry over the same period. Hence, the evidence that higher energy prices caused compositional change is plausible in part, but the extent of this role is not clear, and may ultimately prove to be ancillary.

In the energy-intensive and troubled steel industry, for example, reduced demand for steel output resulted from a variety of factors, of which the rising costs of energy was but one. High interest rates depressed demand for steel products by depressing the business activities of steel's major customers, including construction, automobile sales and manufacture, and investment in capital equipment and durable goods. Competition from industrial nations abroad challenged U.S. markets at unprecedented levels. Foreign steel increased its market share from 15 to 22 percent during this period (**27**). Although some claims of foreign subsidization were confirmed (**28**), most steel imports enjoyed fundamental competitive advantages. Products of high quality were produced by modern and efficient technologies at lower production costs, particularly with respect to labor. The difference in labor costs, for example, was identified by U.S. steelmakers to be the single most important factor contributing to the lack of domestic competitiveness and loss of market share (**29**).

Finally, in the case of the slowdown in overall industrial economic growth, the role of higher energy prices is controversial. Studies in the 1970s suggested that higher energy prices could explain as much as one-half of the slowdown, while later studies suggested a much smaller role for energy and a larger role for the more traditional macroeconomic variables. These include economic and trade policies, interest rates and inflation (caused in part by higher energy prices), declining labor productivity, changing demographics and consumer preferences, lagging technological innovation in some sectors and new technologies in others, and increasing global competition.

The energy and economic developments of the post-1970 period are complex and their relationships to the level and composition of industrial production are not well understood. The increased price of energy, while an important factor reducing growth in energy demand, was surely not exclusive, and may not account for even one-third of the observed behavior. Other factors, including those mentioned above, also contributed to depressed growth in energy demand by fundamentally changing the nature of domestic industrial economic activity and the competitive environment in which it operates.

CAPITAL, LABOR, AND MATERIALS

More insights into the changing nature of U.S. industry may be gained by examining recent trends in those elements of industrial production that are intensive in their use of capital, labor, and materials. Figures 5-13, 5-14, and 5-15 present trend data for industrial production in these sectors, as well as comparative data (shift factors) with respect to the standard index of industrial production introduced in earlier discussions.

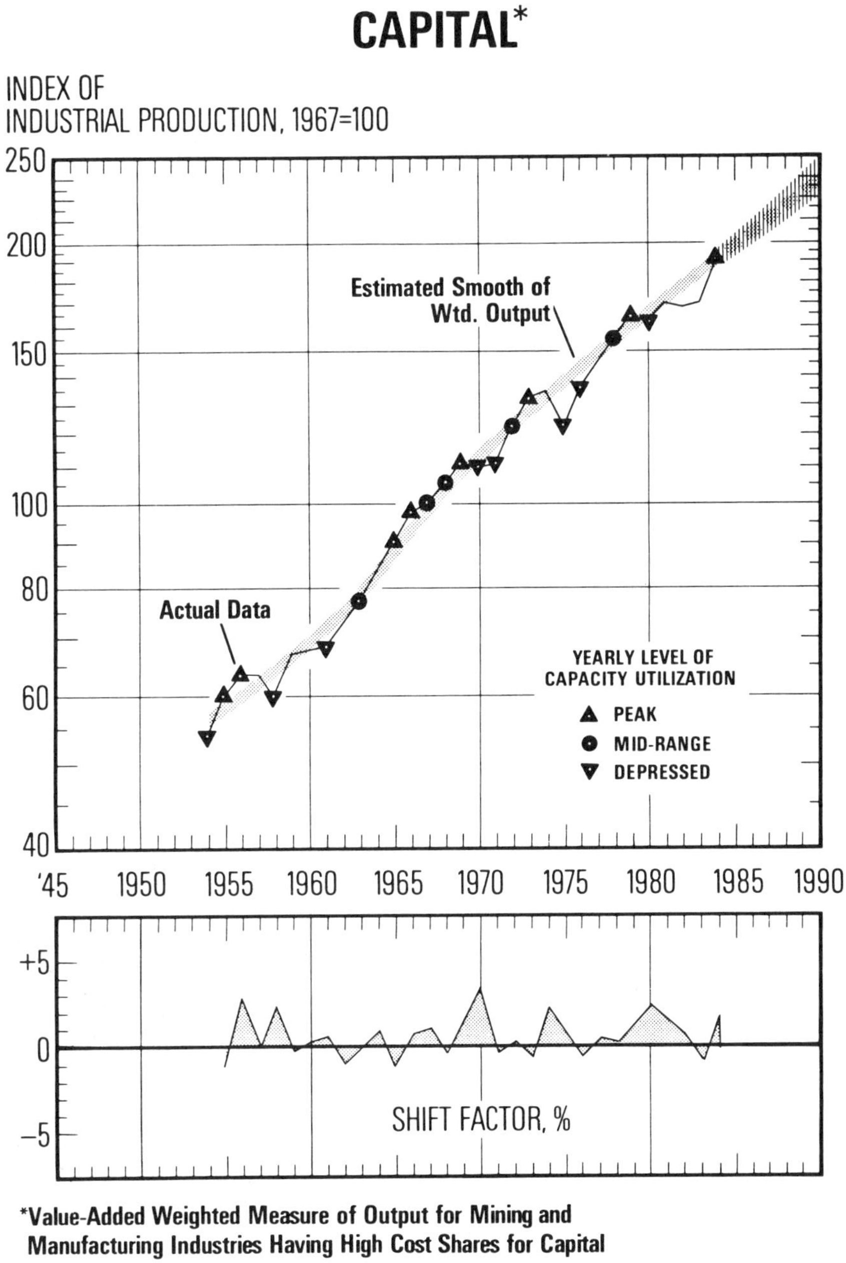

FIGURE 5-13
Index of Industrial Production for Capital-Intensive Industries

FIGURE 5-14
Index of Industrial Production for Labor-Intensive Industries

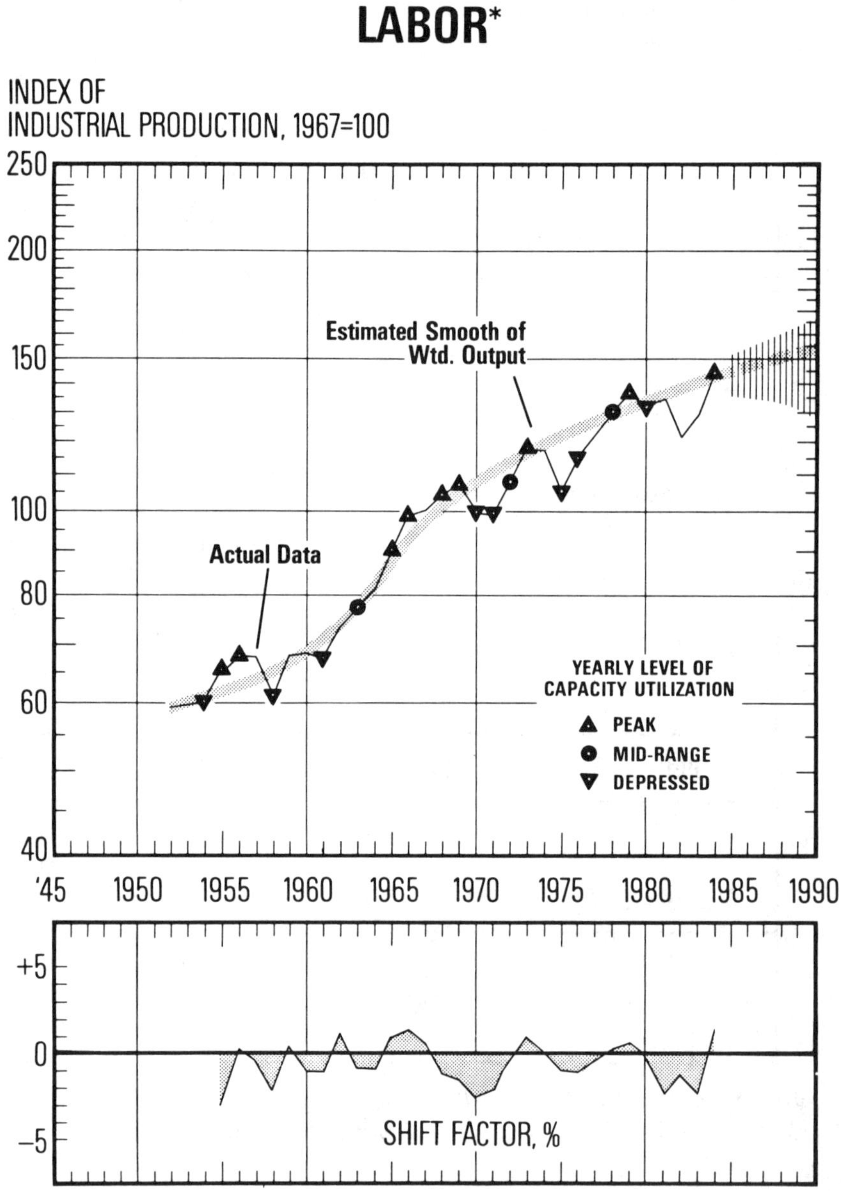

*Value-Added Weighted Measure of Output for Mining and Manufacturing Industries Having High Cost Shares for Labor

FIGURE 5-15
Index of Industrial Production for Material-Intensive Industries

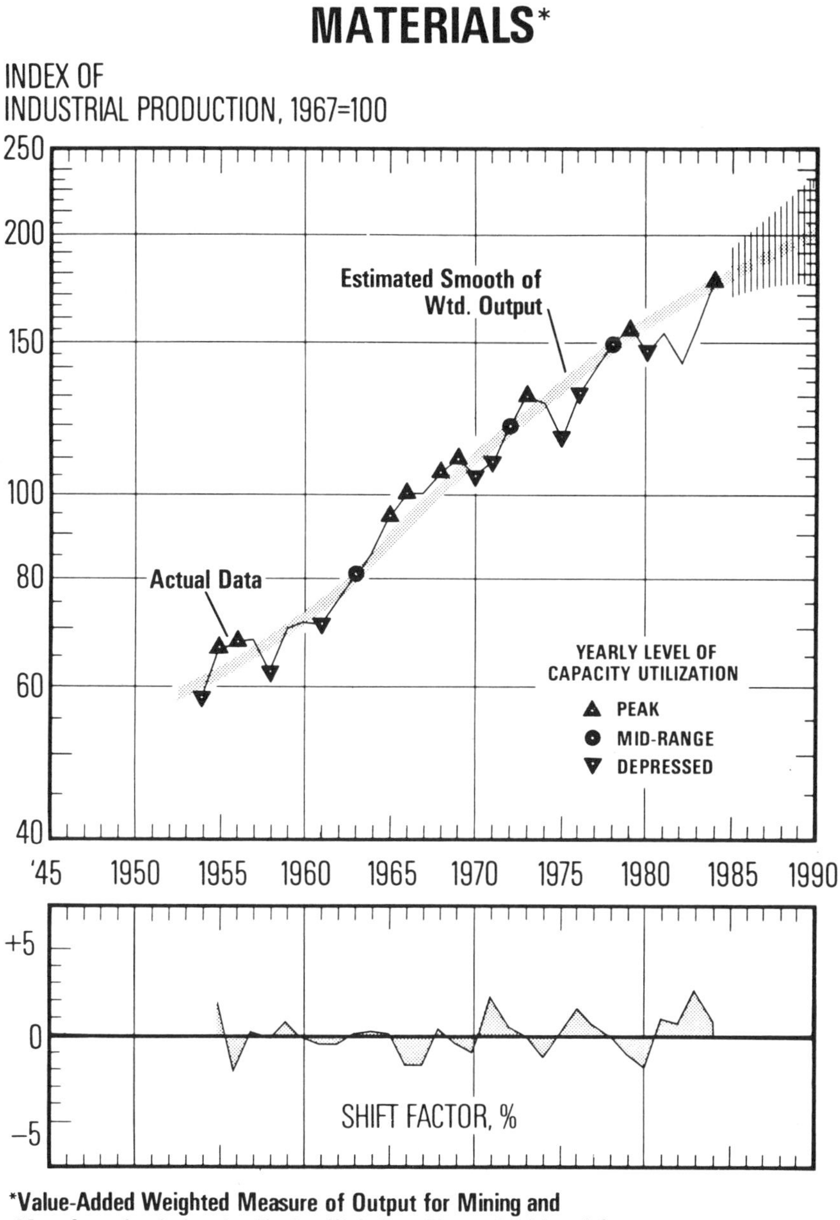

*Value-Added Weighted Measure of Output for Mining and Manufacturing Industries Having High Cost Shares for Materials

From these figures, it can be seen that the long-established trend of labor-intensive industries toward declining relative economic importance accelerated during the post-1973 period. Industries intensive in their use of capital, however, experienced continued and strong growth compared to previous trends, and those intensive in their use of materials posted major gains.

Materials-intensive industries are those characterized by relatively high cost shares (of total production costs) for purchased materials. Such industries are engaged primarily in the latter stages of processing and fabrication. They are often technologically sophisticated. Examples include a host of industries related to agricultural and food processing and others as diverse as aircraft manufacture, ready-mix concrete, computers and electronic equipment, and women's clothing. This latter finding is consistent with and provides further evidence in support of the often-noted trend in the U.S. economy toward services and the lighter end of the manufacturing spectrum, and away from heavy manufacturing and raw materials production.

While changes in the composition of industrial output occurred from year to year throughout the entire post-World War II period, both toward and away from certain industries, no change of like magnitude or duration was observed before 1970. Further, the pervasiveness of the change exended to all sectors, affecting capital, labor, energy, and materials-related activities in different and varying degrees.

These observations and the data on which they are based suggest that U.S. industrial production in both its composition and use of energy underwent and is perhaps still undergoing a transformation of unprecedented proportions, with implications for both long-term energy demand and a range of other socioeconomic issues.

CONCLUSION

In the decade following the 1973 Arab oil embargo, the changing nature and composition of industrial economic activity had a marked negative effect on the growth in industrial electricity demand. Slower economic growth, combined with unprecedented shifts in the composition of output away from industries which are both intensive (relative measure) in their use of energy or labor, **and** consumptive (absolute measure) in their use of electricity, together explain more than 80 percent of the reduced rate of growth of industry's use of electricity.

These two phenomena appear to be the two most significant and least understood mechanisms affecting future industrial electricity demand, and should be addressed explicitly in future theoretical treatments. Although measurable gains were also made toward improved technological efficiency, the negative effect of these gains on the growth rate of electricity demand was less by comparison.

All three developments represent departures from past trends. They occurred quickly, revealing a remarkably rapid facility for change, and all are potentially reversible. The underlying causes of these developments, however, are complex. While they may be linked in theory and apparent observation to the energy price increases of this period, they are more likely to be the result of higher energy prices and a number of other underlying causal factors working simultaneously.

Hence, I conclude that the underlying causal factors of reduced energy demand in the industrial sector from 1973 through 1984 extend beyond higher energy prices. Because the role of increased energy prices was not exclusive, the extent to which future demand would increase should energy prices decline is unclear. Nevertheless, these observations provide important insights into the changing nature of industrial energy demand and suggest new modeling approaches.

REFERENCES

1. E. Hirst, R. C. Marlay, D. L. Greene, R. W. Barnes. Annu. Rev. Energy, Vol. 8, no. 193, 1983.

2. R. Marlay. "Trends in Industrial Use of Energy," Science, Vol. 226, December 1984, pp. 1277-1283.

3. R. C. Adams, D. B. Belzer, J. M. Fang, K. L. Imhoff, R. J. Moe, J. M. Roop. A Retrospective Analysis of

Energy Use and Conservation Trends: 1972-1982. PNL-5026. Richland, WA: Pacific Northwest Laboratory, February 1984.

4. Energy Information Administration. Residential Energy Consumption Survey: Housing Characteristics, 1980. DOE/EIA-0314(80). Washington, D.C.: Department of Energy, June 1982, pp. 9 and 12.

5. Energy Information Administration. Residential Energy Consumption Survey: Housing Characteristics, 1981. DOE/EIA-0314(81). Washington, DC: Department of Energy, August 1983, pp. 16-17.

6. D. Montgomery, D. French, and J. Holte. "Issues in Defining, Measuring, and Forecasting Commercial Energy Use." In S. Schurr and S. Sonenblum (eds.), Electricity Use, Productive Efficiency, and Economic Growth. Palo Alto, CA: Electric Power Research Institute, 1986.

7. Energy and Environmental Analysis. In "Assessment of Trends in Fuel Efficiency and Fuel Use in Transportation: 1972-1982." Washington, DC: Office of Policy, Planning and Analysis, Department of Energy, 1984.

8. Office of Technology Assessment. Industrial Energy Use. OTA-E-198. Washington, DC: Office of Technology Assessment, June 1983.

9. P. C. Gupta and A. Faruqui. Proceedings: Forecasting Structural Change in the U.S.A. Palo Alto, CA: Electric Power Research Institute, 1984.

10. C. Jenne and R. Cattel. "Structural Change and Energy Efficiency in Industry." Energy Economics, Vol. 5, April 1983, pp. 114-123.

11. M. Ross. "Industrial Energy Conservation." Natural Resources Journal, Vol. 24, April 1984, pp. 369-404.

12. P. Werbos. "Industrial Electricity Demand: Prospects and Uncertainties." Proceedings: Forecasting the Impact

of Industrial Structural Change on U.S. Electricity Demand. Palo Alto, CA: Electric Power Research Institute, December 1984.

13. M. Ross. Materials Trends in the U.S. Economy. Ann Arbor, MI: University of Michigan Press, July 1984.

14. M. Ross, E. D. Larson, and R. H. Williams. Energy Demand and Materials Flows in the Economy. Report No. 193. Center for Energy and Environmental Studies, Princeton University, July 1985.

15. E. D. Larson, R. H. Williams, and A. Bienkowski. Material Consumption Patterns and Industrial Energy Demand in Industrialized Countries. Report No. 174. Center for Energy and Environmental Studies, Princeton University, December 1984.

16. W. T. Hogan. World Steel in the 1980s: A Case of Survival. Cambridge, MA: Lexington Publishing Co., 1983.

17. R. W. Crandall. The U.S. Steel Industry in Recurrent Crisis. Washington, DC: The Brookings Institution, 1981.

18. Aluminum Statistical Review, Washington, DC: Aluminum Association, 1981.

19. R. Berk, et al. "Aluminum: Profile of the Industry." Metals Week, McGraw-Hill, 1982.

20. O. S. Yu. Prospects for U.S. Basic Industries 1986-2000: Implications for Electricity Demand. No. P/EM-4502-SR. Palo Alto, CA: Electric Power Research Institute, March 1986.

21. R. Marlay. Industrial Energy Productivity. Ph.D. Dissertation. Cambridge: Massachusetts Institute of Technology, May 1983.

22. National Academy of Sciences. Energy in Transition, 1985-2010. San Francisco: Freeman, 1979, pp. 553-555.

23. J. H. Gibbons et al. *Science*, Vol. 200, 1978, p. 142.

24. *The Census of Manufactures and Census of Mineral Industries.* Washington, DC: Bureau of the Census, Department of Commerce, 1947, 1954, 1958, 1963, 1967, 1972, and 1977.

25. *The Industrial Production* indexes. Washington, DC: Board of Governors of the Federal Reserve Board, from January 1954 through December 1984.

26. E. A. Hudson and D. W. Jorgenson. *Natural Resources Journal*, Vol. 8, 1978, p. 877.

27. R. Marlay. Staff working paper. Washington, DC: Department of Energy, January 1984.

28. Department of Commerce. "Final Determinations in 27 Countervailing Duty Investigations Involving Imports of Basic Steel Mill Products." Washington, DC, August 1984.

29. American Iron and Steel Institute. "The American Steel Problem: Severe but Solvable." Washington, DC, February 1983, p. 16.

6

INDUSTRIAL STRUCTURAL SHIFT: CAUSES AND CONSEQUENCES FOR ELECTRICITY DEMAND

Paul J.Werbos
U.S. Department of Energy

INTRODUCTION

This paper will begin by describing recent projections of industrial electricity demand, and the role of industrial structural change in shaping these projections. After a brief review of the historical studies on the importance of structural change, new work on the causes of structural change is presented. The new work investigates four simple theories for why the energy-intensive industries have declined and shows how these theories lead to alternative projections of energy and electricity demand.

DEMAND PROJECTIONS AND INDUSTRIAL STRUCTURAL CHANGE

In a recent set of runs for the Energy Modeling Forum (EMF), we have projected that industrial electricity demand would grow at an average rate of 3.8 percent per year from 1985 to 1995 **if** two conditions were met:

- Industrial production at the two-digit SIC level follows a reference case provided in the spring of 1985 by the Wharton Long-Term Model. (This case assumes a 2.8 percent growth rate for real GNP.)
- Energy prices follow a "low energy-price case," which assumes prices lower than the Energy Information Administration's (EIA) Annual Energy Outlook 1984 (**1**) and closer to the current expectations of industrial users as expressed at EMF.

This growth is considerably faster than the 1.4 percent per year experienced from 1973 to 1984 (**1**). However, real electricity prices grew at 7.2 percent per year in that period, compared with a 1.1 percent decline in the projections. This difference in price growth, combined with our assumed long-term price elasticity of -0.33 for electricity (**2**), would be enough to raise the growth in demand from 1.4 percent to 4.2 percent by itself.

In actuality, there are several reasons why demand should be lower than this simple calculation suggests. There is the "inertia effect" of past price increases, leading to continued conservation as pre-1973 capital is replaced. There is the competitive effect of lower growth in oil and gas prices than in the previous period. These effects are accounted for in the projections (**3**); however, they are largely compensated for by changes in the growth rates of energy-intensive industries. From 1973 to 1983, the energy-intensive industries **declined** at a rate of 1.1 percent per year, on average. (More precisely, an end-use energy-weighted index of industrial production declined at this rate, as shown in Energy Conservation Indicators 1983 (**4**).) In the Wharton projections, they grow at a rate of 2.6 percent per year, almost as fast as GNP.

To the extent that electricity use correlates with end-use energy, and 1984 with 1983, our projections of electricity growth might have been 0.1 percent per year instead of 3.8 percent if we had assumed the rate of structural change (industrial growth rates) as in the previous decade.

The projections of our model were also compared with those of the other leading models at the EMF meeting in Washington, D.C., in October 1985. In the low world oil price case, our model and the AGA-Tera model were about 2.5 percent higher than the process models in projecting end-use energy in 1995, with all models assuming the Wharton projections; however, when the graph was redrawn to include only heat and power in manufacturing, ours was actually a bit lower than most of the others. (This reflects different treatments of agriculture, construction, and mining, which rely on fossil fuels more than electricity.)

At EMF, the econometric models clustered together in projecting more electricity use, while the process models projected less. However, Samir Salama pointed out that ISTUM2 (the most detailed process model) projected a greater growth in electricity **use** than we projected for purchased electricity; the big difference in projections of purchased electricity is due to differences in assumptions about cogeneration. He also stated that elimination of the PURPA buyback leads to a much lower projection of cogeneration in his model. Even in recent data from DOE/ERA and from trade associations, we see only a minor rebound in self-generation, which is mainly an old technology, and no reason to adjust our forecasts in specific industries; however, if industry should become worried about the **availability** of electricity in the future, it is capable of moving rapidly to protect itself from interruptions in production. Older econometric models, based on the translog, declined to participate at EMF, and are no longer used at EIA for empirical reasons (**5**).

Structural change is especially important to regional projections because individual states tend to depend on particular industries. Even with new state-level economic forecasts (based on the new OBERS projections from the Department of Commerce), our EMF runs project a growth in electricity in the Northwest comparable to that of the United States as a whole. (In theory, we project energy use at a two-digit state level, and have a solid historical data base—available on tape along with the model itself (**3**)—for doing so; however, the projections are only as good as the assumed energy prices and economic forecasts, and are not published because of these and other concerns.) While the biggest source of growth is expected to be in the paper industry, most of the rest is in aluminum, which is assumed to grow at 1 to 2 percent per year. Local experts say that the aluminum industry is "dead," but the Regional Service of Data Resources, Inc. (DRI) projects significant growth, from 50 percent capacity utilization now to 60 percent by 1995; this is in line with a general assumption that primary metals will grow in the United States, and that aluminum will grow a little faster than steel, as in the past. This example illustrates both the importance and the controversial nature of structural change in the future.

HISTORICAL STUDIES OF INDUSTRIAL STRUCTURAL CHANGE

In order to provide a complete breakdown of electricity growth into components like fuel substitution, conservation, etc., one must have reliable data on electricity and fuel use and real output in specific industries. Such data are available only for 1974 to 1981 and a few earlier years from the Annual Survey of Manufactures and Census of Manufactures. In a previous paper for EPRI (**6**), we have shown that the 7.2 percent growth of electricity in that period in manufacturing can be broken down into a 15.9 percent growth related to electrification (fuel substitution), an 11.5 percent decline related to conservation, a 16.5 percent decline due to structural shift, and a 19.4 percent growth due to real GNP growth. Structural change (among two-digit SIC industries) was the most important factor holding down demand, although most of this change represented a change in manufacturing as a whole relative to GNP. That paper also reviewed other work and discussed the methodology and implications for demand forecasting; in particular, it shows why studies using **tons** as an output measure should show more structural change than we do, while studies using numbers such as the Federal Reserve Board indices can be expected to claim less.

Less exact but more comprehensive breakdowns can be done by comparing energy or electricity use against an energy-weighted index of production. Using such an index, Robert Marlay (in these proceedings and elsewhere) has argued that structural change at the four-digit SIC level has been half again as large as structural change at the two-digit level; in our analysis, that would almost wipe out the conservation term. We have also used such an index, in the Energy Conservation Indicators 1983 (**4**) and elsewhere, though not only at the two-digit level (subdividing chemicals and rubber into basic chemicals and other). It appears that electrification and structural shift explain **two-thirds** of the apparent conservation of end-use energy in all of the industry from 1973 to 1983. In the EMF October meeting, Marc Ross stated that structural shift consistently explains about one-third of the reduction in energy intensity in the past decade according to studies done by several authors for several nations. One should bear in mind, however, that this refers to the **end-use Btus** (not counting electric conversion losses), and is only part of the picture shown in Energy Conservation Indicators.

PROJECTIONS IMPLIED BY ALTERNATIVE THEORIES OF STRUCTURAL CHANGE

Starting from four of the scenarios developed for EMF in May 1985 (7), we have developed alternative projections based on four alternative theories of what causes industrial structural change. The theories are:

- The **time-trend** theory. In this theory, energy-intensive industries have declined solely because of long-term factors such as the rise of the service sector or the migration of smokestack industries as an inevitable part of economic development in poor nations.
- The **energy prices** theory. In this theory, higher energy prices lead to higher prices and lower sales for energy-intensive products, above and beyond the effects of long-term trends.
- The **growth and interest** (or cyclical) theory. In this theory, real interest rates and growth in real GNP cause structural change above and beyond the effect of time-trends. This theory assumes that energy-intensive industries depend heavily on the markets for structures, equipment, and automobiles, because half the energy goes to produce the bulk materials used in such products and half of the rest goes to produce such products directly. Investment and automobiles, in turn, respond to real interest rates and the **rate** of

growth of GNP more than the rest of the economy does. (New investment covers the **growth** of GNP, not the **level** of GNP, so that the difference between 2 percent growth and 4 percent growth in GNP leads to a lot more than a 2 percent difference in the level of investment.) Likewise, high interest rates attract foreign capital, which, by raising the dollar, ensures a corresponding level of net imports **regardless** of issues such as quality control, labor practices, etc.
- The **combination** theory, which combines the causal factors of all the others.

To implement these theories, we have done a series of regressions for each of the 18 industries in our data file (**3**). The dependent variable, in all cases, was the logarithm of the ratio between real output (according to the Bureau of Labor Statistics) and real GNP, from 1958 to 1983. The independent variables, in the various versions, were time (the year), the logarithm of real energy prices paid by each industry, the growth rate of real GNP (from the previous year to the present), and the logarithm of the real interest rate (AAA, with a 2 percent minimum). In regressions using energy prices, the data were cut off after 1981.

The results are shown in Tables 6-1 through 6-4. Table 6-1 shows that a pessimistic projection of energy-intensive industries (still assuming the same GNP growth) would lower electricity demand by 4.5 percent at most in 1995, while an optimistic projection would raise it by about 2 percent in all the scenarios. Table 6-2 shows that the uncertainty bounds are smaller for heat and power energy as a whole. The Wharton model **appears** intermediate between the energy prices theory and the combination theory; however, it shows small changes between scenarios, like those of the growth and interest theory, suggesting that the forecasts of that model mainly reflect growth and interest rates but assume a worsening of the historic time-trends due to factors beyond our consideration here (such as the maturing of the chemical industry circa 1970).

EIA has reported (with a few misprints in the table) a similar exercise with the Data Resources, Inc., Interindustry model, in connection with the Annual Energy Outlook 1984 (1). The DRI model followed the growth and interest theory rather closely. An industry-by-industry analysis suggested

TABLE 6-1
Effect on Electricity Demand in 1995 in Billion KWH of Alternative Theories of Structural Change

Structural Change Assumptions	Reference Case	Energy Price & GNP Assumptions (7) High World Oil Price	Gas Prices 20% Lower by 1990	Electricity Prices up 9%
Wharton	1190	1184	1165	1159
Time-Trend	1139	1133	1115	1110
Energy Prices	1170	1154	1171	1124
Growth & Interest	1213	1206	1188	1182
Combination	1203	1188	1201	1157

TABLE 6-2
Effect on Total End-Use Heat and Power Demand in 1995 in Quads of Alternative Theories of Structural Change

Structural Change Assumptions	Reference Case	Energy Price & GNP Assumptions (7) High World Oil Price	Gas Prices 20% Lower by 1990	Electricity Prices up 9%
Wharton	18.54	18.09	18.71	18.49
Time-Trend	17.97	17.55	18.14	17.93
Energy Prices	18.27	17.73	18.77	18.04
Growth & Interest	18.80	18.35	18.98	18.76
Combination	18.66	18.11	19.12	18.43

TABLE 6-3
Effect on Primary Metal Production in 1995 in Billion 1972 Dollars of Alternative Theories of Structural Change (calibrated to BLS data for 1982)

Structural Change Assumptions	Reference Case	Energy Price & GNP Assumptions (7)		
		High World Oil Price	Gas Prices 20% Lower by 1990	Electricity Prices up 9%
Time-Trend	46.61	46.32	46.61	46.61
Energy Prices	51.95	50.81	54.66	50.11
Growth & Interest	56.09	55.49	56.09	56.09
Combination	55.44	54.27	57.83	53.81

TABLE 6-4
Effect on Chemical, Rubber, and Plastics Production in 1995 of Alternative Theories of Structural Change (calibrated to BLS data in billion 1972 dollars for 1982)

Structural Change Assumptions	Reference Case	Energy Price & GNP Assumptions (7)		
		High World Oil Price	Gas Prices 20% Lower by 1990	Electricity Prices up 9%
Time-Trend	145.54	144.64	145.54	145.54
Energy Prices	148.32	145.07	156.45	144.10
Growth & Interest	162.00	160.62	162.00	162.00
Combination	155.65	152.40	163.12	151.75

that the DRI projections were indeed more reliable than the simple ones used here, because of the consistency imposed by their input-output structure. Recent tests by EIA for in-house purposes confirm that the DRI Interindustry model is more optimistic about the energy-intensive industries than is Wharton, which also uses an input-output structure.

EVALUATION OF ALTERNATIVE THEORIES OF STRUCTURAL CHANGE

The Wharton Long-Term Model and the Jorgenson 36-DGEM model are the only well-known input-output models which allow energy prices to change the input-output coefficients; thus, they are the only models capable of doing full justice to the energy prices theory discussed above. The Wharton model shows bigger economic impacts for energy price rises within any year, because it tends to predict that price rises will be passed on to product prices (5). Despite all this, the comparison across scenarios in Tables 6-1 and 6-2 shows that energy prices have little effect. This should be expected, since energy is only about 3 percent of the cost of production in the average manufacturing industry, and its price rises cannot be expected to change relative markets very much. The input-output structure of the Wharton model should fully account for the growth-and-interest arguments above, as well as any first-order effects due to imports (which are projected separately for different industries, as a function of exchange rates projected by Wharton's linked global model).

Table 6-5 shows the quality of fit of the four alternative theories for the five biggest energy users in manufacturing (minus refineries). All four theories **appear** to do poorly in forecasting for the paper industry; actually, the ratio between paper output and GNP has changed very little over the past 30 years, so that all four models averaged between 2 and 3 percent error in predicting paper production. Interest rates and GNP growth consistently improve the fit compared with a simple trend model; this finding is reinforced by Table 6-6, which contains the actual regression coefficients and t-ratios for the growth and interest model (except for the intercept term).

GNP growth and interest rates were highly significant, and had the expected sign in all industries except food; since

TABLE 6-5
Quality of Fit (R^2) of Alternative Theories of Structural Change

Industry	Theory of What Causes Structural Change (%)			
	Time-Trend	Energy Prices	Growth & Interest	Combination
Food	92.0	93.8	94.8	94.8
Paper	0.2	38.9	19.4	46.0
Stone, Clay, Glass	76.8	81.9	95.1	93.5
Primary Metals	66.5	80.7	86.3	85.9
Chemicals, Rubber	62.1	93.6	83.1	96.3

TABLE 6-6
Regression Coefficients for the Growth and Interest Model

Industry	Causal Factors Assumed by the Model					
	Real GNP Growth		Log Real Interest Rate		Time (Year)	
	Coeff.	t	Coeff.	t	Coeff.	t
Food	-.36	2.1	.021	2.9	-.012	20
Paper	.38	1.7	-.016	1.8	.000	0
Stone, Clay, Glass	1.44	6.0	-.073	7.4	-.013	16
Primary Metals	2.24	3.4	-.130	4.9	-.020	9
Chemicals, Rubber	1.34	3.4	-.069	4.4	.013	10

food is not consumed by investment industries or automobiles, this exception is consistent with the underlying theory.

Energy prices, by contrast, appear less important. In three of the industries, the addition of energy prices to the growth and interest theory (yielding the combination theory) actually reduces the overall fit of the model in Table 6-5; in other words, energy prices are statistically insignificant when GNP growth and interest rates are controlled for. In paper, none of the causal factors makes much of a difference. In chemicals, rubber, and plastics, the energy price term appears significant, but is questionable.

The combination theory suggests that a 1 percent increase in energy prices lowers chemical output by 0.12 percent, a statistically significant effect; however, when the same exercise is repeated, using BLS data for the real price of chemical products instead of using energy prices, a 1 percent increase in product prices lowers demand by only 0.39 percent. Given that heat and power energy accounts for only about 5 percent of the cost of chemical products, it seems clear that the success of the energy price term is mostly spurious in this case. The logical discipline of an input-output model, like Wharton's, prevents such spurious results, because energy prices are only allowed indirect effects by way of product prices. (Some readers would be interested in a few further details on this exercise. The automobile and electrical machinery sectors showed product price elasticities greater than one, while the other industries in Table 6-6 all showed lower elasticities than chemicals did; this reinforces the notion that investment and automobiles drive the cyclical behavior of manufacturing.)

In summary, the growth and interest theory seems to fit historical data the best, and the Wharton model provides the best available projections of structural change. In the longer-term, if oil prices rise, electricity prices stabilize, and current research succeeds, one might expect a change in the industrial trend terms which now favor chemicals (plastics) over metals and ceramics; however, this is beyond the scope of the historical data and the near-term future. With lower GNP growth, of course, all the projections discussed here would be lower by more than a proportionate factor.

REFERENCES

1. Energy Information Administration (EIA). Annual Energy Outlook 1984. DOE/EIA-0383(84). Washington D.C.: National Energy Information Center (NEIC), Forrestal EI-20 (Room 1F048), 1985.

2. B. Cohen, J. Holte, and P. Werbos. Demand Analysis Elasticities. DOE/EIA-0475. Washington, D.C.: NEIC, June 1985.

3. P. Werbos. Documentation of the PURHAPS Industrial Demand Model, Volume I: Model Description, Overview, and Assumptions for the 1983 Annual Energy Outlook. DOE/EIA-0420/1. Washington, D.C.: NEIC, 1984.

4. EIA. Energy Conservation Indicators 1983. DOE/EIA-0441(83). Washington, D.C.: NEIC, October 1984.

5. P. Werbos. A Statistical Analysis of What Drives Industrial Energy Demand. DOE/EIA-0420/3. Washington, D.C.: NEIC, December 1983.

6. P. Werbos. "Industrial Electricity Demand: Prospects and Uncertainties." Forecasting the Impact of Industrial Structural Change on U.S. Electricity Demand. Edited by A. Faruqui. EA-3818. Palo Alto, CA: Electric Power Research Institute, 1984.

7. J. Weyant et al. "Final Second-Round Study Design for EMF8: Industrial Energy Demand, Conservation, and Interfuel Substitution Under Alternative Energy Futures." EMF 8.1. Energy Modeling Forum. Stanford University, Stanford, CA: Terman Engineering Center, August 1985.

7

LONG-TERM U.S. ECONOMIC OUTLOOK

Kurt Karl *
Wharton Econometric Forecasting Associates

Forecasting in the short term is easy. I can say that since I forecast the long term. You build a model of the U.S. economy and let the model do all of the work. When the model comes up with the wrong answer, you blame the results on the model. If, by chance, the model results are correct, well, then, you make sure that you tell everyone what a great job you are doing.

*This paper represents a verbatim transcript of the presentation given by the author at the EPRI seminar.

Seriously, short-term forecasting has been very difficult lately. The economic indicators are certainly mixed at best. There are some strong areas in the economy, some weak areas, and some areas that are just in a holding pattern. Strong areas, of course, are in the service sector: service consumption, service employment growth, and the service industry.

Weak areas tend to be the industrial sector, especially for industries that have to deal with international competition. We view the economy now as in a turning point of some sort. It is possible that we'll run into a recession in the next couple of years. It is also possible that we will just simply have slow growth, as we have been having in the past year, followed by returning strength.

The problems in forecasting in the near term and in the long term reflect economic problems. The money supply changes do not seem to affect the interest rates as they once did. Changes in the interest rates do not seem to affect the economy as they once did. The demand for autos, housing, and consumer durables, and business investments do not seem to respond to interest rates as they once did.

Another problem is that the exchange rate outlook does not seem to be influenced as much as it once was by current account deficit. In fact, the current account deficit seems to have no impact at all on the value of the dollar.

We do have explanations for why these factors are occurring currently in the economy. On the money supply side, there has been a lot of deregulation in the financial markets. The Latin American debt situation, agricultural loans, and oil exploration loans have all created a certain amount of uncertainty in the financial markets, making changes and responses to money supply changes slower on the part of the financial institutions.

It is also true that an increase in the money supply now increases inflationary expectations. This was not the case ten years ago. With the increase in inflationary expectations, interest rates do not necessarily fall when the money supply increases.

On the interest rate impact, it is true now that domestic auto sales and housing starts account for a smaller share of the national economy, so that an interest rate change necessarily will have less of an impact.

It is also true that nominal interest rates are higher now,

so, a 50-basis point change in interest rates means less in terms of a percent delta on the interest rates.

And, finally, a factor that has not been examined as much as perhaps it should be is this: Interest income is a much larger proportion of personal income than it was ten years ago, because of the high interest rates, so that, when you have a decline in interest rates, many people have lower income because of the decline. This tends to dampen any kind of positive impact from lower interest rates.

On the exchange rate, it seems now that the value of the dollar is determined more by real factors, such as long-term outlook for growth, real interest rates, and the relative inflation rates between the countries of the world.

However, all of these explanations do not make it easier to forecast what is going to happen next. These things are occurring; they are the results of changes in the economy, changes in the functioning of the U.S. economy. What the outlook is going to be is a different question.

The one view is fairly optimistic. I call it "muddling through" the next couple of years. In other words, we are not looking at a recessionary situation. There aren't sufficient down-side factors to bring us into a recession. However, there is not sufficient strength in the economy to grow at potential growth. It is less than potential growth for the next year.

While we muddle along, interest rates can fall, and we have been trending downward over the next year. Also, with the decline in interest rates, we can have the dollar decline, so that, by late 1986, we expect net exports to be contributing to growth and interest-sensitive sectors of the economy to be strengthening.

There is insufficient strength now to push interest rates up, insufficient strength to push the dollar up, so we have been trending downward, and this will lead to stronger growth of more than 3 percent, late 1986 through 1987 and 1988.

There is a risk of a recession but we don't think it is likely. The reason is that interest rates are falling, and, typically, interest rates rise into a recession. The dollar is falling, and, though this will help the economy with a lag, it certainly will help. We don't envision any shocks on the economy. Oil prices are stable to declining. Fiscal policy doesn't seem likely to have a radical change in direction.

Mostly, we can expect moderate expenditure cuts over the next few years.

As for the monetary policy, we don't expect any surprises from the Federal Reserve Board. With the slow growth outlook that we have, they are hardly likely to push interest rates up. Also, since inflation is low, they don't really have a reason.

Finally, inventories to sales ratio is not high. Typically, in a recessionary situation, inventories are high at the beginning of the recession and the inventory rundown becomes part of the recessionary situation.

How do we get this type of outlook? We have the assumption that the Federal Reserve Board will be accommodative to the slow growth and bring interest rates down. We assume no major tax reform packages. We have oil prices sliding a little bit this year and zero growth through 1990. We have the value of the dollar falling.

As I said, we have had no major changes in the fiscal outlook, and we assume that there will be no constraints met in the labor markets. In the industrial markets, capacity utilization is low now. Unemployment is about 7 percent. There are some areas of employment, especially at the younger ages, where there is some constraint. Fast-food chains is one area that is having difficulty finding workers, but that is not going to have a significant impact on the economy or inflation.

So, we have the Feds pushing down interest rates, in some sense, by their policy of fixed monetary growth, and not a lot of inflation and not a lot of real growth. This should induce a slow decline in interest rates over the next year, and, by late 1986, interest-sensitive sectors of the economy should be recovering. We will have more than 3 percent growth in the second half of 1986.

What we have in our scenario, in the medium term, is the strength from the interest-sensitive parts of the economy—autos, housing, business fixed investment—through 1989 will eventually increase inflation and increase interest rates. So, we have interest rates rising in the latter half of 1986 and through 1987, 1988, 1989, and 1990. And by 1990, we have a correction that is more of a typical recessionary correction where interest rates rise into the recession, inventories are rising, and then they both fall during the recession as the correction occurs. After a strong recovery in 1991, we basically have potential growth from 1992 onward—just less than 3 percent growth.

This brings us to the long-term outlook. Rather than giving a point-by-point summary on the final demand sectors, I will go over long-term issues that we think are probably going to be around for the next ten years, and then we can handle some questions about what is of interest to you. I'm not an electrical expert or an energy expert, so I hope you will confine yourselves to industrial questions and macroeconomic activity.

What are the long-term issues? Obviously the federal deficit is one of the issues that will continue to be discussed over the next ten years. Another issue that we expect to be continuing is tax reform. Productivity growth and international competition are going to continue to be issues even with the decline in the value of the dollar. There are several international trade issues that are going to last a while, from a trade point of view. Developing countries' debt, particularly in the Latin American area, will have to be resolved; the problem of the debt that they have will have to be resolved for us to be able to have more robust export growth to those areas. Growth in Europe is also an issue for international trade. Another international trade issue is, of course, protectionism and the likely impact on the U.S. economy, if protectionism is invoked.

We can't get to the long term without a demographic perspective on things: population growth, household growth, labor force changes.

Finally, energy issues are certainly going to be issues over the next ten years, particularly price movements and demand in types of energy usage.

On the federal deficit, we expect it to be an issue over the next ten years, because it is so large, and, being large, it accumulates a debt that has to be serviced, so net interest to income paid by the federal government is growing rapidly; in the next couple of years, we expect it to be 20 percent of the revenues taken in by the federal government, so that they will always, through the next decade, have a large proportion of expenditures devoted to paying interest on the debt. That is basically why we expect it to remain an issue.

In our forecast, we have the deficit staying close to about $200 billion. This is good news in some senses, in that it means that the deficit falls as a share of nominal GNP, but it does mean that it is large for the next ten years, and, being large, it keeps real interest rates above their historical post-war average.

We do have real government purchases declining as a share of Gross National Product to help out on the deficit. And we also assume that, after the next elections, there will be a 7 percent increase in personal taxes that will also help flatten the deficit outlook, but not enough to balance the budget.

As time goes on, tax increases are increasingly likely, because it will be more and more necessary or clear that the deficit is a continuing problem. In the near term, we suspect corporate tax increases will be more likely than personal, although we don't have any assumed in our base case. We have it less than 50 percent chance, and that will fall out of some tax reform policy if it occurs. Again, we don't expect it to be occurring.

We viewed tax reform as an issue across the decade, because over the past 30 years a lot of special provisions have been built into the tax code and it is complicated. It can be now viewed as an unfair system as an overall package. So, as this unfairness issue and this complicated issue, and the special interest group issue continues to grow, with more and more special tax provisions, the idea of a sweeping tax reform becomes more popular. That is why—and not because of the deficit issue—the tax reform has become an issue. As long as it isn't fully implemented in some form, it will continue to be an issue.

This creates problems for the economy; discussing tax reform causes real changes to occur, as investors attempt to make decisions on their investment projects. We have noticed that there are less vacation homes being sold because of tax proposals, less being built. The indications are, at the beginning of the year, that residential units, the multiple side, were stronger because of anticipated tax changes. And there is bound to be uncertainty on those people making decisions on a business investment about what their write-off outlook is going to be for the investment. That is also a major part of tax reform.

As for productivity growth in investment, in the U.S. competitive position, certainly the dollar must fall to help out, and, fortunately, it finally is. It has been a tough one to forecast, as you probably know. We had it declining before it actually did in our forecast, but I don't think anybody else was doing much better in forecasting its movement. We do not have it falling rapidly, only about 5 or 8 percent per year for another few years, and then slowly over the decade.

We have investments strong in the U.S. They have been strong for the last couple of years, and this will help productivity in the next five years, as the capital improvements become more effective, more utilized.

We also have the labor force slowing down, from a demographic sense. The Baby Boom has entered the labor force. We do have women's participation in the labor force increasing, but this is not expected to be as dramatic as in the '70s. With the slower labor force growth and slowly declining real interest rates, the cost of capital relative to wage cost should be low. More investment in the labor-saving area and in electrical equipment can be expected, helping growth in productivity. That is, in fact, part of our outlook that electrical equipment should do well over the next ten years.

Foreign competition, however, will not go away simply because we are investing and the dollar is falling; this foreign competition will continue to create an incentive to invest. It is also true over the decade that energy conservation investment is less necessary to reduce your total Btu cost, so that labor-saving investment is more likely. We have oil prices very flat, as I mentioned, so it is not as big an issue over the next ten years.

On international trade issues, here again, the dollar must fall. That would certainly help, and it would help export growth and reduce import growth. But it is also true that we must work closely with the Latin American debtor nations, the African debtor nations, and the eastern European debtor nations, to make sure that there is no collapse in their economies, so that we can continue over the next 10 or 20 years to export to those countries.

We have just a slow improvement after 1988 in the value of the dollar, around 1 or 2 percent per year. The slow improvement in the dollar helps to have a slow improvement in exports throughout the decade, and there is a slow improvement in the Latin American debt situation, which has been with us for several years now and has not really been worked off.

So, we have an improvement over the next decade in net

exports. Net exports in 1980 were about 50 billion positive in the real national income accounts in 1972 dollars. They are now closer to 50 billion in the negative side, so we have lost roughly 100 billion in the past 5 years. We have it improving, but we don't have it going positive over the next 10 years, so we still have minus 5 to 10 billion real net exports by 1995; thus, it is helping in growth after 1986, but it is not exactly booming.

Europeans are very pessimistic about growth in their economies. They have 2 percent over the next five to ten years. Wharton is a little bit more optimistic. There seems to be sort of a generic pessimism in Europe, and we have it at closer to two and a half. So, there is room for export growth in that area.

Protectionist measures are likely to be debated for some time, even with the decline in the dollar. There are many industries where the comparative advantage lies overseas, and they are hardly going to let the U.S. Government forget that they are in decline. The steel industry is one. Textiles will have increasing problems, as will apparel and leather goods. There are others as well.

In our analysis of protectionist measures, we can find that selected protectionist measures are beneficial to the economy or a particular industry, provided there can be a clear case made for the protectionism. Reasons might be that some country has been dumping commodities into the United States, or that there are nontariff barriers. This clear case helps to reduce the probability of retaliation; normally retaliation can be sufficiently effective to make the protectionist measure hurt rather than help the U.S. economy, though it may help a particular industry.

As for a blanket tariff, we see a lot of down-side risks on a surcharge, because it is more likely to invoke retaliation, and there are also factors that make it less of a benefit than is typically viewed. A surcharge creates inflation in the U.S., and there is potential overseas to cut the prices in the face of a tariff, so that the international competition could be more difficult with the tariff; we would be in less of a position to compete with our higher prices. All of that is from a macroeconomic point of view. Of course, some industries could do better. It also depends on where the dollar would go during a protectionist situation.

The problem with protectionism is, of course, that a lot

of decisions are made overseas. How much of the tariff is passed through—pricewise—and is determined by foreign exporters? The pass-through issue is this: essentially the decline in the value of the dollar has not been fully reflected in prices that we face. In other words, foreign exporters are collecting greater profits than normally, and they have room to cut their prices. So, if we raise the tariff, they could cut their prices and there would be less than full pass-through of the tariff to the domestic market, and, consequently, less of a price impact on those imported goods. That is the situation where we face greater competition overseas, from a price point of view.

The other risk, of course, is retaliation by foreign governments. And there would certainly be some kind of impact on foreign growth, although the tariff could outweigh that.

The direction of the dollar during protectionism is another big issue. Your older, sort of intuitive point of view, using the current account deficit, would indicate that the dollar should rise in such a situation off of its base case. But our foreign exchange expert at Wharton indicates that this may not be the case, provided the protectionist package increased the revenue of the government substantially and reduced the deficit, the Federal Reserve had the ability to bring down interest rates, and the dollar could go down in such a situation. That is on the optimistic side.

A demographic issue that is having an impact and will continue to have an impact over the next 10 and 20 years is population growth. It is certainly slowing down. The fertility rate is below 2.0 for women of child-bearing age. It's around 1.8 to 1.9, which is not at the level of population replacement. However, population continues to grow at about 1 percent, declining to 0.8 percent by 1995, because there are a large number of Baby Boom women in the child-bearing ages having a greater number of births. Also, there is a lot of net immigration—we estimate it at 750,000 per year—one-third of which is illegal. Furthermore, life expectancy is improving, so people are staying around longer, which is kind of nice to know.

The impact of the slower population outlook is lower growth in nondurable consumption goods; and overall GNP will slow down.

It is also true that the growth in the number of households is slowing down. Most Baby Boomers are in the household age, the large household headship rate ages, and they have formed their households; the growth is not coming from increased growth in those age groups. The divorce rate and the marriage rate have stabilized and this has reduced household growth. Household growth is still greater than population growth—about 1.8 percent per year—but that is because as people age, you find a greater tendency to form a household, or to be the head of a household the older you get, because of death, divorce, or what-not. It is also true that employment ratios relative to population are increasing; this tends to have a positive impact on household formation so that household headship rates are rising.

The slower growth from household formation means that there will be slower growth in consumer durables, slower growth in residential investments, and lower housing starts than in the '70s. That will continue to be the case through the '90s and beyond the turn of the century.

The final issues I would like to bring up are the energy-related issues. We have oil prices weak over the next five years, declining a little bit this year, and then no growth through 1990. Indications are that this may be a pessimistic outlook, that oil prices could be significantly lower, but that is not our base case at this time.

I am hardly going to inform you about what is happening in the utility area, but it is true that nuclear energy is going to be the area of capacity growth over the next ten years, not from new construction or new starts, but completion of existing projects.

We have electricity demand growth at less than GNP growth—about 2.5 percent; GNP growth about 2.8 over the next ten years, and this will have an impact on capacity growth. But, with more efficient utilization of existing capacity, a rise in the capacity utilization rate, we might have fewer investments than would otherwise be the case.

The growth in electricity demand is from the household growth. Electrical appliances are considered to continue to be popular. And it is also true that electrical machinery is likely to be an area where industry is going to invest, and, after putting that in place, of course, their demand for electricity will increase per unit of output. We do not have electricity demand growing from electrical intensity of industry growth. In other words, we don't have a very robust outlook for textiles, aluminum, or steel, but it is true, even with output relatively flat or with slow growth, they can increase their utilization of electricity per unit of output; the basic incentive there is the productivity growth.

Another long-term issue is this: we have a shift toward coal based on a relative price shift, although that won't be as strong as it would have been if the oil prices did not remain flat in the next five years. It would be stronger if oil prices were rising. But, in the long run, there is a lot of coal in the United States, and that is an area which will draw attention.

QUESTION-AND-ANSWER SESSION FOLLOWING PRESENTATION

MR. RAY SQUITIERI, EPRI: You say that the value of the dollar will continue to decline. Now, it seems to me that I have been reading that every quarter, since early 1982, from all of the major forecasters, and the first time it happened was about six months ago. Why should we believe you now?

MR. KARL: Well, it finally happened! That is something. As I was trying to indicate, the past five years have seen a kind of shift in the factors that determine the value of the dollar. So, our models for this, in the short run, at the current account deficit, had an impact on where the dollar should go. As we reestimate these equations and models with the latest data, we find that is no longer the case, or there is a much smaller elasticity.

The case now is that relative international real interest rates have a big impact on the value of the dollar, and the growth outlook in the United States relative to international growth outlook has an impact on the value of the dollar. So, that is our analysis of where we have gone wrong in the last couple of years on exchange rates. Wharton has done no worse or better, I'd like to say, than most of the services on exchange rates.

MR. SQUITIERI: Perhaps I could focus my question a little bit more. This is in the area where everybody went wrong. The classical argument was that foreigners won't continue to pile up dollars at a certain exchange rate, just pile up dollars and not take goods. Sometime they are going to want goods in exchange for these dollars and bank deposits that they are holding.

At that point, they will put more dollars on the market to bid for the things they want. And, putting more dollars on the market will change the exchange rate and drive down the value of the dollar. That is what everyone expected to happen, but it didn't happen because foreigners were content, instead of trading their dollars for goods, to trade their dollars for claims to U.S. assets, like stocks and treasury bills and things like that.

Now, that seems to be the big shift that none of us expected three or four years ago. Do you see that continuing, or changing, or what?

MR. KARL: In fact, we have continual current account deficits throughout the next ten years, the implication being that foreigners will still find U.S. assets attractive, and this is where this potential growth comes in. This is where the real interest rate issue comes in.

As long as our real interest rates are high, relative to the rest of the world, there will continue to be net inflow of capital. And that is, in fact, helping to finance the deficit which is wrapped up in this issue with the real interest rate.

But we still have the dollar declining on a trade-weighted average. The growth outlook in Europe we have better than they expect, so that when we see the growth there, there will be some shifting. The interest rates are falling here now and they are falling more rapidly, I think, than was expected on the international marketplace; this is why the dollar is slipping at this time.

So, provided the interest rates can decline faster than German rates, Japanese rate, et cetera, we will continue to see the dollar decline. It is all of these relative rates that is the issue.

MR. DAVID GOLDFARB, Georgia Power Company: How much more of a decline do you anticipate in the dollar relative to some of the European currencies, like the mark, over the next five years? We have had about a 10 percent decline since February.

MR. KARL: I am not that familiar with all of the different relative rates, unfortunately, but, over the next couple of years, we have a trade-weighted average decline of the dollar, by 1988, of about 20 percent. We'll have a greater decline against the deutsche mark—in the magnitude of 30 percent—and against the yen of 30 percent, and lesser of a decline against some of the European economies that are having more trouble: the U.K, France, Italy; a decline because the inflation there is quite substantial.

MR. EDWARD FISCHLER, Georgia Power Company: This morning, almost the first newscast that I heard indicated that the United States had become a debtor nation for the first time in about half a century. From your perspective, are there any implications about us becoming a debtor nation, and, if so, what are the areas of the economy that might be most adversely or favorably impacted?

MR. KARL: The debtor situation implies that over the next decade we will have a reversal in what we have had heretofore, in what is called "factor income" on the export accounting and import accounting. In other words, we will be sending out interest payments, we will be sending out profit payments or dividends, in a greater amount than we would have otherwise, and receiving less in dividend income and interest income.

So, there is a capital flow system that is being set up—an international capital flow system—that does have real implications for the ability to finance domestic investment. However, provided there is still confidence in United States growth outlook, more capital from overseas will come into the United States, creating a continuation of this factor income problem but sustaining investment growth. And our outlook is that the U.S. growth is good relative to the rest of the world. It is not a high-flier, by any means, over the next 10 to 20 years. The Pacific Basin is expected to be more of a high-flier area, Japan as well.

But, the growth will be good, and it is a safe haven for investment, so we don't have any impact other than these capital flows. It is not going to be a major detriment to the U.S. economy.

MR. RAYMOND E. SUND, Toledo Edison Company: I have a question in regard to your predictions on interest rates and growth of the economy. How much confidence do you have in those numbers? And do you have any alternative to scenarios that would allow you to say the bounds on the interest rates in five years will be between such and such, or one interest rate and another interest rate, with a given level of confidence?

MR. KARL: We at Wharton don't typically do these confidence intervals. Basically, what we try to do is present a wide variety of scenarios and keep our clients informed on potential risks to our outlook. We do have a low-growth alternative which involves, in fact, higher inflation and higher interest rates. We rate it, roughly, at 20 percent probability: 2 percent growth, 6 percent inflation, and interest rates a couple hundred basis points higher. So, you are getting the idea of the magnitude of the variation. Two hundred basis points on interest rates, which is more than our base case outlook, would be perhaps 20 percent probability—perhaps less; up and down about the same symmetry.

We are rating now the risks on the outlook in the near term. We're moving up the risk of an oil price collapse, which is hardly going to push up interest rates here, and it is largely beneficial to the U.S. economy, though not for, of course, all economies in the world.

We see other risks in the outlook. We have been discussing for quite a while a rapid decline in the value of the dollar, and potential inflationary impact, and, therefore, interest rate impact. But that has been moved down from 30 percent, say, to 10 percent risk. It seems to be a slow decline, and that is the indication at this time.

That is basically how we're looking at things now. Offhand, I can't think of another big risk on the outlook that we have or major problems we are discussing, aside from the protectionist issue, which I covered in the talk.

MR. JOE WHARTON, EPRI: Do you have any simple way to define and measure industrial structural change, and, within your moderately optimistic long-term forecast, do you have any insights about it?

MR. KARL: Structural change, the way I tend to look at it, is the relative growth or the relative share in total output of various industries. That is a very simple and straightforward way to look at it, and it is nice to have simple and straightforward ways of looking at things so that we can all comprehend them.

So, we have come around more to the sort of outlook on structural change of what at TVA is appropriately called a sort of a high-tech outlook or high-tech forecast. In other words, electrical machinery as an industry does well. Non-electrical machinery does well. It includes in the SIC classification, computers, of course, and that is where quite a bit of the growth over the next ten years will be as it rises as a share of that industry; but also we have a strong outlook for investment goods in general.

The lower-tech industries where we have less of a comparative advantage—textiles, apparel, footwear, some but not all of the chemical industry, most of the steel, but not necessarily the specialty steel—will have problems. The auto industry is hardly going to make a terrific comeback, because we don't have a robust outlook for total sales. But we do expect to have an auto industry, and the auto industry will be expected to continue to fight to maintain its share, though we expect it to lose a little bit on total sales in units terms.

So, the parameters affecting structural change over the next ten years, basically in the industrial sector, are the international competitive position of the U.S. industries, and also this technology focus. The technology has great implications for productivity growth in our ability to compete, and in those areas where growth—not necessarily electrical but definitely productivity growth—can be increased through some type of machinery; we have a strong outlook. Instruments is another area of strong growth.

As for the manufacturing share in total output, the service versus manufacturing argument, in the long-term perspective—I tend to view it as kind of a bogus or a red-herring discussion. The manufacturing share of value-added output, shared gross national output, has been close to 25 percent since World War II. Now it is declining a little bit because of the fierce competition and what we view as an over-evaluation of the dollar, but we expect it to be roughly 25 percent by the year 2000 as well.

Growth in the service area, which we often tend to focus on, is in employment; it is not necessarily in output. And that is certainly well worth keeping in mind. As for output in the manufacturing sector, we have it roughly constant as a share of GNP, which means that, with this movement within the industries, you can get a structural shift, but not for manufacturing as a whole.

8

IMPACT OF FINAL DEMAND CHANGES ON INTERINDUSTRY GROWTH PROSPECTS

*Stanley J.Feldman**
Data Resources Incorporated

I'm going to talk not only about the impact of the varying changes of final demand on industrial growth rates, but also about the impact of structural change and what industries are going to do in the United States.

Usually, when you talk about final demand, I think what you really want to talk about or think about is what the end

---

*This paper represents a verbatim transcript of the presentation given by the author at the EPRI Seminar.

markets or what U.S. industry is going to do, by detailed industry. As a result of the ongoing interest in structural change, and, in particular, how structural change impacts U.S. industry, I thought I would not only examine the outlook for U.S. industry but combine the outlook, in terms of not only what the industries are going to do but what kinds of structural change are actually going on in these industries. How come the high-tech industries are growing so fast? How come the telecommunications industry is growing so fast? And how much of it is essentially due to final demand changes, that is, the actual size and growth of the U.S. economy and the mix of goods and services that the economy demands? And how much of it is associated with pure structure?

Structure has to do with the kinds of inputs, and the amount of inputs used in the production process of U.S. industry, and, at the same time, what the breakdown of final demand looks like. That is to say: What is the distribution of final demand expenditures? For an example, for a given dollar of consumption expenditure, how much of it is going to be shoes per dollar of consumption expenditure and how much is going to be TV sets?

Both of these changes, the production changes or, we'd like to say, input/output coefficient changes in the A matrix, and changes in the B matrix or the final demand set of coefficients, both combined, give us some estimate of how we think structural change has impacted U.S. industries.

We have done a number of studies—parts of which have been supported in various degrees by Edison Electric Institute, by the Electric Power Research Institute (EPRI), and by the Gas Research Institute. One approach is what I call the distribution of coefficient change (Figure 8-1). The first thing we want to look at is how those coefficients change and how they have changed over time. The distribution that you see here represents the percent growth of the coefficients, on average, over the time period from 1963 to 1981.

The way you should look at this is as follows: This is the way the penetration, if you will, of various outputs of sectors of the economy have penetrated other industries. For example, semiconductor penetration into the computer industry is one. Construction and mining machinery penetrated into the chemical industry, into the blast furnace and

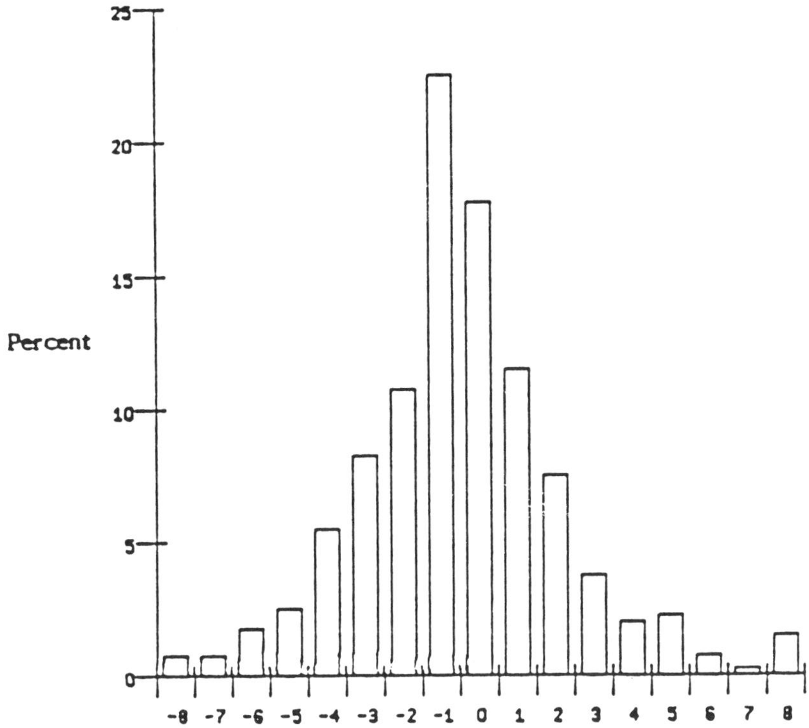

FIGURE 8-1
Distribution of Input/Output Coefficient Change (1963–1981)

Source: Data Resources Interindustry Service.

steel industry, for another example. These represent the recipe or the way the input/output coefficients have changed over time.

The reason why this is important—and certainly what is important about it is the distribution, as opposed to any one of the specific sectors—is that the distribution is normal. Figure 8-1 indicates that the percentage change in input/output coefficients is somewhere between minus 2 percent and plus 2 percent over this whole period. They change very slowly. About 60 percent of the sectors that make up this model—and it is very disaggregated; there are 400 sectors that make up the model—have very slight coefficient change, but they are changing. Figure 8-1 indicates that changes also take place at the tails: On the right side, the industries are growing very rapidly because of penetration;

on the left side, the industries are dying—the leather industry, for an example.

The thought is that structural change in the United States tends to occur rapidly. In any one particular sector, it may occur rapidly but for sectors taken as a group this is not the case.

We think there are two reasons for this. First, there is a basic infrastructure in the U.S. economy that doesn't change quickly over time. For example, if you look at the energy area, the first oil shock generated great increases in relative energy prices, with very, very little impact in terms of the distribution of usage. It was very difficult for an industry which had been using oil, traditionally, to move to electricity because electricity happened to be cheaper.

The second reason has to do with whether relative price changes are going to be permanent. Do you expect, for example, in the energy area, that the prices of electricity and oil are going to continue to rise forever, relative to other goods and services? Certainly, in the mid-1970s, most of us thought that. In reality, this has not happened. But the fact is that getting industries to change their production processes, or getting individuals to change their buying or their purchasing habits, is extremely difficult. For changes to take place relative prices have to change in a way that offsets the costs of making the change.

Another factor is the state of knowledge. Some industries simply cannot change the way they produce their output. Simply, the state of knowledge is not brought to the point where they can react very quickly to changing relative input prices. The result in terms of structural change is slow and gradual; very slow, in fact, and very gradual.

If you examine the coefficient distribution change between 1963 and 1973, and then 1973 and 1981, you will note there are some differences (Figure 8-2). But I think you will find, when you look closer, that, basically, the coefficient changes over the time frames are about the same. They are both normal distributions, and the growth rates that you witness are about the same. That is to say: Most sectors grow or change. The structural changes, the input/output changes that are going on are gradual, and they are gradual over the latter period as well as the earlier period.

The table that was used to do all these calculations was

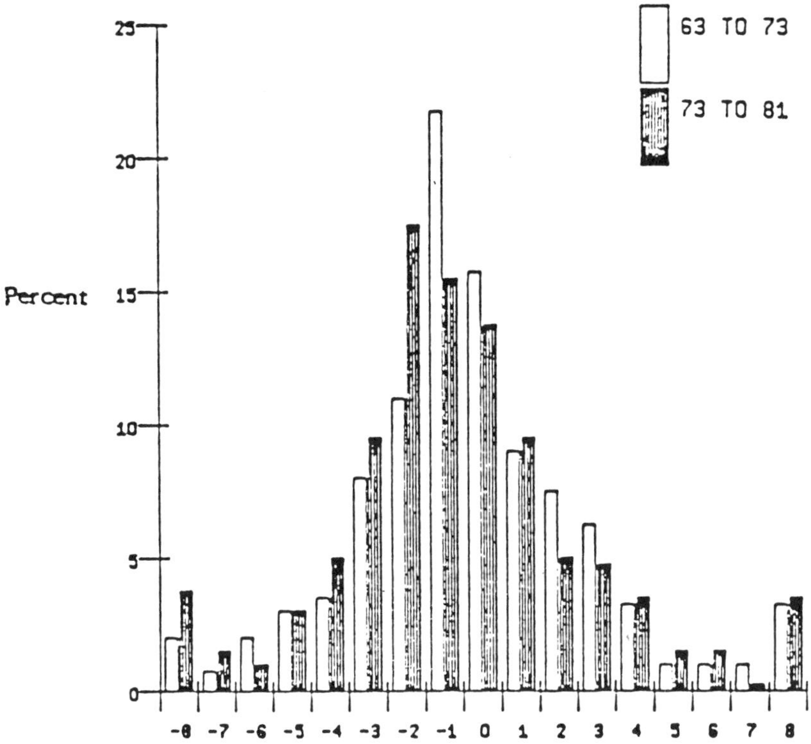

FIGURE 8-2
Distribution of Input/ Output Coefficient Change (1963–1973 and 1973–1981)

Source: Data Resources Interindustry Service.

the 1978-based input/output table. The Bureau of Economic Analysis recently released the 1977 table, and we're in the process of doing this kind of analysis using the 1977 table. Our 1978 table is a table that we put together, benchmarked in 1972, and we have a new table, which is based on 1981, benchmarked to the 1977 table; this kind of analysis will be done again to be sure that the structural change implications of our research are consistent and not dependent upon the fact that we altered a table or that we are using a different benchmark table.

But I think most of the work that I have done in the past, both in the energy area and other areas, has suggested quite strongly that structural change is simply something that goes on gradually.

The outlook for industries, that is, how an industry does,

therefore, will depend on two things: how its end markets will grow and what its penetration rate happens to be into its end markets. Now, the penetration rate has to do with the way input/output coefficients change. And the popular thesis, at least up until a few years ago, was that input/output coefficients change in ways that are nonpredictable. Therefore, input/output tables in that kind of framework are not useful. The fact of the matter is they change in a very predictable way. There are other legitimate concerns with input/output tables, like how much of the change that you witness, for example—and this is an issue—is associated with the way the table is actually put together and how much we are witnessing actual structural change.

But, in general, when you look at the evidence, you will find there are two aspects of the way an industry performs. One is its penetration rate into its customer markets, and one is how fast its customer markets are growing.

Now, in the next part of my discussion, I will talk about several key growth industries, and what I'd like to do is talk about not only what their growth prospects happen to be, but give you a feeling for the way these industries have performed in terms of the penetration rate, and then make some comments about final demand, impacts on output growth.

One of the things that we will notice (Table 8-1) is that there are two sectors whose penetration grows quite strongly. These are the service sector—wholesale trade actually is the one that is growing very rapidly—and the durable equipment sector in the U.S. economy. That is, the two areas that have penetrated their customer markets are wholesale trade and durable goods, which includes things like semiconductors, computers, construction and mining machinery—industries like that. And their growth rates, in terms of penetration, are in the neighborhood of 1.5 to 2 percent a year. So it is quite significant.

Now, one of the industries that I think is of particular interest to us is the construction and mining machinery industry. First, let me tell you what I think that industry is going to grow at, and then we can look at what the determinants of that happen to be. This is an industry that has been heavily competed against by the Japanese and by the Taiwanese. It is also an industry that has tended to grow and be profitable, although it has not been particularly

TABLE 8-1
Average Aggregate Rowscaler Growth Rates

	1963-1973	1973-1981	1963-1981
Wholesale Trade	1.7	0.7	1.3
Durables	0	2.8	1.2
Transportation, Communication and Utilities	0.2	1.1	0.6
Finance, Insurance and Real Estate	0.5	0.1	0.4
Services	-0.3	1.0	0.3
Nondurables	-0.1	-0.2	-0.2

Note: In order to limit the analysis to those groups of sectors for which the rowscaler concept has the most meaning, agriculture and mining, the two extraction aggregates, are excluded from the table. Fluctuations in these aggregate rowscalers are biased by insufficient data on inventory change in these industries.

productive. That is to say, price increases have been significant relative to their cost of production, because American industry or American companies in this industry have had a strong market position up until about 1980. Very large profits were generated in that industry as a result of price; because of lack of competition, penetration could occur despite rising relative prices. Now, what you will find is that, when you look at the data, the penetration rate has been significant, and the biggest growth part of this has come from the actual export market.

This industry's growth will slow in the 1980s. That is, there are companies outside the United States who are more price competitive in world markets for this particular product. As a result, the domestic construction and mining equipment industry will slow to about 3 percent between 1985 and 1990. That is a tremendous slowdown relative to what has happened historically.

The situation in agricultural machinery is similar, in that agricultural machinery is being produced cheaper outside the United States than it is produced inside the United States. Hence, both domestic users of agricultural machinery and foreign users are going to foreign producers as opposed to U.S. producers.

So, in this case, the actual penetration rate of agricultural machinery is into its end markets, which are essentially agriculture, all over the world, both in the United States and other countries, but the price of U.S. goods is high relative to its competition. And here is a case where there is no proprietary technology involved; it is just as easy to buy a piece of agricultural machinery from Komatsu, for an example, as it is from International Harvester.

To reiterate my point, there is a market for construction mining and agricultural machinery, but more of that market now is being served by exports of foreign producers, because these industry prices are rising relative to the prices of other commodities, and those commodities are essentially no different than those produced in the United States. So, there isn't any reason, other than price, in terms of the purchase decision, and that is a key item for a number of these industries.

Now, when I look at computers and semiconductors, here is something I think most of you have seen or certainly have noticed in the past several years. Part of the growth in electronic components is simply due to the fact that more components are used per dollar of customer market output. There are more components in automobiles per dollar of automobile output. There are more components in computers per dollar of computer output. There are simply more components per dollar of end market output in all sectors.

Now, the high-technology aggregate, of course, includes not only semiconductors and computers, but also resistors, capacitors, and items like that. And, again, the penetration rate has been phenomenal. If you looked at what the demand for computers would be and assumed no penetration at all, the growth in computers would be something like 2 percent or 3 percent a year. That's it. It would grow very, very slowly. As it is, the computer industry is growing somewhere in the neighborhood of between 6 and 8 percent between now and 1990, and a very large percentage of that, more than half, is due to penetration. So computers are penetrating all markets.

Now, you might ask: How many of the computers that are actually being purchased are produced in the United States, and how many are produced outside the United States? This most recent downturn in electronic components doesn't mean that electronic components aren't continuing to penetrate customer markets. They are. However, what it means is that more of electronic components are being produced and supplied by foreign producers or American companies that have production facilities outside the United States that are selling or importing electronic components into the United States.

Some of you might have a view that somehow the Asian Basin countries are taking over the market for semiconductors and American manufacturers are no longer competitive. I can assure you that this is not the case. In fact, that is not even in the data; it is a myth.

If you segment the semiconductor industry into what I will call commodity semiconductors and high value-added semiconductors, the high value-added market is still being served by American companies. High value-added market means the applications market. Take a commodity chip, put an application on it, and sell it to an end-user. The United States still has an edge in that marketplace.

The market for the commodity semiconductors—for example, the 256K-RAM type, which is the state-of-the-art semiconductor in the marketplace—is pretty much owned by the Japanese, though Americans produce about 30 percent of the output of 256K RAMs.

Now, it turns out that there is a view that, while semiconductor penetration will continue, and we think it will, American companies won't participate in it, and, in fact, if we don't participate in the 256K-RAM market, we will eventually lose our competitive edge in the semiconductor industry. That is, the state of our knowledge will dissipate relative to the rest of the world, and over time, we simply won't produce semiconductors.

Now, the fact is that, if any leapfrogging is going on, American companies are doing it now. Last week, AT&T announced the production and shipment of a 1000K chip. It is the only company in the world that is producing it. I will give you an idea of what that means.

What it means is that you can take an IBM 360 mainframe—and for those of you that remember that big machine, it took up the size of a room—and you could put a PC on your desk, that has got the same power as that IBM 360 with that 1000K chip.

Now, AT&T has produced a batch and has actually shipped it for customer use. If, in fact, this chip can be commercialized and AT&T thinks it can, then we have leapfrogged the Japanese in the semiconductor market. And this is a market of leapfrogging. If you get there first, you make a lot of money; others enter; prices come down and profits are reduced or even disappear.

In the computer market, the United States is the major producer of computers in the world. If you look at the guts of the computer, there is the box—the plastic box, now, as it turns out—and there is the operating system of the machine. The operating system is really the computer. The box is just the way it is delivered. Operating systems are pretty much the sole market of U.S. producers. There is very, very little operating software and application software of any significance that is produced outside the United States. Most of the software that goes in the computers that the Japanese produce and sell in the United States is produced by U.S. companies.

Now, what does this mean in terms of what we should expect in the high-tech area? I think what it means, basically, is that American companies are growing; they are thriving. There may be downturns in the semiconductor industry and in the computer industry, mostly due to cyclical changes as opposed to secular changes, but don't think for a moment that American companies have stopped competing. In fact, it is just the other way around (Figure 8-3).

If you look at what the causes of semiconductor growth have been, one of the things you need to think about is its relative price change. Relative prices, in this case, represent the price of semiconductors relative to the price of all other goods. There is a definite relationship between changes in relative prices and the change in the penetration of semiconductors.

Now, if we forecasted (Figure 8-4) the solid line, straight out, the semiconductor market becomes immense. Therefore, we assume that the future path of penetration is

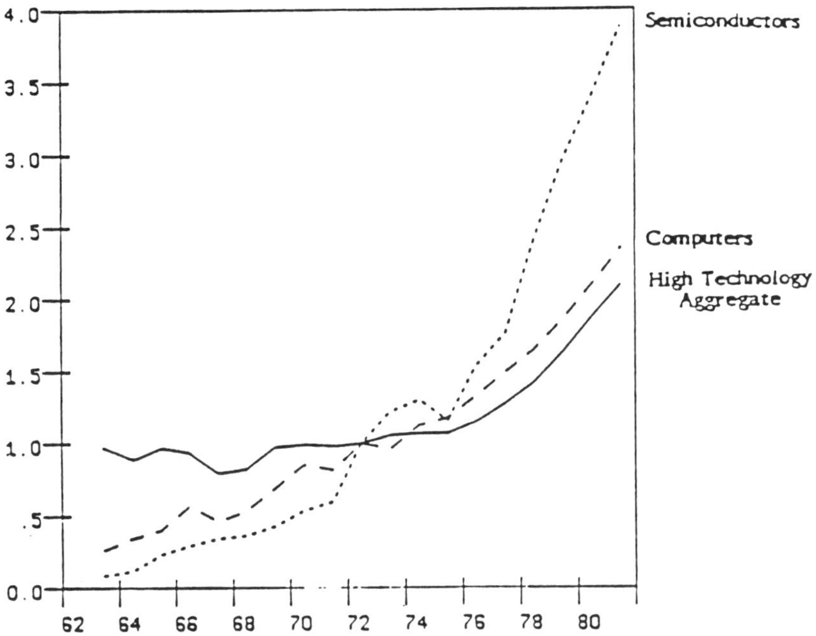

FIGURE 8-3
High-Technology Equipment and Instruments Rowscalers

Source: Data Resources Interindustry Service.

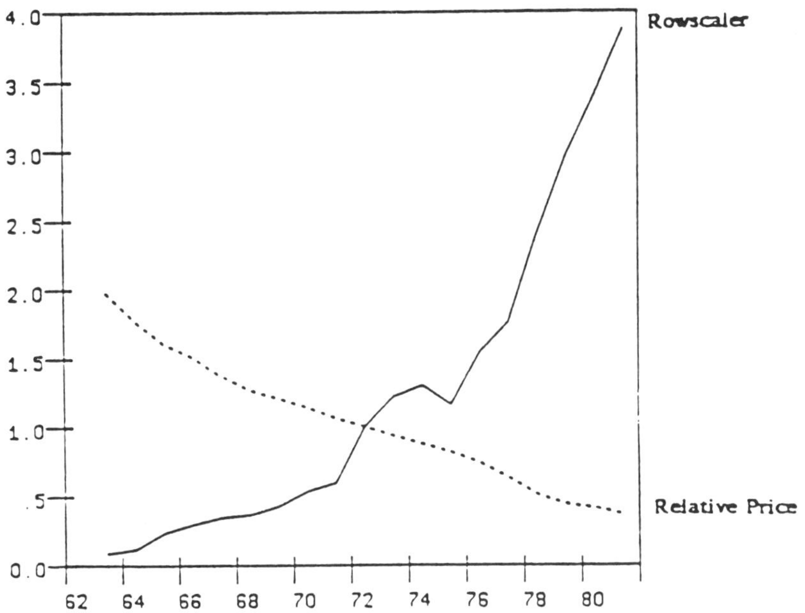

FIGURE 8-4
Declining Relative Price and Market Penetration by Semiconductors

Source: Data Resources Interindustry Service.

slower than in the past for semiconductors. Thus penetration will increase by about 3.5 or 4 percent a year between now and 1995.

Plastic materials is another area that is of great interest, and it is a fast-growing market. I think it is of tremendous interest to energy-producing sectors. One, energy is used as a feedstock. Two is the tremendous penetration of plastic materials into customer markets and substituting for metals, substituting for glass, substituting for ceramics, and it represents a really important growth industry for U.S. industry.

Now, the plastic materials market is broken up into two distinct markets. The first is actually the plastic resins themselves, which a large part of you are probably familiar with. The second is the miscellaneous plastic products, which take plastic resins and make fabricated products out of them.

The plastics category also includes advanced composite materials. Now, here is an industry where relative price makes a difference, but the state of knowledge makes at least as much difference; in fact, I think it makes an even greater difference. That is, the state of knowledge is developing in such a fashion that, by 1995, large parts of automobiles—I mean the operating parts of automobiles, the engines, for an example—will be ceramic or plastic. The bodies themselves will use more and more plastics or plastic composites. The space shuttle, today, for example, has a large element of advanced composite materials in it, which represents essentially plastics woven with synthetic fibers, and also some, what is termed, metal matrix materials. They are very light; they are very strong; they are heat-resistant; and they are corrosion-resistant. And we're in the process now, in U.S. industry, of actually producing—I don't mean, when I say "producing," lab-size amounts—but actually running production runs of significant amounts of plastic products that meet certain types of specifications in terms of weight and strength.

By 1995, we think, it wouldn't be uncommon to see in automobiles plastic engines—not ceramics—plastic engines. I guess it was in the movie "The Graduate," in 1968, it was said, "Plastics is the industry to be in." He said that tongue-in-cheek. But the fact of the matter is it is certainly true.

The United States has proprietary technology in the

production and development of plastic materials. There is no other country in the world that produces advanced composite materials or knows that technology the way we do. Now, that is not to say plastics aren't produced outside the United States. In fact, they are. Plastic products—for an example, you know, paper clips made of plastic—that kind of stuff is being imported into the United States all of the time. But plastics are penetrating and they are penetrating significantly.

The rowscaler or average penetration rate for plastics is an index of penetration of plastic materials into customer markets. And you can see from Figure 8-5, not only is plastics a better material than lots of metals, it is also cheaper. And that has increased the penetration. The amount of plastics in automobiles is supplanting metals because of this relative price effect. It also has some nice properties as well. It makes the car lighter, more mileage per gallon, and it also doesn't corrode.

Now, even if you look at plastics relative to ceramics (Figure 8-6), which is a very, very close substitute for plastic products, what you will find is that, again, relative prices make a big difference. Here is an industry, therefore, that produces a product that penetrates other markets as a result of relative price change. The reason why relative price change here is so important is that it has all of these nice qualities associated with the fact that it is cheaper. It is as strong as metal, in many cases, or can be made as strong as metal. It is light weight. It has a number of properties that go beyond the fact that it is cheaper, that make the substitution process continue.

This story is interesting because it gets down to what structural change is all about, and that is, in this particular case, that plastics are substituting for steel, in durable equipment, construction mining machinery, and automobiles: one, because they are cheaper, and, two, because they are light weight.

A few years ago we did a study for a steel company, and what they wanted to know was: Given that imports have done damage to the U.S. steel industry—this was a statement made by, I guess, an Under Secretary of Commerce—how much of the damage done to the steel industry can be associated with exchange rate movements? How much of it is due to dumping and things like that?

FIGURE 8-5
Plastics Materials Rowscaler Versus Price of Plastics Relative to Price of Steel

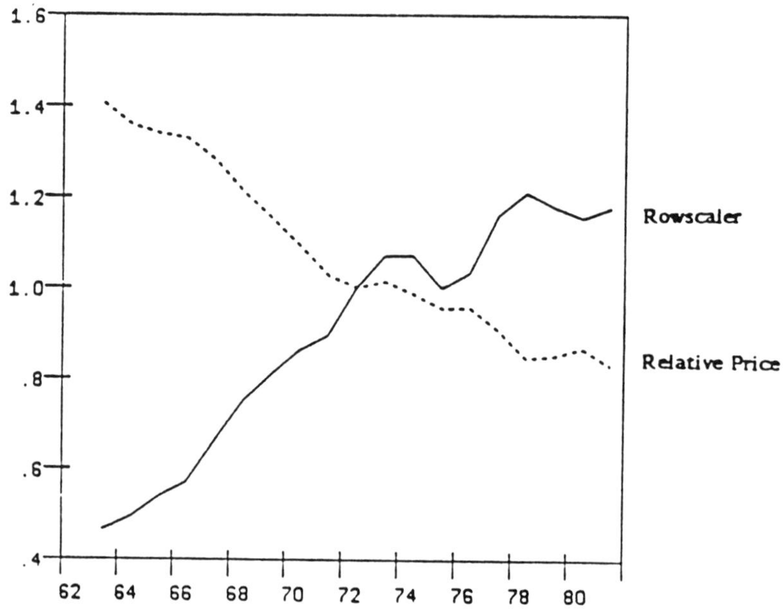

Source: Data Resources Interindustry Service.

FIGURE 8-6
Plastics Materials Rowscaler Versus Price of Plastics Relative to Price of Ceramics

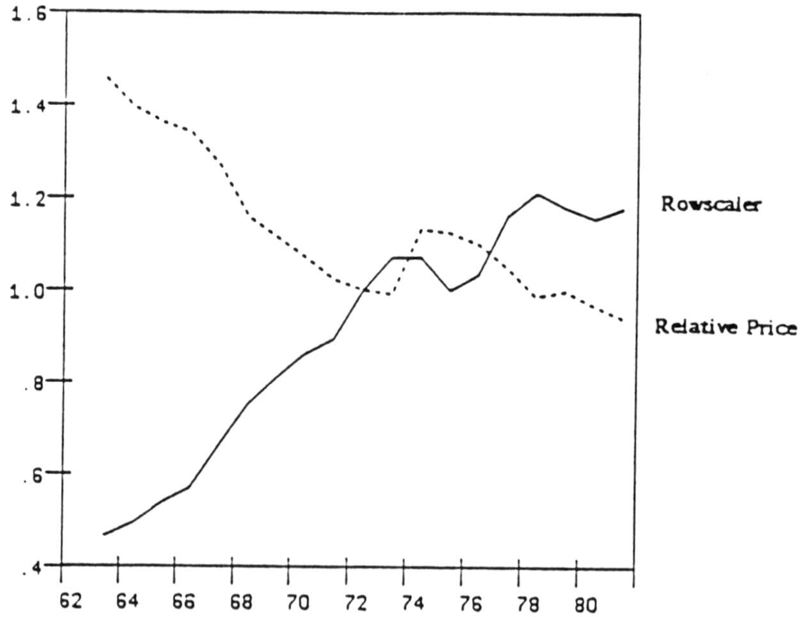

Source: Data Resources Interindustry Service.

We found that exchange rates had very little to do with the story on steel, but a dominant determinant of reduced steel usage in the United States was the fact that technology was going on in the end-use industries that precluded the increased use of steel and encouraged the substitution of plastics for steel. Clearly, in the steel industry, plastics were taking over a large part of the end-use markets served by metals, thus reducing the growth rate of metals.

The telecommunications industry (Figure 8-7) is growing by leaps and bounds. Its growth will be about 6 percent a year between 1985 and 1990. And what is interesting here is that the penetration of telecommunication services to every other industry is associated with a major decline in its relative price. That is, the price of telecommunication service has come down so quickly, there has been an inducement of firms to substitute away from other resources toward telecommunications. Industries that have suffered would be advertising, postal use, for an example, and other forms of information transfer.

Now, the argument, typically, has been that somehow there is a relationship between telecommunications and computers which increases the demand for telecommunications services, and that is true. But the fact is, when you run a historical simulation, most of the penetration for telecommunication services results from the fact that the industry itself was extremely productive. Its productivity rate was in the neighborhood of 4 or 5 percent a year, had a basis for reducing its relative price, and encouraged increased use. And here is a case where an industry is growing fast and most of it is due to penetration.

Finally, I want to make one statement about electricity. Again, the case here is that electricity actually penetrated its customer markets historically between 1963 and about 1978-1979, and then has remained flat, and is actually declining now (Figure 8-8).

Now, most of that has to do with two things. Most of it has to do with the relative price story. During the 1970s, the relative price of electricity was declining and there was an inducement, where possible, to switch from other fuel sources to electricity.

If you look at what is going on now, the relative price of electricity is rising, and major energy-using industries—for example, plastics, chemicals, and others—are now switching,

FIGURE 8-7
Declining Relative Price and Market Penetration by Telecommunications

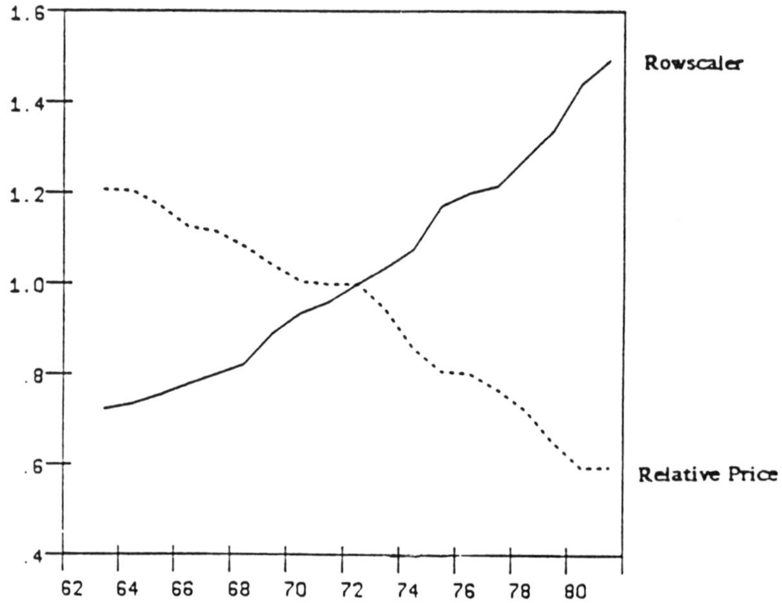

Source: Data Resources Interindustry Service.

FIGURE 8-8
Electricity Rowscaler and Relative Energy Prices

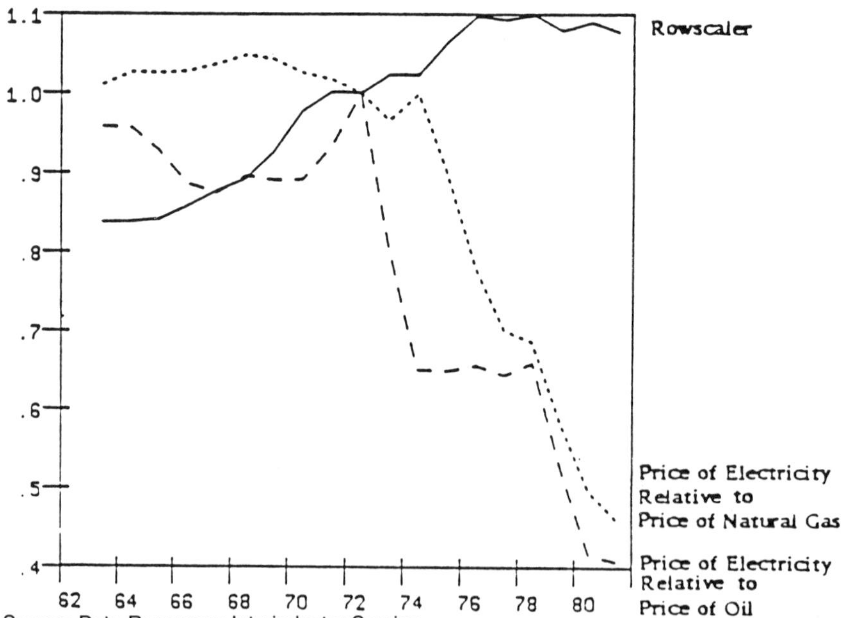

Source: Data Resources Interindustry Service.

where possible, from electricity to things like natural gas and oil. Now, as a result, the growth rate of electricity usage in the industrial sector will probably be at most about 1.5 percent a year. That is due to two things: (1) the fact that overall energy productivity in end-using sectors is growing, which means they are economizing on energy resources, and (2) electricity won't get its share of that productivity because its relative price is rising.

So, I think the bottom line from all of this is that, from a utility perspective, what it needs to do is limited in some sense. Unlike other industries, the utility industry is not mobile. It can't export energy. What it needs to do, I think, is ask the question: How can I possibly attract industries into my service territory? Because, for many electric utilities, they are involved in the sections of the country that are not growing, or are growing but very, very slowly. The fact that the price of electricity is rising relative to energy sources creates a demand for alternative fuels.

The real issue is this: How can you attract industries that are growing that tend to be electricity users? In terms of increasing the demand for industrial electricity, the industrial load, those kinds of issues have to be looked at and seriously evaluated.

QUESTION-AND-ANSWER SESSION FOLLOWING PRESENTATION

MR. BOB CILIANO, Dun and Bradstreet: Stan, I find your approach to analyzing structural change refreshing, in the sense that it does consider interindustry linkages and the fact that it is dependent on considering things like changing product design definition, the process steps that a product goes through in its manufacture: materials, labor, inputs.

An overworked example is the automobile in the 1980s being something that is comprised of special composite materials or microprocessors, and in terms of materials inputs in size and weight is something quite different than it was 15, 20 years ago.

The thing that bothers me about the approach, though, is this. How does one rigorously go about forecasting changes? The scenario approach that was introduced to us leaves me sort of cold in the sense that it seems to confound cause and effect and there doesn't seem to be any way in my mind—I hope I will learn more about it in the next day or two—to impose rigorous constraints on the consistency of these assumptions.

Aside from going to a Herman Kahn type approach, is it relative price? What factors can we really lean on to make consistent forecasts?

MR. FELDMAN: There is a host of literature, which has been developed since about 1979, on the econometric side, pure economics—Jorgenson and Bernt—that concentrates on what the production structures of the industries happen to be, and really looks at the role that relative prices play for a given capital stock, and what I call exogenous technological change, which is what is going on qualitatively in that industry that is going to change that production process above and beyond changes that would occur as a result of relative input price changes.

We have been extremely successful in using that model for about eight or nine industries to give us an idea of what role relative prices play in energy usage. Now, that is one area. The second thing is that there is a lot of work done by the Department of Energy that I have used on energy use by industry, and there is a wealth of information, which has been produced by EPRI and also under contract from DOE, that examines where the energy, in fact, is being used, just qualitatively. Can electricity substitute for natural gas in parts of the chemical industry? What we found is that the econometrics, when appropriately done, mirror what our qualitative assessment of those industries happens to be. And there have been eight industries like that that I have done recently, plastic materials being one.

When the elasticities of substitution that are coming out of the econometrics make some sense, given what we know about the industry, the actual production structure of the industry—and there are great descriptions in the literature of what that production structure looks like and where in the production process energy is actually being used—I mean, there is just a wealth of information. What you need is somebody to cull through it and put it together, which is what we did for about eight industries. I could offer a suggestion. There are a number of text search programs that make that a lot easier to do than simply going to every article and doing it. But there is a lot of work that is being done.

Now, how do you forecast it? Well, on the forecast, there are fairly rigorous econometric models on relative prices produced by DRI and others. That is number one. And, number two, if you buy my story about structural change being, in general, kind of gradual, there are even uses, for lack of anything else, of time trends in some of these models that will give you an idea of what the penetration rate ought to be or might be. And I think you get a good idea in separating out penetration rates and their causes from the fact that these end markets are growing.

So, there has been a lot of work in production theory that I don't think has been levered off in any serious forecasting context. And when I say "forecasting," I mean analytical context, not just push a button and the model solves and throws out some output. I mean some serious analysis.

For example, some of the scenario material that I listened to before might be very relevant in terms of specifying the parts of these models or the input you need that really can't be quantified.

MR. ADAM KAHANE, Lawrence Berkeley Laboratory: I am very interested in this link between econometric discussion of individual industries and end-use discussion, and one of the comments you made at the very end I found surprising in that context, when you said that some of your work showed potential shift from electricity to fuel—I think you said from electricity to gas.

Work that EPRI has done in different industries on electrotechnology has identified technologies where the shift is the other way, if at all, from gas to electricity, and it is the first time I've heard anybody suggest that you might see a shift caused either by technological change or by relative prices from electricity to fuels.

MR. FELDMAN: The plastic materials industry is one that comes to mind readily, where there is a switch. Motor vehicles and parts is another industry where natural gas and electricity are both used, both in the heating context; and electricity, of course, is used in the operation of the machines, where there is some substitutability between those two in the production process. And, for a given capital stock, it tends to be sensitive to changes and relative prices.

MR. KAHANE: But when you take into account the turnover in the capital stock, do you still see a trend?

MR. FELDMAN: Yes. It turns out that in the motor vehicle and parts industry, the capital stock tends to be biased to two energy types: electricity and natural gas. When I look at the embodied technology in that stock, it tends to be either electricity-using or natural gas-using. I mean that is just what the numbers say. It is not oil-using, for example. So, what I have is that, when relative prices change for a given capital stock, there is some switching that goes on. What the models are saying is this: How much of that can I actually quantify? And what I am saying is that, as a result of the relative price increase in electricity relative to natural gas, the growth rate of natural gas is increasing relative to electricity in that particular industry. It is also happening in plastic materials. It is not happening necessarily in all industries. But what the model does is separate out the pure technology effect or the embodied technology in the capital stock from the pure relative price effect. Kind of the switching issue.

MR. KAHANE: And what time period are you referring to?

MR. FELDMAN: Actually the forecast goes from 1982—because the data stops at 1981—through about 1990.

MR. SHISHIR MUKHERJEE, EPRI: Could you explain something about the international perspective of the structural change? I just want to explain it this way. Two things have been happening, especially, I would say, in the past ten years. One thing is that, to take advantage of the labor cost, many of the U.S. manufacturers are sending things abroad, starting from the garment industry to the electronics. The second thing is the technology shift; I believe very advanced technology is moving to the other parts of the world very quickly, within two years. Apple computers are produced in Taiwan within one year.

So, what is the effect of that, because most of the analysis is very much domestic still? Isn't it time to really look at what is happening internationally? Because most of the products, including IBM-PCs, are now an international product.

MR. FELDMAN: The computer is, I think, a poor example of what you are alluding to or what you want to say. The reason I say that is because it is true that the box that it comes in may be produced somewhere else but the operating systems definitely aren't. They are produced here in the United States. That's number one.

Number two is that many of what we call imports are production facilities of U.S. companies located outside the United States that import, that sell their goods back into the United States. Now, I know that, in terms of pure economics, that is simply saying, "Well, if that's the case, then U.S. labor and other resources aren't being used in the production process." But there are some benefits from the equity side, namely, stockholders of these companies presumably benefit from that.

Finally, what we are exporting is high value-added product, but that has always been the case for the United States, and what is coming into this country are commodity-type items that are very sensitive to changes in relative prices. And, when you look at, for example, components coming into the United States from Japan, for an example, what goes out are components. For example, I think last year our export of components to Japan was in the neighborhood of 18 percent growth, 1984 over 1983. What we sell them are very high value-added components, and the applications associated with them. Custom chips. We do that in the computer industry. And so what we are witnessing is that the commodity is coming in, you know, just produced and unfeathered; we do something to it and send it back out.

What we are witnessing in the past couple of years is that these commodities are coming in at a faster rate than we can sell the high value-added product, which is what is worrying people. But if you look at where the high value-added products go, they tend to go to parts of the world that have been growing very slowly in the last year and a half or two years. Number one, if their growth rates actually increase, our exports to those parts of the world should increase significantly.

Number two is this. There is a lot of technology transfer. I mean, the United States, at the moment, is the largest exporter of what I call proprietary rights and knowledge. It is an actual item in the balance-of-payments accounts. It is a service item. The neighborhood of our exports is about $8 billion. But most of that goes to subsidiaries of U.S. companies outside the United States, so it is not getting out into the public domain.

The high value-added niche business in high-tech, for example, is something which is guarded, very closely. Up until recently you couldn't get a measure at all of what the size of the chip produced by AT&T happened to be—that is, the volume of their production. Certainly, you know, the major producer of semiconductors in the United States is IBM. But we don't know how much they produce a year. It is very, very secretive.

I think what you are seeing is that those commodities that can be produced very efficiently either with labor or different types of production techniques—automobiles, machine tools—that is what is coming into the United States. We do not have a comparative advantage in that any more. But we do have a comparative advantage in the value-added products that go out.

9

IMPACT OF TRADE, TECHNOLOGY, AND MANAGEMENT FACTORS ON U.S. MANUFACTURING

*Barry Bluestone**
Boston College

In the next 20 minutes I want to look at the question of structural shifts in the economy, paying particular attention to the manufacturing sector. At the end of my talk, I will discuss what the broader shifts in the economy toward a so-called post-industrial service economy mean for your industry.

---

*This paper represents a verbatim transcript of the presentation given by the author at the EPRI seminar.

This morning you looked at capital and raw materials. As a labor economist, I tend to look at employment. Together with Ben Harrison at Massachusetts Institute of Technology and Alan Matthews at Boston College, I recently examined structural shifts versus cyclical and trade effects in what we call the decline of American manufacturing.

The question over deindustrialization—the idea that we have endured and may continue to endure major employment losses in basic industries in particular regions of the country—is one that became quite important beginning in 1980, if for no other reason than the publication of a rather somber special issue of Business Week called the "Reindustrialization of America." In that issue, Business Week pointed out that during the 1970s the American economy had hit the wall and lost its ability to produce and compete, and that, increasingly, the newly industrialized countries of Taiwan, South Korea, Brazil, and Mexico were following closely the lead of Japan and were catching up very rapidly.

During the next few years through 1984, there was a great deal of debate about the impact of foreign competition, of U.S. outsourcing of production, and, of course, also of multinational investment. The question was: Have we really lost our ability to compete, and what are the implications for American enterprise?

However, there is a smaller question within this broader one. There is no doubt that employment declined in key sectors rather dramatically. There is also no doubt that overall employment in the manufacturing sector remained approximately constant. It did not grow, despite the enormous growth in employment across other sectors of the economy during the entire '70s. But the real question and the one that I want to discuss today is this: What was the source of that stagnation in the manufacturing sector and the absolute decline in a significant number of individual industries?

The Brookings Institution responded to the discussions raised by Business Week and the work done by Ben Harrison and myself, Felix Rohatyn, Robert Reich, and Lester Thurow, with a four-point offensive against the claims of deindustrialization. Number one, deindustrialization is simply a myth; it doesn't exist. We are experiencing a cyclical decline because of the slow economic growth rate in the United States, exacerbated by the high value of the dollar.

The second point was that if deindustrialization is not a myth, that is, if the United States faces a serious problem that is structural in nature, there is really not much that can be done about it. Attempts to develop industrial policies along the lines of Japan's Ministry of International Trade and Industry (MITI) will not work. According to the Brookings analysis, MITI is nothing more than a nice little discussion group that has no power whatsoever over the economy. Therefore, we can't look to Japan, or any other country for that matter, for a way out.

The third point was that even if MITI works in Japan, and even if deindustrialization is not a myth, there is not much we can do about it here in the United States. American bureaucrats are simply not smart enough to outguess the market in picking winners and losers.

Finally, the fourth point of the Brookings Institution was that even if American political bureaucrats are smart enough to pick winners and losers, the 435 Congressmen and the 100 Senators will not let their choices survive.

I'm not going to debate the last three issues. What I want to do is focus on the first point: Is deindustrialization a myth? In particular, are the changes in employment within manufacturing industries basically due to structural causes having to do with our loss of competitiveness, multinational investment, and new technology, or are the changes fundamentally cyclical in nature, presumably adjustable through enlightened fiscal and monetary policy? In short, are the losses that we find in key sectors of the economy structural, presumably requiring something beyond a macroeconomic solution, or are they cyclical?

The analysis that I want to present today is just a couple of months old and is still not complete, but I think it is complete enough to show it to the public and get some response. The analysis is based on a set of regression analyses of 74 manufacturing industries over the period 1958 through 1984. These 74 industries represent roughly 83 percent of all employment in the manufacturing sector.

The model attempts to investigate how much of the change in employment in each of these 74 industries can be directly related to changes in the business cycle measured as deviations from GNP trend; how much can be related to changes in the value of the dollar measured as the trade-weighted exchange rate with our major trading partners; and then how much can be related to a change in long-run trends after controlling for these two cyclical factors.

To do this, we had to select a set of years over which we expected some change to occur. Bob Marlay looked at the year 1973, as did Bob Lawrence. We looked at the years 1967 to 1973 as a "crisis period" in the American economy. It was the period in which we first began observing inflationary pressures following 4 straight years of unemployment rates below 4 percent. It was the first planned recession in over a decade—the 1970-1971 recession—planned to try and weed out inflationary pressures. It included August 1971 and the advent of the new Nixon economic program that imposed wage and price controls for the first time since World War II. It included 1973, the first oil embargo, and, in the political/economic context, the beginning of our forced withdrawal from Vietnam. The period 1967 to 1973 marks the era when America moved away from the go-go economy of the 1960s to the stagflation of the 1970s. The question is: How did manufacturing industries respond to that shift?

The data for this analysis come from the historical output and employment data base generated by the Bureau of Labor Statistics using Bureau of Economic Analysis input/output categories. Essentially, we project out to 1984 the employment trends that would have occurred in each of the 74 industries if there had been no structural shift in the industry. We then compared this projected employment level with actual employment in each industry. Clearly, you can get various employment pictures. One picture would be of an industry that was growing very rapidly and then, following the crisis period or beginning within the crisis period, began to collapse. Another would be of industries that were declining during the early '60s and turned around. In fact, we found a number of industries that seemed to have positive structural adjustments in employment after a relative decline. Still another picture would show industries that were in long-term decline, and whose decline accelerated following this period. In this sense, the methodology was highly flexible in providing a taxonomy of the types of shifts that occurred in the economy, again, after controlling for cycle and exchange rate.

The results of our study break down into three groups (Table 9-1). The industries that underwent no significant

TABLE 9-1
Trend Employment in Industries Exposed to the Possibility of Structural Shifts During the "Crisis" Period 1967-1973 (thousands)

Industry (1)	BEA Code (2)	Actual 1966 Employ (3)	Actual 1984 Employ (4)	Break Year T* (5)	Change in Employ. T^*_{-1} to 1984 (6)	ESTEMP minus PRO-JEMP 1984 (7)	Ind. Type (8)
I. Industries that underwent no significant structural change during 1967-73							
Fabricated textiles, n.e.c.	19.01-19.03	175	180	NA	NA	NA	IA
Printing & publishing	26.05-26.08	513	632	NA	NA	NA	IA
Screw machine products	41.01	108	95	NA	NA	NA	IA
Fabricated metal products, n.e.c.	42.04-42.11	297	331	NA	NA	NA	IA
Elec. transmission equipment	53.01-53.03	190	221	NA	NA	NA	IA
Elec. industrial apparatus	53.04-53.08	214	213	NA	NA	NA	IA
Telephone & telegraph apparatus	56.03	128	151	NA	NA	NA	IA
Electronic components	57.01-57.03	389	685	NA	NA	NA	IA
Aircraft	60.01-60.04	754	636	NA	NA	NA	IA
Railroad equipment	61.03	61	36	NA	NA	NA	IA
Fabric, yarn, & thread mills	16.01-16.04	611	431	NA	NA	NA	IB
Hosiery & knit goods	18.01-18.03	236	207	NA	NA	NA	IB
Logging	20.01	129	82	NA	NA	NA	IB
Paints & allied products	30.00	67	62	NA	NA	NA	IB
Leather tan. & indust. leather	33.00	32	18	NA	NA	NA	IB
Manufactured products, n.e.c.	64.05-64.12	242	177	NA	NA	NA	IB
TOTAL		4146	4157		NA	NA	

TABLE 9-1
Trend Employment in Industries Exposed to the Possibility of Structural Shifts During the "Crisis" Period 1967-1973 (thousands) (continued)

Industry (1)	BEA Code (2)	Actual 1966 Employ (3)	Actual 1984 Employ (4)	Break Year T* (5)	Change in Employ. T^*_{-1} to 1984 (6)	ESTEMP minus PROJEMP 1984 (7)	Ind. Type (8)
II. Industries that underwent a positive structural shift during 1967-73							
Floor covering mills	17.01	45	53	1969	2.1	+11.6	IIA
Furniture & fixtures	23.01-23.07	150	196	1971	48.9	+48.3	IIA
Newspaper printing & publishing	26.01	360	443	1973	53.0	+13.4	IIA
Drugs	29.01	128	201	1967	72.7	+42.7	IIA
Fabricated struct. metal prod.	40.04-40.09	408	454	1967	45.8	+92.2	IIA
Cutlery, handtools, & gen. hard.	42.01-42.03	162	147	1973	-15.8	+13.7	IIA
Service industry mach.	52.01-52.05	125	181	1969	44.7	+64.0	IIA
Motor vehicles	59.01-59.03	866	867	1973	-9.8	+90.4	IIA
Ship & boat building & repair	61.01-61.02	183	205	1967	21.8	+63.1	IIA
Scientific & controlling instru.	62.01-62.03	189	225	1971	44.0	+30.2	IIA
Medical & dental instru.	62.04-62.06	62	176	1967	113.5	+73.7	IIA
Meat products	14.01	331	356	1967	25.0	+54.0	IIB
Grain mill prod.	14.14-14.17	133	130	1967	-3.5	+21.1	IIB
Household furniture	22.01-22.04	307	289	1970	-27.5	+44.2	IIB
Petrol. refining & rel. prod.	31.01-31.03	185	188	1967	2.6	+93.0	IIB
Cement & concrete prod.	36.01,.10-.14	223	228	1970	-0.2	+42.4	IIB
Pottery & related prod.	36.06-36.09	44	41	1968	-1.9	+15.7	IIB
Heating appar. & plumbing fixt.	40.01-40.03	71	65	1969	-7.1	+25.7	IIB
Engines, turbines, & generators	43.01-43.02	99	114	1967	14.5	+51.0	IIB
Electrical mach. & equip. n.e.c.	58.01-58.05	116	160	1967	44.2	+97.8	IIB
Optical & ophthalmic equip.	63.01-63.02	64	73	1972	8.2	+27.1	IIB

TABLE 9-1
Trend Employment in Industries Exposed to the Possibility of Structural Shifts During the "Crisis" Period 1967-1973 (thousands) (continued)

Industry (1)	BEA Code (2)	Actual 1966 Employ (3)	Actual 1984 Employ (4)	Break Year T* (5)	Change in Employ. T*-1 to 1984 (6)	ESTEMP minus PRO-JEMP 1984 (7)	Ind. Type (8)
Jewelry & silverware	64.01	71	73	1967	2.2	+28.9	IIB
Textile mill prod., n.e.c.	17.02-17.10	78	62	1967	-15.8	+13.5	IIC
Sawmills & planing mills	20.02-20.04	246	209	1968	-28.2	+119.8	IIC
Tires & inner tubes	32.01	107	101	1068	0.3	+33.1	IIC
Rubber products	32.02-.03,.05	170	150	1972	-5.1	+36.7	IIC
Structural clay products	36.02-36.05	70	39	1967	-31.1	+10.0	IIC
Metal stampings	41.02	237	222	1969	-25.5	+48.8	IIC
Household appliances	54.01-54.07	180	153	1969	-25.9	+24.3	IIC
TOTAL		5410	5801		346.1	+1330.4	
III. Industries that underwent a negative structural shift during 1967-73							
Sugar	14.19	36	26	1973	-10.8	-12.6	IIIA
Apparel	18.04	1243	1022	1970	-222.5	-183.8	IIIA
Paperboard	25.00	210	196	1969	-26.8	-51.5	IIIA
Synthetic fibers	28.03-28.04	118	176	1973	51.8	-48.0	IIIA
Iron & steel found. & forg.	37.02-37.04	309	191	1971	-108.1	-80.9	IIIA
Metal containers	39.01-39.02	79	61	1969	-22.1	-11.2	III
Radio & TV receiving sets	56.01-56.02	163	90	1967	-72.9	-41.4	III
Musical instru. & sporting goods	64.02-64.04	144	137	1973	-13.7	-20.5	IIIA
Dairy products	14.02-14.06	285	163	1971	-87.7	-18.2	IIIB
Bakery products	14.18	292	213	1972	-57.8	-13.4	IIIB
Tobacco	15.01-15.02	84	67	1971	-16.2	-5.4	IIIB

TABLE 9-1
Trend Employment in Industries Exposed to the Possibility of Structural Shifts During the "Crisis" Period 1967-1973 (thousands) (continued)

Industry (1)	BEA Code (2)	Actual 1966 Employ (3)	Actual 1984 Employ (4)	Break Year T* (5)	Change in Employ. T^*_{-1} to 1984 (6)	ESTEMP minus PROJEMP 1984 (7)	Ind. Type (8)
Leather prod. incl. footwear	34.01-34.03	335	184	1971	-115.3	-50.5	IIIB
Blast furn. & basic steel prod.	37.01	653	337	1971	-290.3	-185.4	IIIB
Watches & clocks	62.07	36	16	1973	-13.6	-6.0	IIIB
Paper products	24.01-24.07	460	486	1971	3.1	-47.7	IIIC
Periodical & book print. & pub.	26.02-26.04	184	262	1971	52.5	-38.7	IIIC
Agricultural chemicals	27.02-27.03	64	61	1971	-2.9	-20.8	IIIC
Cleaning & toilet prep.	29.02-29.03	109	148	1972	25.1	-26.7	IIIC
Plastic products	32.04	237	545	1971	231.2	-640.8	IIIC
Glass	35.01-35.02	179	165	1969	-11.9	-8.3	IIIC
Farm machinery	44.00	154	115	1970	-25.7	-33.7	IIIC
Construc., mining, oilfield mach.	45.01-45.03	191	194	1971	-10.7	-55.6	IIIC
Material handling equip.	46.01-46.04	88	82	1971	-11.3	-36.6	IIIC
Metal working mach.	47.01-47.04	343	310	1971	-13.9	-75.1	IIIC
Special industry mach.	48.01-48.06	205	168	1971	-28.6	-48.3	IIIC
General industrial mach.	49.01-49.07	285	277	1971	-8.1	-75.9	IIIC
Nonelectrical mach., n.e.c.	50.00	232	268	1971	33.7	-78.6	IIIC
Electric lighting & wiring	55.01-55.03	196	204	1971	7.0	-40.2	IIIC
Photographic equip. & supplies	63.03	98	126	1967	28.0	-16.4	IIIC
TOTAL		7012	6290		-738.5	-1972.2	
GRAND TOTAL		16568	16248		-392.4	-641.8	

structural change and are labeled IA in the far right column are ones that are growing rapidly or were growing before the period 1967 to 1973; they continued, after cycle and exchange rate, to grow at approximately the same rate. The ones labeled IB had just the opposite scenario; they were declining before the crisis period, and they continued to decline afterward.

The industries that underwent positive structural shifts during the 1967 to 1973 period make up Group II. One of the more interesting is motor vehicles. As a matter of fact, following 1967 to 1973, the motor vehicle industry grew a little bit faster than it had been growing before, and its crisis does not occur until after 1978. Part of the reason why the industry still continued to be above trend is not because of macroeconomic variables but because of the protection offered the industry through the voluntary export restrictions and also, in part, because of the Chrysler loan guarantee.

All of these industries had positive structural growth. The group labeled IIC were ones that were declining before the crisis period. They are declining at a statistically significantly lower rate today, and, therefore, we call that positive structural adjustment.

The industries that we are most concerned about are in Group III. These industries underwent negative structural change. There is a wide array of them—29 out of 74—and they include a series of industries that were growing (for example, apparel), and are now declining. The set of industries that were declining, but are declining even faster than before, includes the steel industry and footwear. Finally, the bottom set of industries, which includes a whole set of machinery production industries, was growing before but has had a significant decline in growth rate. The outstanding one here is the plastic products industry, which no doubt is continuing to grow, but at a much lower rate than its previous trend.

Table 9-2 presents a summary of these results. For now, I will just point to the Type III industries which represent about 42 percent of all manufacturing employment. Across these industries, there was a loss of about 1.9 million jobs, due to structural factors. Another 500,000 jobs—484,800, if you want a precise estimate—were lost as a result of the business cycle, and roughly 200,000 jobs were lost as a result

TABLE 9-2
Summary Statistics (thousands)

Industry Type	(No. of Industries)	1966 Employment	Percent of Total (74 Industries) Employment	Total Job (+) or (-) Due to Struct. Shift (T^*_{-1} to 1984)	Total Job (+) or (-) Due to GNP Cycle (T^*_{-1} to 1984)	Total Job (+) or (-) Due to Exchange Rate (T^*_{-1} to 1984)
I. No Structural Change						
IA	(10)	2,829	17%	NA	-288.1	-51.9
IB	(6)	1,317	8	NA	-83.0	-45.9
TOTAL I	(16)	4,146	25%	NA	-371.1	-97.8
II. Positive Structural Change						
IIA	(11)	2,678	16%	+543.3	-427.5	-43.4
IIB	(1)	1,644	10	+500.9	-197.8	-44.9
IIC	(7)	1,088	7	+286.2	-201.0	-0.5
TOTAL II	(29)	5,410	33%	+1330.4	-826.3	-88.8
III. Negative Structural Shift						
IIIA	(8)	2,302	14%	-449.9	-216.1	+25.8
IIIB	(6)	1,685	10	-278.9	-33.3	-45.5
IIIC	(15)	3,025	18	-1243.4	-235.4	-166.0
TOTAL III	(29)	7,012	42%	-1972.2	-484.8	-185.7
GRAND TOTAL	(74)	16,568	100%	-641.8	-1682.2	-372.3

of the rise in the value of the dollar.

For that set of 29 industries representing 42 percent of the manufacturing labor force in this country, it is clear that the dominant factors responsible for the decline in employment are structural variables, that is, variables **not** due to business cycle or dollar.

One last point I need to make is that these numbers are all **national** numbers. They all have to do with what is happening nationally in the economy. Of course, the concern of many people has to do with the fact that these industries are specialized to certain areas of the country. The Bureau of Labor Statistics recently released a report to the Joint Economic Committee on employment losses in manufacturing between March 1979 and March 1985. The states that have lost the largest number of jobs in the manufacturing sector are ones that are no surprises. In order of absolute number of jobs lost the top ten are: Illinois, Pennsylvania, Ohio, Michigan, New York, Indiana, Wisconsin, New Jersey, Iowa, and Kentucky; basically the Northeast, excluding New England, and the industrial Midwest.

What this suggests is that a continued regime of dealing with the structural and regional shifts in our economy, exclusively through national macropolicy—through an expansionary fiscal or monetary policy—will probably not have much effect on at least this set of industries or regions. Furthermore, with 7 million workers in these structurally impacted industries (out of approximately 20 million workers in the entire manufacturing sector) this is not a trivial problem. In fact, if we try to regenerate growth in these industries and regions simply through rapid fiscal and monetary expansion, we would probably find ourselves developing inflationary pressures in other parts of the economy before we did much about what was happening in the structurally impacted industries.

The implication is that we are going to need some specific policies to deal with these industries, particularly industries like machine tools. That means specific domestic industrial policies, and it may mean certain forms of short-term trade protection.

Let me conclude by suggesting how structural shift affects your industry. I studied the national input-output model for 1977, which was just released by the Bureau of

Economic Analysis earlier this spring, and looked at the total dollar requirements from public utilities (which unfortunately includes gas, water, and sanitary service as well as electric in the 1979 industry categorization) for every dollar of output of final demand in the manufacturing sector.

The range over industries from food to textiles, paper products, chemicals, plastics, glass, motor vehicles, and radio/TV communications equipment is roughly from 2.9 percent in radio and TV communications to about 10 percent in the chemicals industry. And, of course, in nonferrous metals, such as aluminum, it would be above that. Overall, the average is somewhere in the neighborhood of 4 to 4.5 cents of expenditure on utilities per dollar of output in those industries.

We have at SWRI (our institute) the largest microsimulation model in the world of the American economy. It breaks out electric utilities into a separate category. Using it, I looked at the following scenario: What would happen if we had a 10 percent cut in the value of imported automobiles offset by an equivalent increase in domestic production? In 1977 dollars, this would mean an increase of about $2 billion in increased demand for domestic vehicles, or a boost of about 3.75 percent in final demand in 1977. That would tend to increase the demand for electric utilities, per se, by approximately $45 million. If you multiply that, however, over the whole auto industry, this would have suggested, at least in 1977, about $1.2 billion worth of electric service requirements in order to produce American vehicles.

The amount of electric utility use in other sectors of the economy comes to about 3.5 to 4.5 percent in manufacturing with closer to 3.5 percent in automobiles and about 10 percent in things like chemicals and aluminum smelting. But look at other sectors of the economy. For every dollar of final output or final demand in wholesale and retail trade, one of the most rapidly growing sectors of the economy during the 1970s, total utilities, **including** gas, water, and sanitary, only account for about 2.8 cents per dollar of final output.

In finance, banking, and insurance, only about 1.86 cents worth of electric utility input is used for every dollar of final demand. In real estate, another growth industry, only about 1.4 cents of utilities is used for every dollar of final output. In business services, everything from advertising to public relations to putting on conferences like this, only about 1.7 cents worth of total utility input is needed for every dollar of final output. The big ones: hotels, about 5 cents, much of that because of heat and also water and sanitary services; and eating and drinking establishments require about 4 to 5 cents per dollar of final output.

This suggests that the structural shift that we're seeing is twofold: one, a continued erosion of major manufacturing sectors which continue to be large-scale users of electric utilities despite efficiency gains in that area, and that poses a problem for your industry; and, two, the shift from manufacturing to the service sector, which also suggests, in the long run, as we move to a postindustrial society, a decline in the relative intensity of use of utilities, including electric, gas, water, and sanitary services.

Overall, models that only look at the cycle, models that only look at macro-variables and fail to get down to the disaggregated level of studying sector versus sector, region versus region, give very poor forecasts of electric demand, something I presume you already know.

Unless we look at the basic vitality and the fundamental competitiveness of the American manufacturing sector and see how we fare versus not only Japan but, increasingly, the newly industrialized countries, we will not have a complete picture of what the forecast looks like for your industry. Indeed, we may look down the line and still be making forecasts that are too high rather than too low, relative to what a straightline macroforecast would give you.

QUESTION-AND-ANSWER SESSION FOLLOWING PRESENTATION

MR. AHMAD FARUQUI, EPRI: When you mentioned structural shifts, were you referring to factors such as the saturation of the market and perhaps international competition, in terms of the cost comparisons and other issues?

MR. BLUESTONE: The analysis so far is just to see whether, in fact, there is a structural shift going on, for whatever reason, outside of cyclical characteristics. This would include major structural shifts in consumption behavior, but it also includes productivity increases as it affects labor, international competition, and the investment decisions of firms, particularly those that are moving abroad.

The next step of our research is to look within this structural shift and identify which specific factors have been most important. What we will find, on the basis of what we have done already, is that, for certain industries, it is almost all international trade. The footwear industry is getting blown out of the water primarily because of international competition. In other industries, like textiles, almost all of the declining employment is due to tremendous productivity increases, because of the introduction of a whole new set of high-tech machinery into the textile industry. Labor productivity increases are responsible for the employment declines there, whereas trade is most important in the footwear industry.

In auto, it is a combination of both. The Commerce Department recently reported a projection that, by 1986, imports will make up approximately 35 percent of all automobiles sold in this country, up from about 25 to 26 percent now, and up from about 20 percent in 1978. More importantly—and this is the one that I think will be most important in terms of your industry and the kind of factor that is often ignored—the Commerce Department suggests that somewhere in the neighborhood of 30 percent of the value of domestically produced cars—basically GM, Ford, and Chrysler cars—will be made up of imported components, which, of course, use electricity generated in Japan, Brazil, and South Korea, not electricity generated near Detroit. All of those factors are the ones that we need to look at much more closely.

MR. SAM SCHURR, EPRI: This period of crisis that you referred to, beginning in 1967 and running until the early '70s, was sort of a transition period. There were systemic factors at work presumably.

I wonder if you would venture a hypothesis as to what the systemic factors were that began to have this effect around 1967. As I recall, this was in the Lyndon Johnson presidency. We had just gotten through the Golden Years of the Kennedy Administration into the Johnson years, the New Frontier, the War on Poverty, things like that. What is it that happened at that time? What are the underlying factors that began to produce this slowdown in the American economy, a slowdown which shows up, incidentally, also in rate of growth of productivity? You dated productivity slowdown 1973, but it actually began around 1966.

MR. BLUESTONE: That's correct. I have to give a longer-term historical answer to that in capsulized form.

During the period 1946 to, roughly, 1970, the American economy had what could only be described as hegemonic power in the world economy. The other economies of the world, of course, had been devastated by World War II. And, during that period of time, it was very clear that the American economy could grow at very rapid rates, not only because there was no competition from abroad, but because we could also be a major contributor to the economic growth of the countries that had been devastated during the war.

We had also had tremendous pent-up demand during World War II and a pent-up supply of savings. The result was a period of dramatic economic expansion during the 1950s and 1960s only marred by a couple fairly major recessions—1953 and 1958—and then continuing right through the '60s. During the '60s, of course, the demands for sharing the wealth created during that period of time led to the Great Society programs, the growth of the so-called social welfare state in the public sector. Similarly, there was a tremendous growth in the strength of the trade union movement, also attempting to get a share of the great wealth that had been created during the '50s and the '60s. The result was—and I am saying this as somebody who happens to be fairly left-wing politically—a great number of pressures put on both the private sector and the public sector during the 1960s going into the 1970s to maintain relatively fast-growing standards of living in the private sector and a relatively high social wage in the public sector.

What happened, of course, however, during the same period of time, was dramatic increases in international competition. At the same time that the demands for increased sharing of the wealth were being felt in the private sector, that sector was also feeling the increased competition from Europe and then from Japan, and, as I said, now from the NICs, the newly industrialized countries.

The question, therefore, is this: How does the system respond to that? How do managers respond to it? My argument is that there are a number of ways the corporate sector could have responded. One, managers could have redoubled their efforts at building better quality products. Two, they could have tried to find new forms of labor management relations that would have increased productivity and increased job satisfaction.

However, in general, during the 1970s, coming off of the halcyon 1960s, management did not choose either of these strategies. Instead, it chose strategies of investing abroad, of importing components from abroad, of moving production overseas, and, in the period in the late '70s, divesting themselves of manufacturing capacity in order to make short-run profit gains in the nonmanufacturing sector—U.S. Steel buying Marathon Oil, Mobil Oil buying Montgomery Ward, etc.

The result was that, by the end of the 1970s, many managers had failed the management test of maintaining a strong competitive manufacturing sector, and it was only by the end of the 1970s and the early 1980s that management began to wake up. I think if there is promise in the 1980s, it is the promise that our management skills have become better honed, that we understand that to keep a competitive economy we need to work with long-term horizons, that we need to develop new forms of labor/management relations, and that we need to develop higher-quality products. I think the one really hopeful sign for the '80s and the '90s is that we have learned that lesson.

I know that is a long-winded answer to your question, but it suggests that the environment has changed dramatically, both domestically and internationally. We have come through that period and we may do a little bit better in the next decade ahead.

10

INTERNATIONAL COMPETITIVENESS AND ITS IMPACT ON INDUSTRIAL STRUCTURE

*Robert Z. Lawrence**
The Brookings Institution

My topic today is "International Competitiveness and Its Impact on Industrial Structure." The only problem with my topic is the difficulty in defining exactly what I mean by "competitiveness." Indeed, I have been working on this topic for a number of years and, as far as I'm concerned, "competitiveness" is a word that is like "love" or "democracy." It really means what people want it to mean.

---

*This paper represents a verbatim transcript of the presentation given by the author at the EPRI Seminar.

The word competitiveness has several distinct meanings which depend on how you use the concept.

I think the concept does have important implications for the topic of this conference. What I'd like to do today is to explore what we know about some of these concepts of competitiveness and then to try to spell out what their implications are for changes in industrial structure. This is extremely difficult to do, needless to say.

In fact, I am reminded, as an economist, of the story about the two men who went out in a balloon. They went out over the countryside and were really enjoying themselves when, all of a sudden, a cloud storm came up, and the clouds came together, and there they were totally lost. They didn't know where they were, and they started to panic. But, all of a sudden, the clouds parted, and they looked down and, fortunately, there was a man standing below them in a field. And they looked down and they cried out to him, "Where are we? Where are we?"

And he looked up and he said, "You are in a balloon. You are in a balloon." Whereupon, the wind blew; the clouds came together again; and they were still totally lost.

And the one turned around to the other, and he said, "You know that guy down there—he must have been an economist."

And the other one said, "Why?"

He says, "Only an economist could have given us an answer with such great precision and of so little use."

Well, what I'm going to do today is to try deliberately to be imprecise. In a sense, my approach is going to be how to think about these questions rather than to give very detailed answers. And so, when I hear people talking about competitiveness, I think they use that term at least in two very distinct ways. And within those ways, you can start to make qualifications.

In one sense, when they say "Can America compete?" and "How well are we doing?" they are really asking a question: How well do we compare with other nations? And really this doesn't necessarily have anything to do with our interaction through trade or through other means with other nations. We ask how well we are competing, in terms of the various performance standards by which we rank economies. One criterion, the one that first comes to mind, is: How productive are we? And, here again, one has to distinguish between levels of productivity and rates of change of that productivity growth.

It is clear that, when you look at those stories, they are very, very different. The United States, by most measures, remains number one. Among the industrial worlds—let's say among major industrial economies, excluding certain oil-producing countries—the United States remains the world's most productive nation in terms of output per employee. (We can measure that best by a variety of purchasing power parity calculations.) Also, as best we can measure it, we remain the world's most productive manufacturing nation. Our output, again, per man-hour, appears to be higher than anywhere else in the world. But the rate of change of our productivity growth has been much slower than in other economies, and that was also true for the manufacturing sector. It is also not true, I think, that we remain number one across the board, in terms of our productivity capacities. Indeed, the Japanese, although they are measured at something like 90 percent of our productivity levels absolutely in manufacturing, have exceeded our productivity levels, certainly in steel, apparently in automobiles, and in a variety of other areas. If we look back over the post-war period, too, we find that our productivity growth has increased much slower than productivity growth in other countries.

Now, insofar as I am concerned, when we stand back and ask about the reasons for that development, I think, first and foremost, that these two facts are related: The fact that we were on the technological frontier, and the fact that it is easier, of course, to copy than it is to push out that frontier, is an important explanatory factor as to why productivity growth abroad has been much more rapid than in the United States. And indeed, one would expect, and we have observed, to some degree, that as nations come to the technological frontier, as they exhaust those benefits of relative backwardness, their productivity growth tends to slow down as well.

The Japanese, though, have maintained a remarkable productivity growth performance, particularly in their manufacturing sector. We have seen some slowdown, but productivity growth remains remarkably rapid. But, nonetheless, as a broad generalization, I think it remains an accurate explanation.

So, what we find, then, is a post-war period in which our productivity growth grows, for the most part, particularly prior to 1973, slightly better than it had historically over the long-term, but nonetheless, much more slowly than in other countries. We have, therefore, a convergence of international productivity levels, particularly among developed economies. That is a condition that will be with us for the next decade or more. I see a continued catching-up in productivity levels among developed countries, and I see the newly industrializing countries also accelerating their productive capacities relatively more rapidly than the United States, and, as I say, converging toward our levels.

That means the global environment to which we look differs particularly from that in the 1950s and 1960s, primarily because we cannot take our leads for granted, and, indeed, we are just like the British Prime Minister, one among equals, rather than an American President towering over other power authorities in the United States. And that means an economy in which pressures, because of a variety of factors—be they exchange rates, relative wages, regulatory tax policies—are all transmitted and have international implications to a much greater degree than they had in the past.

I still believe, though, that our wealth, our absolute living standards are not contingent upon our being number one; our wealth, in a sense, does not depend on others' poverty. In fact, we would face both opportunities and costs living in an environment as one among equals. Certainly, the competitive pressures are greater, but there are also the benefits: the benefits from being able to sell to fast-growing markets, and the benefits from being able to learn as others take over some of the burdens of pushing out that technological frontier. These will be clearly features of the environment into which we are moving.

Let me turn to the second notion of competitiveness, which I would like to address in much greater depth and devote most of my remarks today. That second notion has to do with the actual performance in international trade. How well or how badly are we performing in international trade? And here I'd like to make a second redivision, which concerns our aggregate trade performance. Let's take it as measured by our aggregate trade balance and ask: What are the implications of that trade balance? And then we must try to deal with the composition of trade, which is really what we are mainly interested in. That first question concerning our aggregate performance is, I think, of some significance, certainly in explaining the short-term fluctuations in our relative competitiveness.

Nations can perform well in international trade because of a variety of factors, and that is why, again, it is so difficult to appraise and to answer the question: How well are we competing? You can increase your sales, because your productivity growth runs faster than that of other countries. That is one part of the equation. And your costs, therefore, rise less rapidly than those of your competitors.

It is not necessary for a country which has more rapid productivity growth to experience an improvement in the price of its products. The cost of productivity can be reflected in higher wage growth and in higher profit growth, and, consequently, costs need not decline at all. So, you cannot simply jump from a statement about productivity growth to a statement about the price of the product, and hence, to a statement about performance in international trade. Indeed, you have to work on both sides. You can achieve productivity improvements while wage costs and profits remain constant, and therefore, the price will come down. Alternatively, you can have no productivity improvement while the wages and prices decline for some other reason, particularly, in the world in which we live, an exchange rate change. An exchange rate change may well lower our costs and improve our ability to sell without having any improvement in our relative productivity performance.

Indeed, over the 1970s, in my view, that was how we competed. There were a series of devaluations of the U.S. currency over the 1970s, so that, even though our relative productivity performance certainly declined, as compared with that in terms of productivity growth, our goods remained competitively priced in international markets. Over the 1970s, from the point of view of purely performance volume of sales—and that is what is important if you are concerned with industrial structure—the United States performed rather adequately. By that I mean, our trade balance—if we look at those criteria in manufactured products—moved from a rough balance in 1973 to about an $18 billion surplus as of 1980. If you look at the volume of our manufactured exports over the 1970s, they increased by something like 100 percent. By comparison, the volume of our imports of manufactured products increased by 72 percent.

So although U.S. productivity growth was slower than productivity growth in the rest of the world, our exports grew more rapidly than our imports. In my book, Can America Compete?, I found that an employment increase from those export flows offset the decrease from the import flows. On balance, therefore, our manufacturing sector wasn't much affected in its overall size by international trade.

That is not to say the composition of the manufacturing sector remained unchanged. Clearly, we were rapidly engaging in a process of international specialization, but, in the aggregate, the impact of our trade performance was not enough to reduce the size of the manufacturing sector. And, indeed, what I conclude from that is, in the environment in which we live, the United States can compete.

Since 1980, needless to say, the story is very, very different, but, again, in my view, it reflects not the contribution of the environment but the role of the exchange rate. The exchange rate, in turn, reflects our own macroeconomic policies, and the change in the exchange rate is the major structural change in terms of the United States' relationship with the rest of the world: Our large trade deficit is not, in my view, the result of policies abroad, but overwhelmingly the result of our own policies.

Basically, there is a direct link between the two deficits which we are experiencing today: the budget deficit and the fiscal deficit. After all, another way to think of a trade deficit (the difference between exports and imports) is the difference between what the nation earns—its income—and what the nation spends. The only way you can spend more than you earn is to borrow from the rest of the world, and, indeed, that is what a trade deficit is. Technically more correct, a current account deficit tells you that a nation's expenditures exceed its income. And that deficit can, in turn, stem from two factors: Either the country's government spends more than its income or the private sector spends more than its income.

What has happened since 1980 is, of course, that the government has vastly changed its spending patterns relative to its income or its taxation patterns. A large government deficit has resulted. The private sector hasn't really changed its spending patterns much and, therefore, unless the private sector is prepared to change its spending patterns by increasing its savings and buying those government bonds, the government deficit has to be financed from the rest of the world. And that is why our trade deficit today is essentially a macroeconomic phenomenon.

Of course, our trade deficit has structural implications, and one implication has been the dramatic strength of our exchange rate, which has, in my view, priced our products virtually out of international markets. Now, the issue looking out into the medium- and long-term is: What are the implications? In my view, in the short run and the long run we see very, very different implications from the posture of our macropolicy.

The short-run implications are to provide a source of downward pressures on the size of our manufacturing sector. Now, it is actually quite remarkable that, in the aggregate, we have experienced, for the most part, a normal economic recovery. Trade has negatively influenced our manufacturing sector. We have got about a $90 to $100 billion decline in our trade balance in manufactured products, and yet the manufacturing sector has recovered very normally, given the growth rate of the overall GNP. If you take a simple regression of manufacturing employment and relate it to Gross National Product, you will find you are dead on track throughout this whole economic recovery. The same is true of aggregate industrial output. It is also on track in the course of this recovery. So, we have got a strong negative influence coming from the traded goods sector, yet we have got a manufacturing sector that is doing almost exactly as we would have expected. That creates a kind of a puzzle.

In my view, what has happened is that there have been offsetting switches in the domestic demand that have compensated the manufacturing sector in the aggregate for the slow growth of its international markets. One stems from defense procurement, a relatively small factor. It was important in giving us a kind of a normal recession, given GNP. In the recovery, it has been the strength of equipment expenditures. This recovery is noteworthy for the dramatic rise in equipment investment. So, what we have got is an aggregate story, certainly not for every industry in the manufacturing sector, but an aggregate story that has, peculiarly enough, tended to cover up the impact coming from the traded goods sector.

Now, I know that, in the past year, when the economy slowed down, manufacturing employment started to taper off and has declined somewhat, and industrial production has also remained flat or slightly downward, but that is more or less expected, given a GNP growth rate of 2 percent or below. Manufacturing is supersensitive to the cycle.

But, be that as it may, those industries which are price-sensitive, in terms of their products, have been severely affected in an adverse way, by the strength of the real exchange rate. And that runs the gamut from the agricultural sector, whose products are most price-sensitive, through the basic mining and the metals industries. Even the high-technology sectors of the economy have been negatively affected, less severely, but, nonetheless, clearly present.

In my view, the United States, now a debtor nation, is on a path in which we will accumulate a very, very significant debt to the rest of the world that will bring about a reversal of these current circumstances, and the question is not whether but when. What will the timing of the reversal be?

One possibility is that the foreigners will continue to have faith in this country and they will be willing to lend to us more or less at the current conditions. Under those circumstances, we borrow $100 billion or $120 billion every year, and we pay 10 percent interest. So, what we know is: A 12 percent interest flow starts to build up for each year we borrow, moving the other way. Eventually, the cumulative impact of our borrowing will be to build up a tremendous negative outflow in terms of interest payments. And that, in turn, will start to bring down the exchange rate, gently if foreigners are still willing to lend to us for some purpose or another. But, nonetheless, what we will see is a slowly declining exchange rate.

Another possibility is that, at some point in time, they will lose confidence in the United States, and therefore, we will get a more severe set of circumstances: an exchange rate which falls rather rapidly; the need to raise interest rates in order to attract the capital; a slowdown in our economy perhaps as a result of rising interest rates. But, in either scenario, the exchange rate will come down. Let's take 1980 as a sort of benchmark for the real exchange rate, and assume that, as I believe, it was a rough sort of equilibrium. We may ask: Are there any real factors that have changed? Are there any reasons for us to believe that the dollar should remain permanently higher than that level? In my view, the only reason is what has happened to the oil price and our energy dependency. That is on the positive side.

But, on the negative side is the legacy of this experience which has transformed us from a net creditor nation with positive net inflows from interest income to a net debtor nation, which will gradually accumulate outflows of interest payments. Therefore, I think, broadly speaking, this latter effect is going to dominate. And what we are looking at is a development over the next decade or decade and a half in which this country has no choice but to export, and, indeed, in which the relative price effects which we've seen move so negatively over the past four or five years reverse themselves. That is relatively good news for the manufacturing sector in the aggregate.

I would not expect the same industries which bore most of the burden of international competition, or rather all of those industries, to restore themselves. Some of them, say steel, were hanging on by the skin of their teeth, and they have now been tipped over past the point of shutdown. When the dollar now falls, the new facilities that have to come on line to service our international requirements will come from other areas, and I would expect them to be in the areas where we have experienced growing comparative advantages. Let me now turn to that second aspect of the international competitiveness, which is the changing industrial composition of our trade performance.

Broadly speaking, one of the interesting features over the past two decades is how strong the trends in the aggregate have been. If you look at a chart of the U.S. net exports divided up by technology, just as one split, in terms of technological intensity, you find two almost mirror images. On the one hand, we have a growing surplus in high-technology products, while on the other hand, we have a growing deficit in products which, for want of a better word, could be termed as low-technology products. That picture is extremely striking.

Now, over the past few years, both those balances have

declined. As a result of the dollar's strength, our high competitive performance at the aggregate level hasn't been too good either, and the balance has been shrinking, but the decline in the balance, if you look, has been much, much lower in the high-technology sector than in the "low-tech" sector. And, indeed, the patterns of U.S. international specialization conform quite broadly to what we would expect in terms of the various theories that economists have dreamt up to explain trade performance.

The most famous of those theories is the Heckscher-Ohlin explanation for comparative advantage, which looks at changes in relative factors of production. That theory predicts that a country will specialize in exporting the products which use the factors of production which are found in that country in relative abundance. Let me stress that that is only one theory explaining trade patterns. And, indeed, if you look at statistical estimates, that theory can explain maybe 22 percent of the variance of trade performance. We are talking about something that is significant but not determinative. There are many other factors as well.

But, broadly speaking, that makes sense, in terms of what we've seen in the U.S., growing specialization in the high-tech part of production, and, indeed, in agricultural products. In fact, the relative abundance of land in this country has actually increased compared to our trading partners, over the 15 years from 1960 to 1975, and the relative abundance of skilled labor has also increased, surprisingly. Unskilled labor, on the other hand, is the key determinant of our imports.

Capital, although it fits into that theory, has not been as powerful an explanatory variable. There is some debate, actually, as to whether the United States remains relatively capital-intensive. Certainly, it is true that, over the post-war period, investment abroad was much more rapid than in the United States. So, therefore, in terms of changes, we have experienced a declining relative capital intensity in our economy. And, indeed there is some sense, too. If you look at the pattern of trade, what you find are the trade balances in those capital-intensive products—steel, for example—beginning to turn negative in the late 1960s, early 1970s, automobiles, of course, turning in the 1970s as well. So, that theory offers us some kind of an explanation.

In my view, the theory has been very useful in explaining the trade performance, particularly between the United States and the developing economies. But, increasingly, that theory becomes less and less useful in explaining competition, particularly between the United States and other developed economies, where we are becoming increasingly similar to one another.

The other striking feature of the post-war period is the tremendous growth in intra-industry trade. If we took the simple theory of comparative advantage very literally, and countries were continually specializing, we wouldn't have expected to see interpenetration of markets of the kind that we have seen in the differentiated products. Indeed, once others converge to our levels, it appears as if the more technologically sophisticated the products, the more intra-industry penetration we get.

We can sell computers to Japan. IBM can sell computers to Japan. Fugitsu can sell computers to the United States. That implies that it is much tougher to forecast an overall pattern in terms of the profile of that outcome. But I would expect these trends to continue and, looking into the future, in terms of our trade with the developing economies, comparative advantage theory, which stresses the labor intensity of the LDC production processes, will remain a powerful explanation. The statistical evidence, though, suggests that all of the major industrial economies are already all specializing in what is termed as high-technology products, and I think that they will increasingly follow this pattern of specialization.

One of the theses of the book that I wrote was, in fact, that Americans who discovered the importance of the rest of the world only in the 1970s, now actually exaggerated the role of international trade as a source of structural change in the U.S. economy, and I tried to explain why that is. In fact, what is happening to us is that the changes that are taking place for domestic reasons are tending to reinforce the changes that are taking place because of international trade.

Since 1973, as I'm sure you are all aware, those industries which we call basic or low-tech, are experiencing slow economic growth, because of factors internal to the U.S. economy as well as because of international trade. Demand patents and productivity growth remain dominant sources of structural change, and what we have seen since 1973 is a switch in domestic demand and technology factors toward these high-tech sectors. On top of that change is a continual pattern of international specialization which has reinforced it. We are increasingly specializing in high-technology products and we're becoming major importers of labor-intensive products and more standardized commodities. That pattern again, I think, will continue out into the future.

So, to sum up, let me say that I see a manufacturing sector, contrary to what others have suggested, whose output performance is really quite similar to what it was in the past. Production of manufactured goods, I would expect, would increase roughly as fast as or maybe even slightly faster than GNP, over the next decade and a half. Employment in manufacturing will continue to decline as a share of total employment in the U.S. economy, although I think that aggregate employment in the U.S. manufacturing sector could remain constant or even slightly grow—maybe half a percent a year, on average, looking out over the medium term—provided the economy can sustain its overall economic growth. However, within that, the changes towards a high-technology-based manufacturing economy will continue, driven both by domestic demand factors and technological change as well as by the process of international specialization.

QUESTION-AND-ANSWER SESSION FOLLOWING PRESENTATION

MR. AHMAD FARUQUI, EPRI: You indicated that during the 1970s when the changes were occurring in the industrial structure—considering the sector as a whole—the employment effect wasn't particularly significant, but there were offsetting changes.

Would you hazard a guess as to what the implications might have been for energy consumption, given that several of those industries where heavy losses occurred were also energy-intensive in one sort or another?

MR. LAWRENCE: Well, since we know what happened to energy consumption, I don't think you have to hazard a guess, although there remains the question as to how much was due to structural change and how much was due to relative price changes. Clearly, these were the two—let's put it this way—and those two issues are related. And so it is very difficult for me to disentangle the relative contribution of the fact that high energy prices got passed through; in time, high materials costs, in turn, led to substitution away from those materials and, in turn, led to a greater economizing on energy. And it is noteworthy that most of the price-sensitive energy substitution took place in the industrial sector.

But I would say that, generally, I do think that this transformation away from what I term basic industries is also a transformation away from heavy energy-using sectors, and, in particular, I would say that we are moving to a world in which scale is much less important than it used to be, and I see this partly as a result of the electronics revolution.

If you were asked, for instance, to tell me how many automobile companies we are going to have competing in the United States over the next ten years, I think you will come up with a number like eight, nine, or ten—something like that. Yet, in the period we have just been through, only a few could survive. And you will notice, in a variety of indicators, that it is really the large capital-intensive industries who have experienced particularly slow growth. So, I think that that is partly related to energy, but it is also related to sort of longer-term structural transformation toward high technology and that, in turn, has a negative impact on underlying energy use.

Do you have a follow up?

MR. FARUQUI: Just a follow up. To what extent is the evolution of the economy toward a high-technology orientation related to the United States having a comparative advantage in the production of those types of products that use technology more intensively? In other words, I am trying to reconcile that observation with a statement I think you made earlier that trade has not been as important an influence in determining structure.

MR. LAWRENCE: Yes. Let me give you a number, and this is roughly correct. Over the 1970s, our total employment in manufacturing was about constant. Our employment growth—actually, this is since 1973—in high-tech was about 15 percent. Of that, my estimate is that roughly a quarter of the employment growth could be ascribed to greater export growth and lower import growth, relatively.

So, I would say, looking over the 1970s—and these numbers are actually quantified in my book—about a quarter of the change was due to international factors, and about three-quarters to domestic factors. So that would be my rough sense of what the relative contributions were from the point of view of the manufacturing sector.

MR. A. J. NORTNEY, a visitor: Would you have any comment on the impact of the world communication network on your economic equation? In other words, with the global satellite communication network, the world becomes one huge network of communication. This portends both more education, among the lesser-developed countries, and more knowledge of industrial processes all around.

Do you have any comment on networking?

MR. LAWRENCE: Well, I think, clearly, what we have are major opportunities. I see two trends. Like most economists, I have two hands. And I don't know how they quite come together, and they are both sort of the result of this electronics revolution in both telecommunications and in production technologies.

The first leads you to believe that we are going to see shifts toward increased international specialization. Primarily because of these networks, it becomes increasingly possible to rationalize production processes.

Now, what that says is: Don't be misled into looking at industries as units, because, at the disaggregated level, the processes are going to be rationalized. That is why a lot of the discussion about whether we are going to lose this industry or whether we are not going to lose that industry really to me is much too aggregative. We are going to lose parts of industries and we are going to specialize in other parts of industries. Those will reflect a comparative advantage. And you see that today. With industries like the textile industry, remarkable amounts of specialization exist even within that industry, in terms of knowledge-intensity as compared to labor-intensity.

So, I see that as increasing international specialization and rationalizing it and, therefore, to some degree, accelerating the pace of change.

The other side of it, though, is that the capacity to automate and to tailor production runs, rather than a need for large economies of scale, can potentially lead firms to locate quite close to a market and to tailor product needs to the specific demands of that market.

To use autos as just one example, where economies of scale were terribly important, if you needed to make 500,000 autos because the refueling process was extremely expensive, well then basically, you had to make your small car in Japan, and you had to have a big domestic demand so you could get, say, 300,000—just to dream up some numbers—and then you could export 200,000, and you could get down to those scale levels.

But, if 150,000 was all you need, and if those costs of transforming your production system went down, well then, you might be able to locate closer. And so that, in a sense, may lead us to see, in certain areas, where product differentiation matters; we may find products coming back on shore. So, I see the technological revolution as meaning different things. I think it is difficult to come down one way or another. You can get plausible arguments that take you both ways.

MR. SHISHIR MUKHERJEE, EPRI: My question is related to structural change in relation to what we call industrial policy. Now, we know that, after oil prices rise and after some of the new industrial countries move in, structural changes will have to take place in the industry.

But, I'd say, during the past three or four years, we should consider the sort of uncertain economic involvement with growth and no-growth, high interest rates, and all these things. How much has that contributed in, let's say, the absence of an industrial policy for mobilizing rationalization or mobilizing capital investment?

It seems that the United States is much slower to rationalize the industries as compared to Japan or some of the industries, even in Europe, in terms of new technologies to increase productivity.

MR. LAWRENCE: Well, that is a very big question. Let me just give you some broad indications of how I would answer that.

First, in terms of the nature of the structural changes that we have experienced, I try to indicate that I see those as overwhelmingly due to our own macroeconomic policies. And, in fact, if you ask yourself what an industrial policy would have done, in my view, it would have been the wrong instrument. For instance, we have created an aggregate deficiency of resources in this economy, because our spending is exceeding our production at home. Therefore, we have to get those resources by importing.

Now, industrial policy advocates would put on quotas or tariffs, and try to keep that production in certain sectors of the economy. But we have an aggregate deficiency. So, in my view, what they would do is to shift the burden. They would not deal with the fundamental problem. It is like treating the wrong disease. So, overwhelmingly, I think that our difficulties, to the degree that we have experienced them, and some of our successes in economic growth, stem from our macroeconomic policies, not from a failure in micro.

Now, if you do compare the performance of the United States in adjusting to international change, I would respectfully disagree with you, fundamentally, in terms of the flexibility of this economy, certainly as compared with the Europeans, and possibly even with the Japanese. It is true that we haven't had deliberate government policies designed to facilitate structural change, but perhaps, because we haven't, and more importantly, because of the nature of our economy, we have been remarkably flexible. And one of my great fears about an industrial policy is that it would represent a rigidity in adjusting to the changes from whatever source.

Now, I believe that there is a lot of scope for improved structural policies in this economy, for aiding the adjustment of workers, for increasing research and development expenditures, for what you can call generic kinds of policies. But I am skeptical that industry-specific policies are likely to be either efficient or even effective in facilitating international adjustment. In fact, I think they are more likely because they will be implemented through our particular political system. Given our political system, which has great strength in its responsiveness to a variety of interests, I think that it would be a force for preventing structural change rather than for accelerating it.

I think that, if you study industrial policies around the world, it is really striking that, where they have succeeded, they have been tailored to reflect particular political and institutional realities. I personally think that the Japanese industrial policy has made a contribution to their industrial performance, but I am skeptical that that kind of approach will work, given our political system. And that is why I think we should improve our structural policies, but they won't do anything until we get our fundamental macro-policies in line.

11

ASSESSING THE FUTURE COMPETITIVENESS OF AMERICAN INDUSTRY IN WORLD MARKETS: SOME PRELIMINARY RESULTS

Pradeep Gupta
Electric Power Research Institute

Ahmad Faruqui
Electric Power Research Institute

International trade has become an important factor for the U.S. economy at the same time that the U.S. economy has become less important in world markets: see Figures 11-1 and 11-2 (**1**). This paper examines the predictability of U.S. international competitiveness. It divides the problem into two parts: (1) forecasting the aggregate trade balance and (2) forecasting the commodity composition of the trade balance.

FIGURE 11-1
Declining Share of U.S. Manufacturers in World Trade, 1953-1975

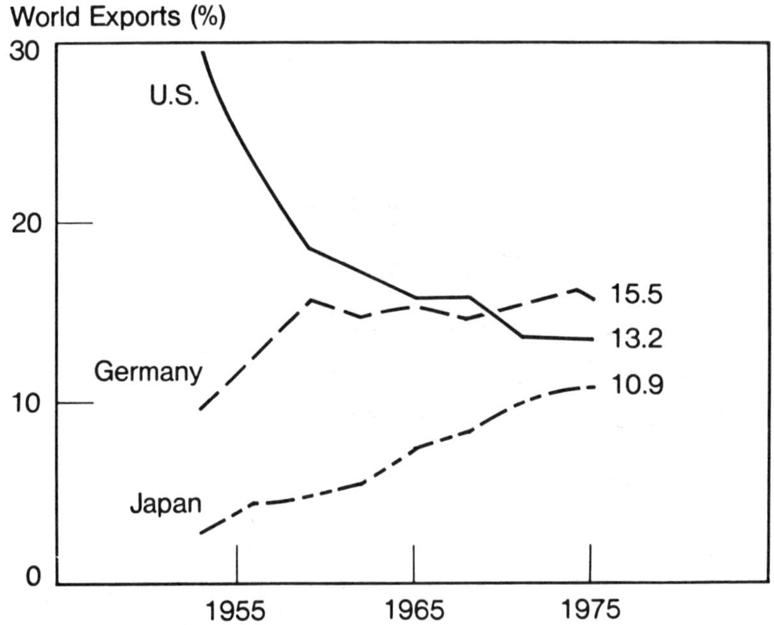

Source: William Branson, *Trends in United States International Trade and Investment Since World War II, National Bureau of Economic Research Monograph*, 1981, p. 196.

AGGREGATE TRADE BALANCE

Several factors determine the aggregate trade balance: the currency exchange rate; the costs of labor, capital, energy, and raw materials; growth of productivity; and export pricing policy. Each of these factors is reviewed briefly below.

EXCHANGE RATE

Figure 11-3 describes the relationship between the U.S. merchandise trade balance and the dollar exchange rate. The observations for 1982, 1983, and 1984 show a pronounced negative trend: As the dollar exchange rate goes up, the trade balance becomes more negative. While the earlier data points do not support the negative trend, the apparent negative relationship witnessed during 1982-84 appears to be continuing. The merchandise trade deficit in

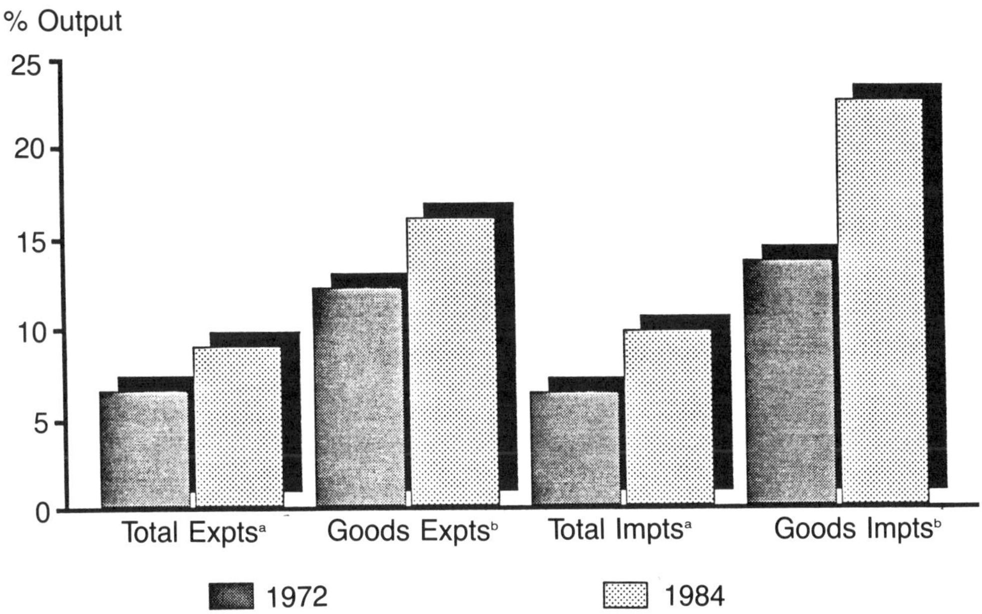

[a]Percent of GNP.
[b]Percent of goods component of GNP.
Source: *U.S. Industrial Outlook*, 1986.

FIGURE 11-2
Increasing Role of International Trade in the U.S. Economy

1985 was $124.3 billion, not much higher than the 1984 figure of $123 billion, while the dollar fell by about 19 percent since its peak in March 1985. Prospects for the future are uncertain. Even if the dollar continues to decline—it had reached 168 yen at the time of this writing—it is not likely to dampen the rising industrial prowess being displayed by several of the Newly Industrializing Countries (NICs). The U.S. trade deficit with the Asian NICs (Taiwan, Korea, Hong Kong, and Singapore) increased from about $7.5 billion in 1980 to nearly $127 billion in 1984. To quote from **(3)**: "A decline in the dollar may not significantly reduce their competitiveness, since **most of their currencies are managed so as to maintain a desired relationship with the dollar.** Thus, their wage costs have remained about 10 to 20 percent of those in the U.S. These much lower labor costs also hold down indirect manufacturing costs. Building a new plant costs less than in the U.S., as does the multitude of services that support production operations. In sum, the

FIGURE 11-3
Aggregate Trade Balance and Exchange Rate

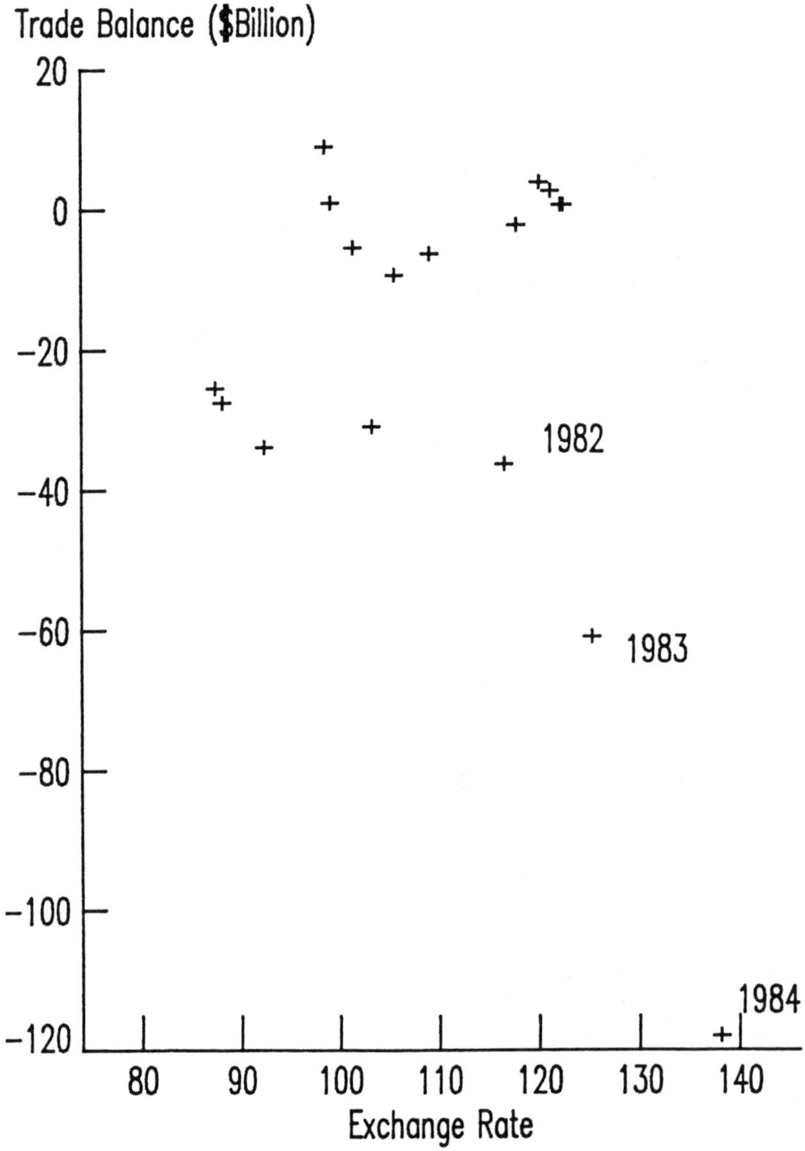

Source: *U.S. Statistical Abstract* and *Federal Reserve Bulletin*, various issues.

Asian NICs, already a competitive force in world trade of manufactures, should increase their global market share . . . even if the dollar declines substantially" (emphasis added).

Additional skepticism about the beneficial effects of a declining dollar on the U.S. trade balance is provided by Michael Boreksky, a senior economist with the U.S. Department of Commerce. In a September 1985 interview, he said: "All things considered, I would attribute no more than 20 percent of the entire trade deficit and no more than 10 percent of the deficit in technology-intensive manufactured products to the surge of the dollar's exchange rate. This implies that if for some reason the dollar's current external value plunged during the next year or so some 25 percent—that is, to roughly what I estimate to be its equilibrium level—the U.S. trade situation would most probably improve, but still remain far from satisfactory" (**4**).

All of this discussion points out some basic flaws in the macroeconomic school of thought regarding the aggregate trade balance. Their views are displayed in the central track of Figure 11-4. It can be concluded that the exchange rate is only one of several factors that will determine future industrial competitiveness.

COST OF PRODUCTION

Figure 11-5 compares U.S. steelmaking costs in 1976 with those in Japan. Similar data are presented in Figure 11-6

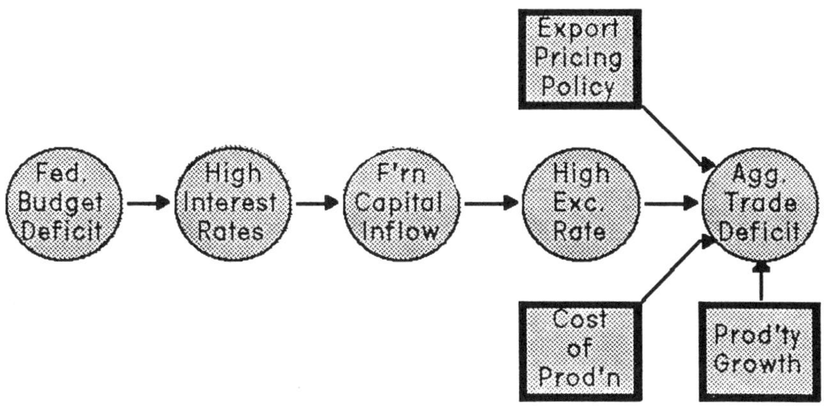

FIGURE 11-4
Role of Exchange Rate in Determining the Aggregate Trade Balance

FIGURE 11-5
Steelmaking Costs, 1976

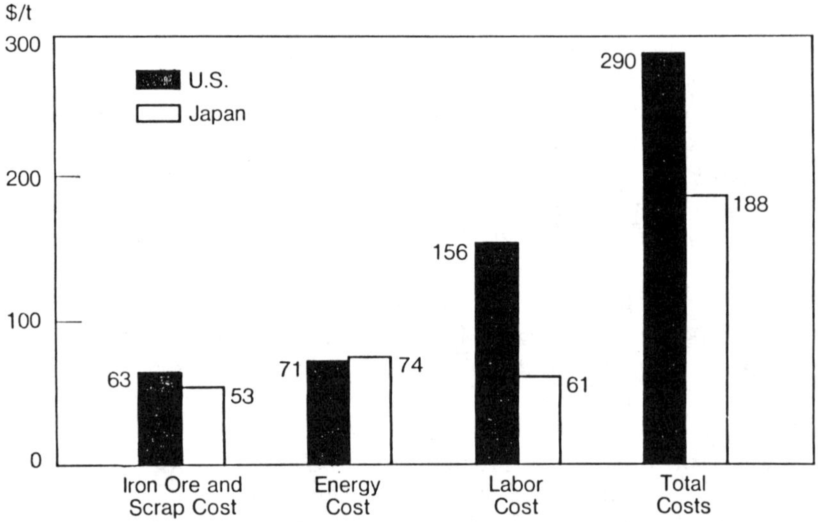

FIGURE 11-6
Automobile Cost Comparison

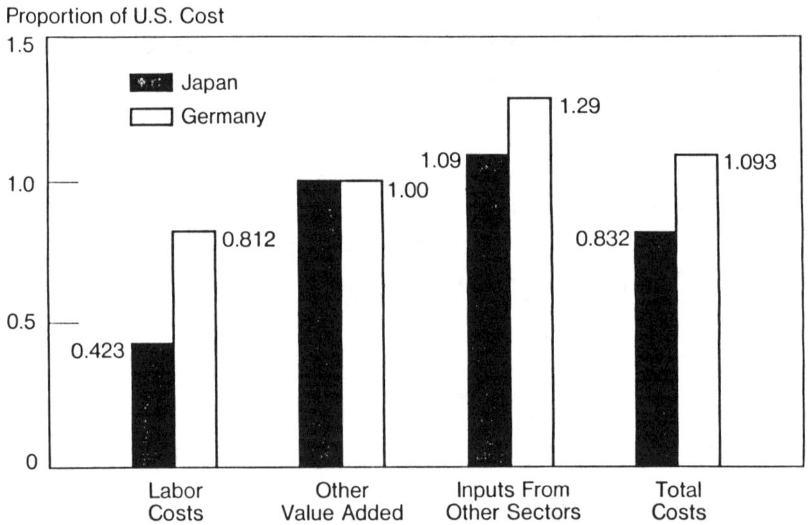

for the automobile industry. In 1976, the Japanese were able to make one ton of steel for $188 compared to $290 in the United States. The big difference is in the labor costs: $156 per ton in the United States versus $61 per ton in Japan. Many steel industry analysts feel that this difference in labor costs is not because the American steel industry is using antiquated technologies and the Japanese are using super-advanced technologies. Indeed, some studies indicate that U.S. output per man-hour is much higher than Japanese or at least comparable. One product study on hot-rolled band estimates 3.6 man-hours per ton in the United States and 3.4 in Japan, 5.9 in Brazil, and 5.5 in Korea. But at U.S. labor costs in the $20-plus range, this works out to $84 per ton in the United States, against $16 for Korea, and $37 for Japan (**5**). The main source of the labor cost difference is the wage rate.

During 1973-84, unit labor costs measured in dollar terms rose at an annual rate of 6.7 percent in the United States, 3.1 percent in Japan, and 4.4 percent in Germany (**3**). The U.S. position vis-à-vis Japan deteriorated annually by 3.6 percent, vis-à-vis Germany by 2.3 percent. The relative position of the United States was made worse off by the appreciation of the dollar. However, it is interesting to note that even if the dollar had not appreciated, U.S. labor costs would still have moved adversely: Japanese costs measured in yen increased only .8 percent and German costs measured in marks increased 4.1 percent. The U.S. deterioration vis-à-vis Japan would have been 5.9 percent, and 2.6 percent vis-à-vis Germany.

The cost of capital is another element affecting international comparative advantage. As shown in Figure 11-7, there is a consistent difference in the cost of capital in the two countries. Several factors are responsible for this difference, including higher rates of savings in Japan—20 percent of disposable income versus 5 percent in the United States (**6**)—and permissibility of higher debt/equity ratios in Japanese financial markets.

PRODUCTIVITY GROWTH

Disadvantages in the cost of capital and labor can be offset by advantages in the growth of productivity. However, on this front, U.S. manufacturing has lagged behind; see Figure 11-8.

Several explanations for the slower growth in U.S. productivity have been offered. Many of these can be traced to the lower rates of investment experienced in the United

FIGURE 11-7
Comparison of U.S. and Japanese Capital Costs

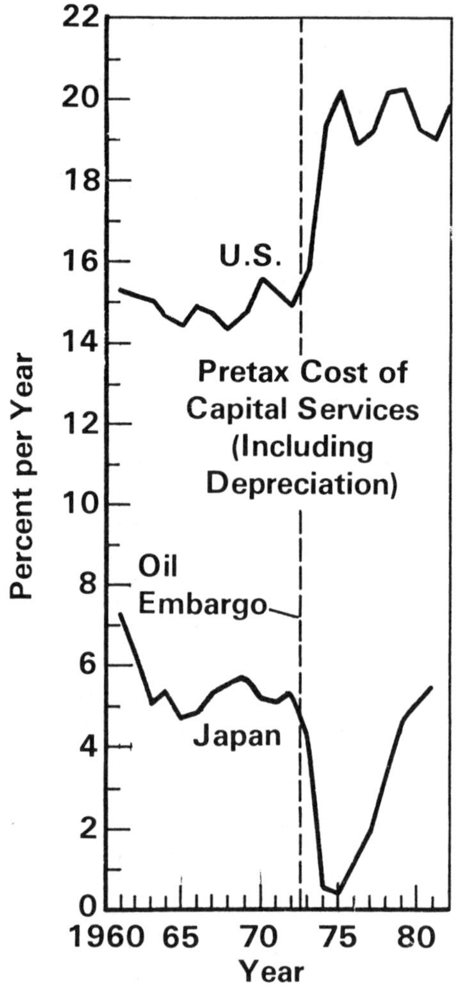

Source: Thermo-Electron Corporation, *High Cost of Capital: Handicap of American Industry*, 1983.

States: 17 to 19 percent of the GNP in the United States versus 31 to 35 percent in Japan, and 20 to 26 percent in Germany (**6**). The slower investment rate has resulted in the U.S. capital stock becoming technologically outmoded and economically obsolescent. A survey conducted in the mid-1970s estimated that almost 60 percent of U.S. industrial

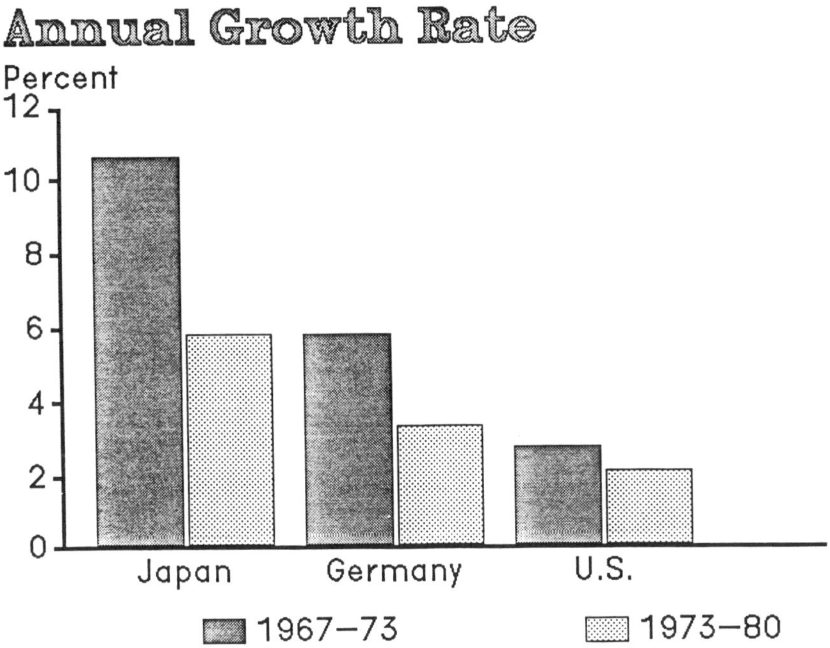

FIGURE 11-8
Comparison of Productivity Growth, 1967-1980

Source: *Monthly Labor Review.*

equipment was 25 years or older. In the machine tool industry, more than 60 percent of Japanese equipment was less than 10 years old, as compared to about 30 percent of U.S. equipment. Figure 11-9 presents a graphical picture on the age structure across the spectrum of manufacturing activities.

EXPORT PRICING

Our major trading partners include Japan, Germany, the United Kingdom, France, Canada, Italy, and several NICs such as Korea, Taiwan, Mexico, and Brazil. These countries have penetrated American markets, and they have competed with the U.S. manufacturers in world markets to capture some of their share. A major reason why many of them have succeeded is their cheap labor: A worker in Brazil costs about $3/hour, including benefits, whereas a Japanese

FIGURE 11-9
Age Structure of Equipment in U.S. Manufacturing, 1980

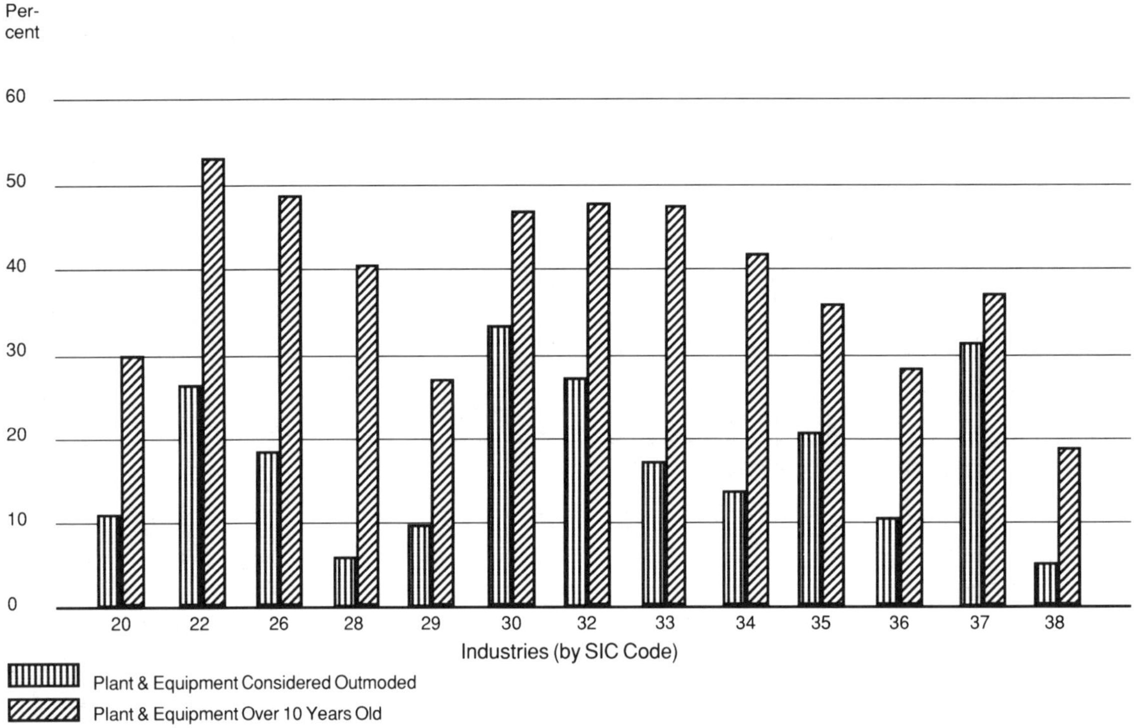

Source: U.S. Department of Energy, *The Industrial Energy Efficiency Improvement Program*, 1980.

worker costs $12, and an American worker $24. However, even beyond the real cost advantages, there is the phenomenon of dumping, or below-cost pricing. This is a variable that has not been adequately quantified and fully integrated in past analyses.

Many of the export-oriented trading partners have initially priced their products at below their costs of production to get a foothold in American markets. Then they ride the cost curve downward to complete the takeover of the targeted market segment. This phenomenon, variously known as forward pricing or dumping, raises moral as well as economic issues (**7**).

We will not go into the issue of whether dumping is good or bad. We are, however, going to look at what leads to such aggressive pricing policies and whether we can predict such behavior on the part of our trading partners in the future.

STAGE I: CONCEPTUAL SUMMARY

Having identified the major determinants of aggregate trade competitiveness, we can now formally develop an approach to forecasting international competitiveness. This consists of two stages. In stage I, we look at the aggregate trade balance, and in stage II, we look at the commodity composition.

In terms of stage I, we focus on the determination of export price (not just cost), since this variable is a key measure of international competitiveness. We assert that the export price, Px, is a function of actual production costs, C, in the country of origin, the transportation costs to the market, T, and, in addition, a third variable called the "Need to Export," Nx. Symbolically,

$$Px = f\ (C, T, Nx) \tag{1}$$

Production costs typically reflect the resource endowments of the producing country. Transportation costs are related to the distance from the market; the third variable, "Need to Export," is based on the need to import, i.e., the "import elasticity."

The rationale for this assertion follows. A country with hardly any indigenous resources can provide for growth in its standard of living only by maximizing exports, especially if most of the goods consumed domestically are imported directly or manufactured by first importing the raw materials. If the imports include food and other necessities of life, it becomes even more important for that country to push exports in order to pay for the import bills. Such a case of extreme need to export can be said to be due to a low elasticity of import, i.e., imports do not move with the level of economic activity, perhaps because many imports are needed to meet basic necessities of life.

STAGE II: EMPIRICAL EVIDENCE

What is the evidence on the import elasticity of different countries? Do they fit into any type of evolutionary pattern? Can we say that the countries that we think are most aggressive in their pricing policies also have low import elasticity? We analyzed the trade balances of nine countries and one regional group over 1960-1981. The results are shown in Figure 11-10.

The specific import elasticity shown in the figure is the GNP elasticity of imports, which has the following interpretation: If the GNP declines for any one of these countries, does the corresponding import volume also go down? If it went down by the same percentage, the elasticity will be one; if it went down more, it will be more than one; and if it went down by less than one, it will be less than one.

The several countries can be grouped into three clusters: (1) the first group involving the United States, Germany, France, and the rest of Europe, all with import elasticity of 1.6 or more; (2) the second group of countries involving

FIGURE 11-10
Import Elasticity

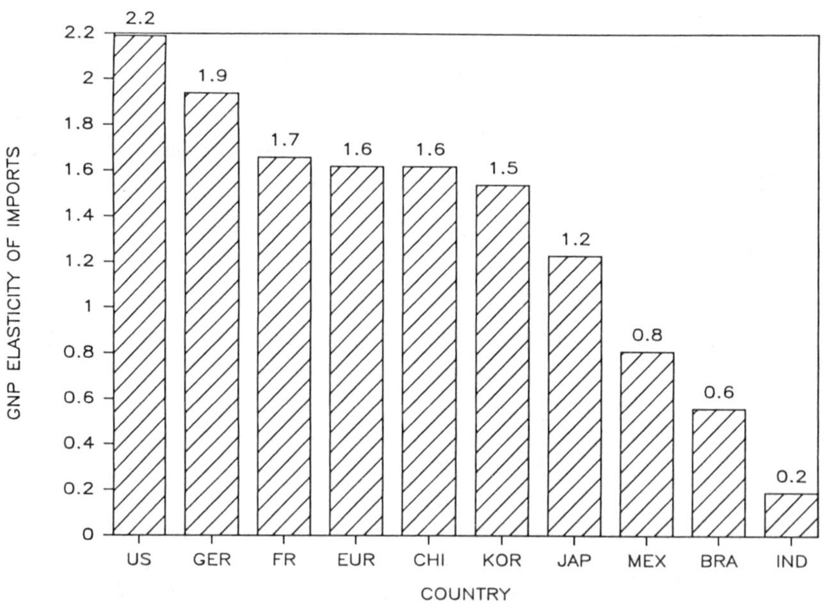

Source: Estimated by the authors.

Taiwan, Korea, and Japan, with import elasticity between 1.2 and 1.6; and (3) the third group involving Mexico, Brazil, and India, with elasticities between 0.2 and 0.6.

This grouping points to a predictable interpretation of the concept of import elasticity. Brazil, Mexico, and India, the least developed economies in our sample, have the lowest import elasticity, while the most developed economies, such as Germany, France, and the United States, have the highest import elasticity. The positive relationship between import elasticity and per capita GNP, a measure of development, is shown in Figure 11-11.

Do countries with the lower import elasticity have a greater need to export? Data relevant to this question are displayed in Figure 11-12. The vertical axis measures the relative export performance of each country in the past 20 years. Export performance is measured as the growth rate in each country's share of world exports, normalized by that country's share of world GNP. This variable is similar to the various indices of comparative advantage used by the U.S. Industrial Development Organization (**8**).

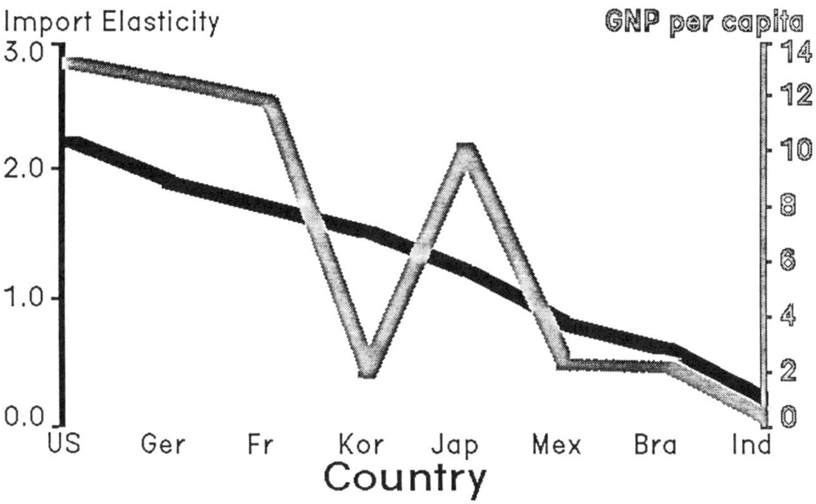

FIGURE 11-11
Import Elasticity and GNP Per Capita for Eight Countries

Source: World Bank, World Development Report, 1984.
Note: GNP per capita is in $\$ \times 10^3$, 1982.

The point on the extreme left in Figure 11-12 is India. It has a very low import elasticity and also very poor export performance. Other countries in the third group—Brazil and Mexico—are also located close by. In the middle, we have the second group countries, and toward the right, we have the first group countries.

The plot is a bell-shaped curve, which says that Korea has performed extremely well in terms of its export performance and also has a lower import elasticity compared to the advanced, industrialized countries. Taiwan, right now, is going up the export performance curve while Japan is just past the peak of the curve.

FIGURE 11-12
Export Share Growth Versus Beta

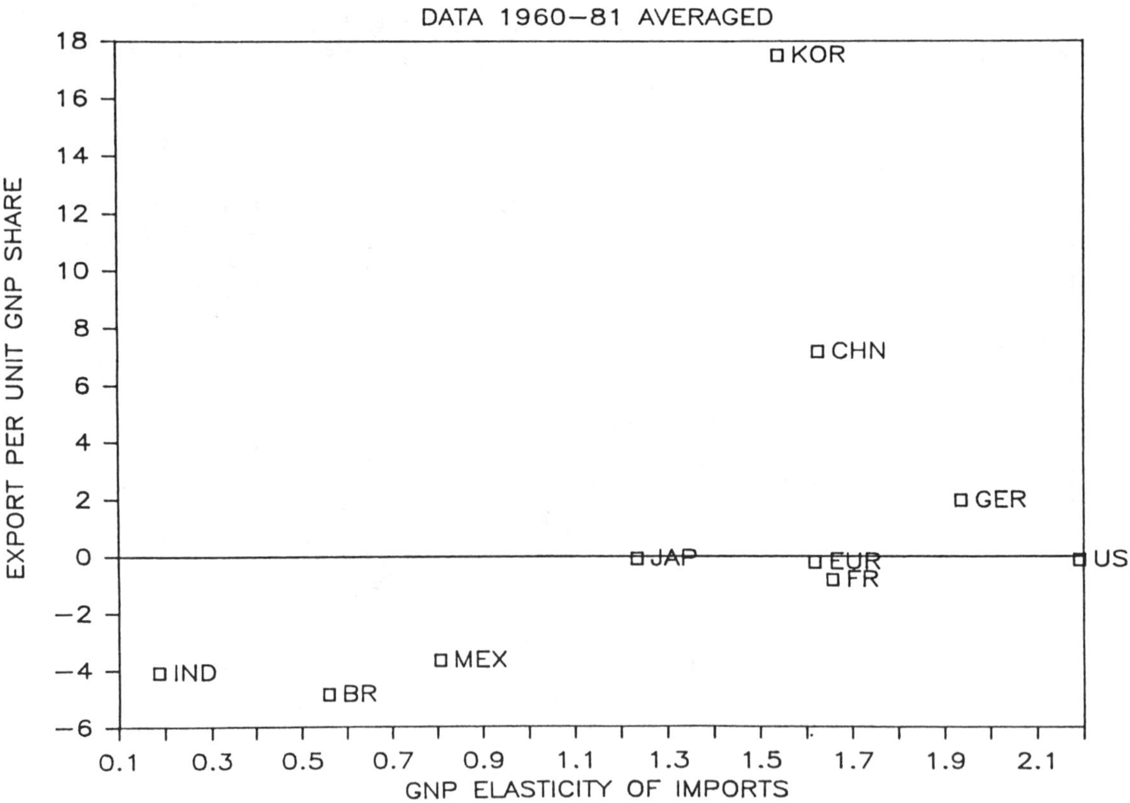

Source: Estimated by the authors.

The Asian NICs are very competitive in the world manufactures market; they have low wage rates, rising educational levels, strong work ethics, increasingly well-developed infrastructures to support an expansion of manufacturing, and a proven ability to diversity into new products. These advantages will continue to be important to them as they face growing challenges in traditional, labor-intensive product exports from other developing countries. These NICs will certainly become more competitive with the industrial countries in more and more capital- and technology-intensive products.

Why are the countries in the third group, Mexico, Brazil, and India, which have the lowest import elasticity, not higher up on the curve? Our hypothesis is that although their need to export is strong and they probably perceive a need to price their products aggressively, yet the weak infrastructure of their economies does not allow them to develop a flourishing export industry, at least in the manufacturing area. However, all countries in this group need to be watched carefully because once the infrastructure comes into place, they will enter the high-performance region, becoming the Koreas and Taiwans of the next decade, and mounting a major offensive on the United States trade balance. Of course, we also have to keep a very close eye on today's high performers, Japan, Taiwan, and Korea. Our traditional trading partners, Germany and France, have slid down the scale, and it is unlikely that their pricing policies would ever be as aggressive as those of countries in the other two groups.

Based on the discussion so far, one can make a plausible case for using the trading partners' import elasticities as a measure of their propensity for aggressively pricing their exports. This information, coupled with the more traditional information on natural resource endowments and production cost advantages, will yield a framework for predicting the aggregate trade balance.

COMMODITY COMPOSITION OF TRADE

The second part of competitiveness forecasting involves the mix of traded products. As shown in Figure 11-13, the United States trade balance for R&D-intensive industries is

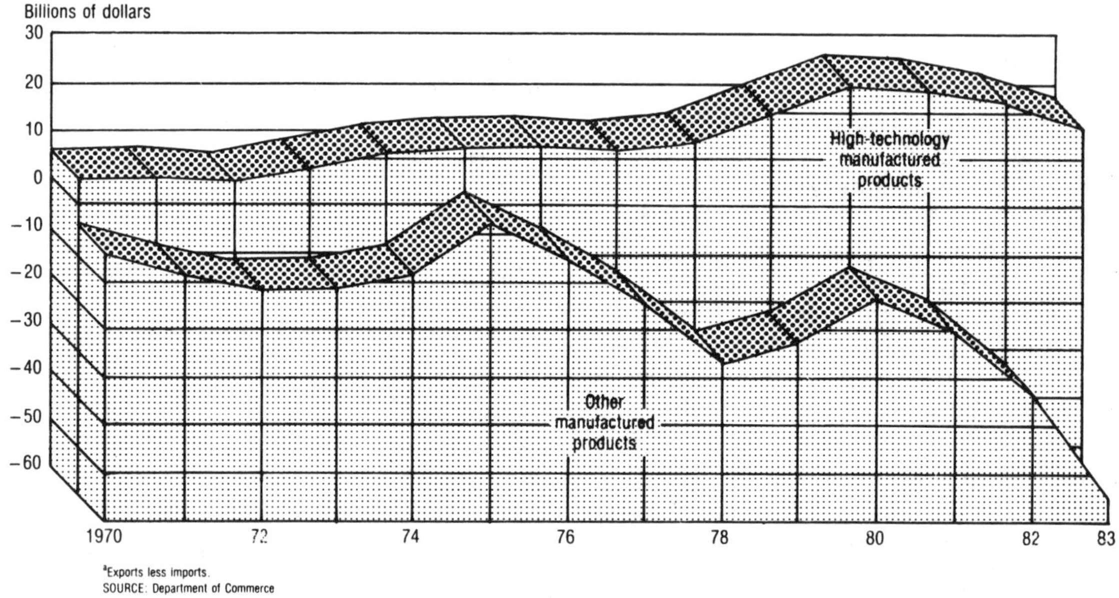

FIGURE 11-13
U.S. Trade Balance in High Technology and Other Manufactured Product Groups

quite positive. Examples of these industries include specialty chemicals, aerospace, scientific instruments, and advanced machinery. Trade in other manufacturing industries yields a negative balance. We contend that similar statements hold true for other countries, and the trade mix can be forecast based on the relative industrial developments of the trading partners, as explained in the following.

Using the product life cycle hypothesis (**9**), we have specified a trade composition matrix in Figure 11-14. Following (**10**), we have divided traded products into four commodity group areas:

1. labor-intensive industries, like apparel and leather;
2. resource-intensive industries, like lumber and wood products and pulp and paper;
3. capital-intensive industries, like iron, steel, and petroleum refining; and
4. R&D-intensive industries, like computers and aerospace.

The hypothesis is that as countries climb the ladder of economic development, progressing from the preconditions

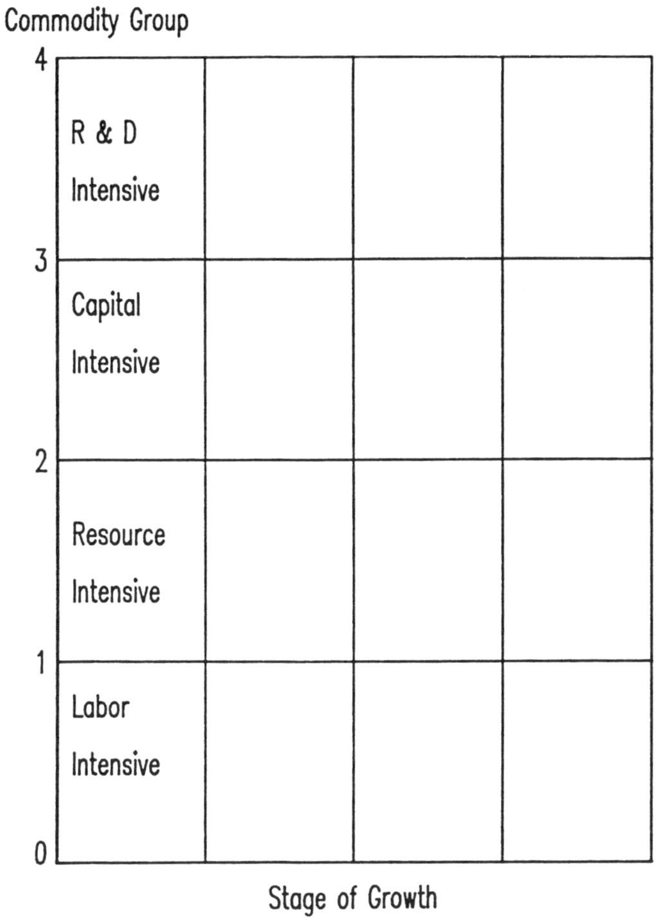

FIGURE 11-14
Ladder of Development and Commodity Composition of Trade

for take-off, through take-off and the drive to maturity, to high mass consumption (**11**), their industrial product mix undergoes a predictable transition from labor-intensive industries through R&D-intensive industries.

Figure 11-15 portrays the industrial evolution of the Japanese economy using empirical data on Japan's export mix during 1952 to 1979. As we can see, the trend is positive, consistent with the ladder of development hypothesis. As is well known, Japanese firms are now competing directly with United States firms in what was up to now

FIGURE 11-15
Ladder of Development Weighted Score for Japan

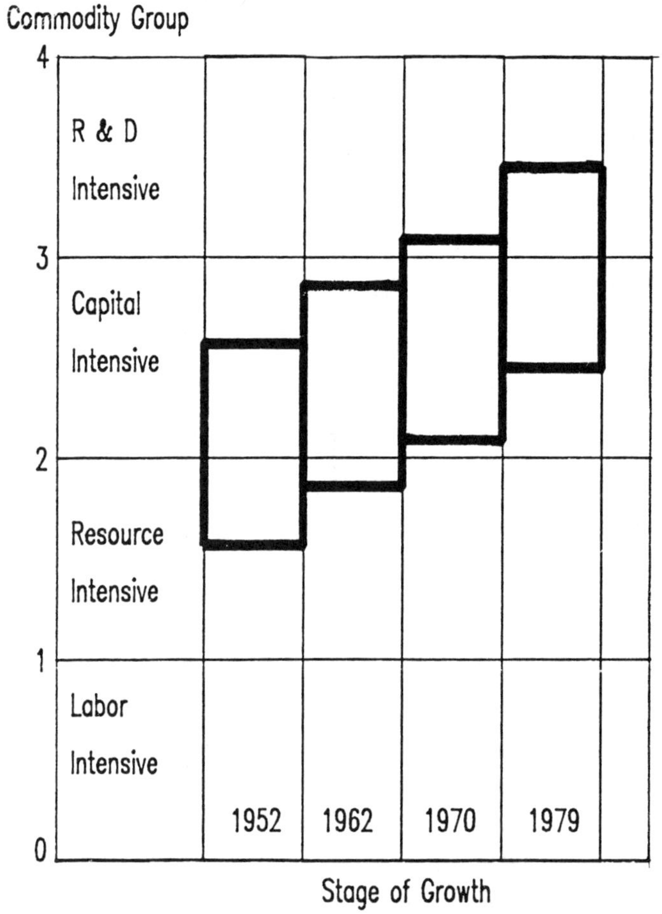

Source: *United Nations Yearbook of International Trade Stats.*

exclusively an American preserve: high-technology industries. If we represent export mix by Zx, and the state of industrial development by D, then we are saying:

$$Zx = F(D) \tag{2}$$

Based on the ladder of development concept, we can develop equations such as (2) for predicting the composition of trade based on the position of various trading nations on the development spectrum. This empirical development will be addressed in future research.

CONCLUSION

Our approach consists of two stages. In stage I, the aggregate trade balance is expressed in terms of productivity growth, comparative production costs, and the import elasticity. Empirical relationships are derived between aggregate export performance and import elasticity.

In stage II, the commodity composition of trade is expressed in terms of the level of industrial development. This relationship is developed empirically for Japan.

Further research needs to address the development of stage II empirically for other countries besides Japan. Additionally, conceptual work is needed on a third stage to analyze policy-oriented issues, such as trade barriers, tariff policies, and technology transfer.

REFERENCES

1. Ahmad Faruqui et al. "Ten Propositions in Modeling Industrial Electricity Demand." In Adela M. Bolet (ed.), Forecasting United States Electricity Demand: Trends and Methodologies. Boulder: Westview Press, 1985.

2. San Jose Mercury. January 31, 1986 and March 19, 1986.

3. U.S. Department of Commerce. U.S. Industrial Outlook 1986.

4. Chemical and Engineering News. September 1985, p. 13.

5. Forbes. March 10, 1986, p. 84.

6. U.S. Department of Commerce. International Economic Indicators. Quarterly.

7. William R. Cline (ed.). Trade Policy in the 1980s. Cambridge: MIT Press.

9. Raymond Vernon (ed.). The Technology Factor in International Trade. New York: National Bureau of Economic Research, 1970.

10. Robert Z. Lawrence. Can America Compete? Washington, DC: The Brookings Institution, 1984.

11. W. W. Rostow. The Stages of Economic Growth: A Non-Communist Manifesto. Cambridge: University Press, 1965.

12

ABSOLUTE AND RELATIVE TRENDS IN U.S. RESEARCH AND DEVELOPMENT INVESTMENT

*William L. Stewart**
National Science Foundation

For those of you who are not familiar with our activity, the Division of Science Resources Studies is the unit in the Foundation responsible for collecting, analyzing, and disseminating information on the nation's human and financial resources for science and technology. We also look at the output of science and technology and its impact on society. That is a congressionally mandated activity for the

*This paper represents a verbatim transcript of the presentation given by the author at the EPRI seminar.

Foundation, and we have been doing it since the Foundation began in the early 1950s.

The topics I am going to cover today start with the overall picture of the national R&D investment; some international comparisons thrown in; some detail on the federal portion of the research and development investment in the nation; a look at what officials in high-tech industries expect to spend on R&D through 1985 and 1986; and then I will follow up with some final comments on academic R&D.

Let me start with a quick breakdown of the nation's R&D effort by source of support, performers, and type of activity, just to put the information to follow into perspective (Figure 12-1). Industry accounts for the largest single source of R&D funds in the nation. In 1985, we are projecting that national R&D will exceed $100 billion for the first time—$106.6 billion, if our estimates are correct.

FIGURE 12-1
The National Research and Development Effort (1985—$106.6 billion)

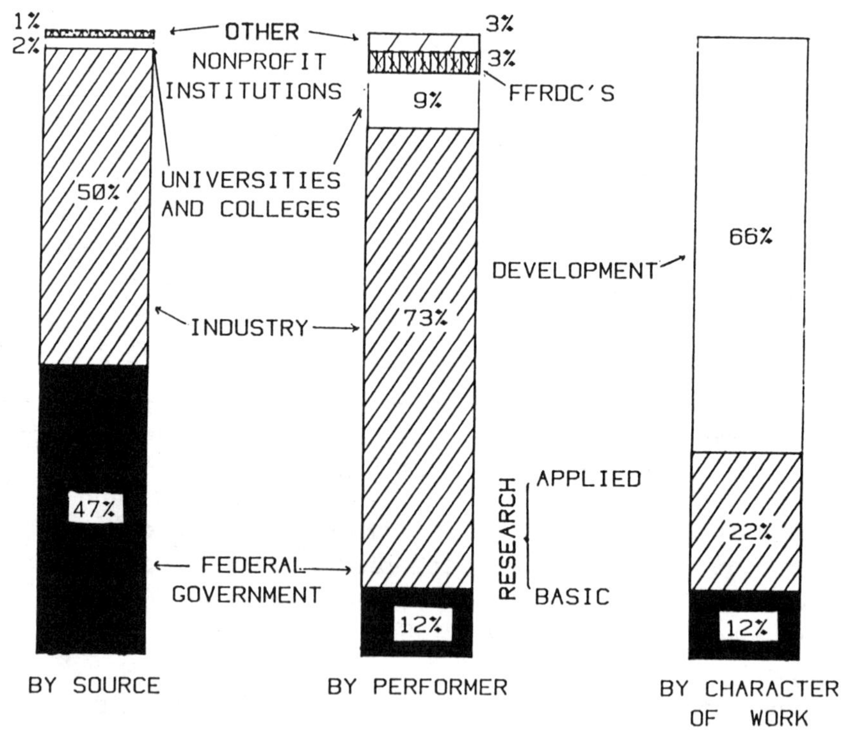

Source: National Science Foundation

That is three times the amount spent in 1975, just ten years ago.

The federal government provides the second largest amount of money for research and development. Industry, as you can see, performs almost 75 percent of all of the R&D performed in this country. And two-thirds of the R&D activity is in development. Applied research leads basic research by a ratio of about two to one. All of these ratios have remained fairly constant over many, many years. The major change that we have seen has been in the source of funds, where industry has started exceeding the federal government back in the late '70s.

Before we celebrate too much about passing the $100 billion mark, I should tell you that this translates into a real increase, over the ten-year period, of 63 percent after adjusting for inflation, instead of 300 percent that I spoke of in current dollar terms.

Between 1975 and 1982, the nation increased its R&D support at an average annual rate of 4.4 percent in constant 1972 dollars (Figure 12-2). By the way, all of these figures are based on constant dollars, and they are computed using the GNP implicit price deflator. Since 1982, this rate has been 6.3 percent.

The 1985 estimate represents a 7 percent increase over 1984. During the past decade, we have seen no increases below 3 percent, and one as high as 8 percent. The federal government is the source of about 47 percent of the national R&D total. In 1985, we estimate the government will account for about $50 billion. That, too, is well above the 1975 level in current dollars and 50 percent over the 1975 level in constant dollars. Much of this increase can be traced to the rapidly growing defense-related programs since 1982. I will touch a little bit more on that later.

Throughout the period 1975 to 1985, the private sector—mainly industry—has recorded higher increases than federal sources every year except 1983 and 1985. Nonfederal support jumped ahead of federal support in 1978 and the gap has been widening.

I will come back to industry's R&D investment in a few minutes. But, for now, let's concentrate on the national level. On the surface, this level of growth in the nation's R&D investment looks pretty good. But if you look at it in light of what has happened to GNP, does it look as good? As

FIGURE 12-2
National Research and Development Expenditures (1972 constant dollars)

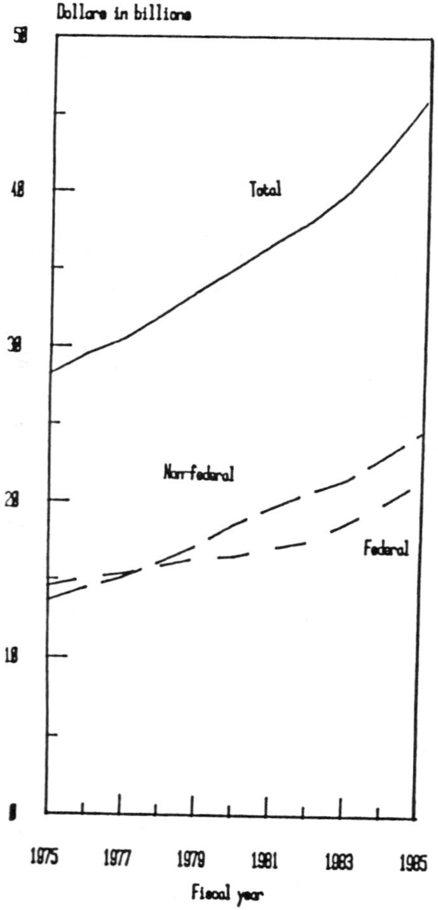

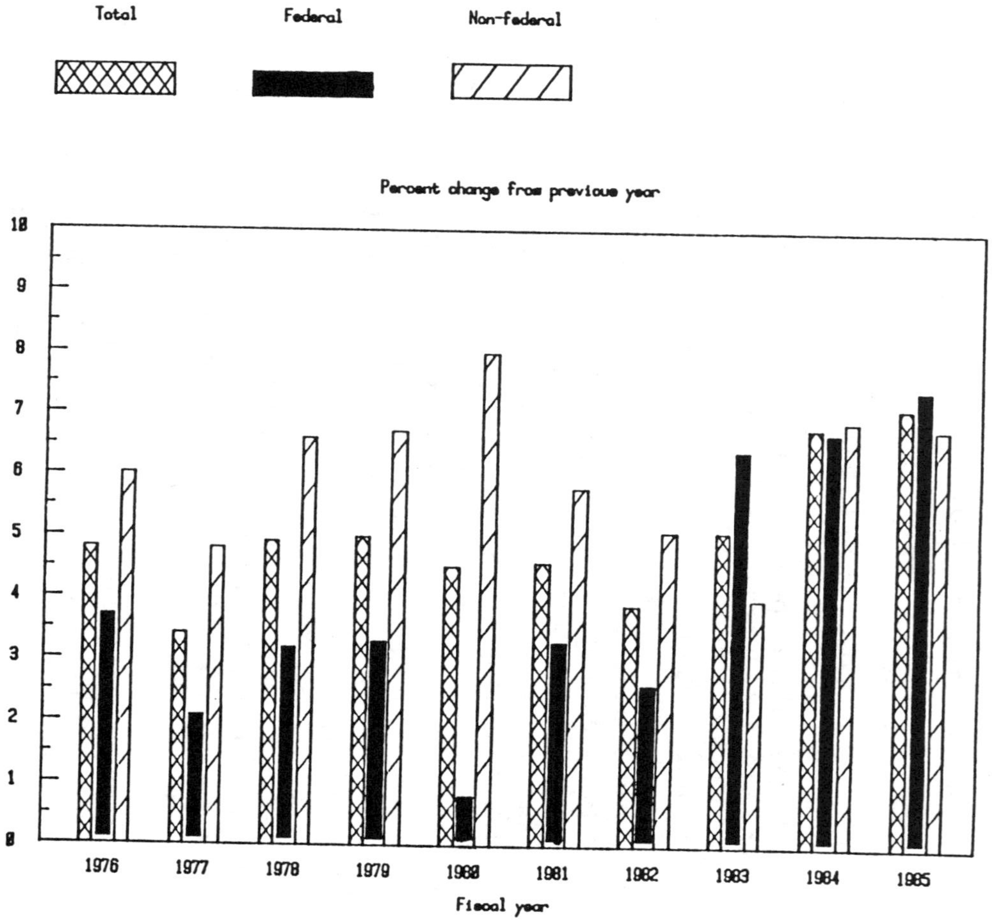

Source: National Science Foundation

it turns out, yes, it does. In 1978, this country's R&D spending amounted to 2.2 percent of GNP, the lowest ratio since we started keeping track. In 1985, we estimate the R&D GNP ratio will be 2.7 percent. Three percent is the highest level we have recorded. That was back in the '60s.

How does this compare with our major foreign competitors (Figure 12-3)? The United States spends more on research and development than France, West Germany, and Japan combined. You can add the United Kingdom to that list also. The 1983 data for the U.K. are not available, so it is not on the chart. Their expenditures usually run about even with those of France.

FIGURE 12-3
National Expenditures for Research and Development for 1983 (constant 1975 dollars)

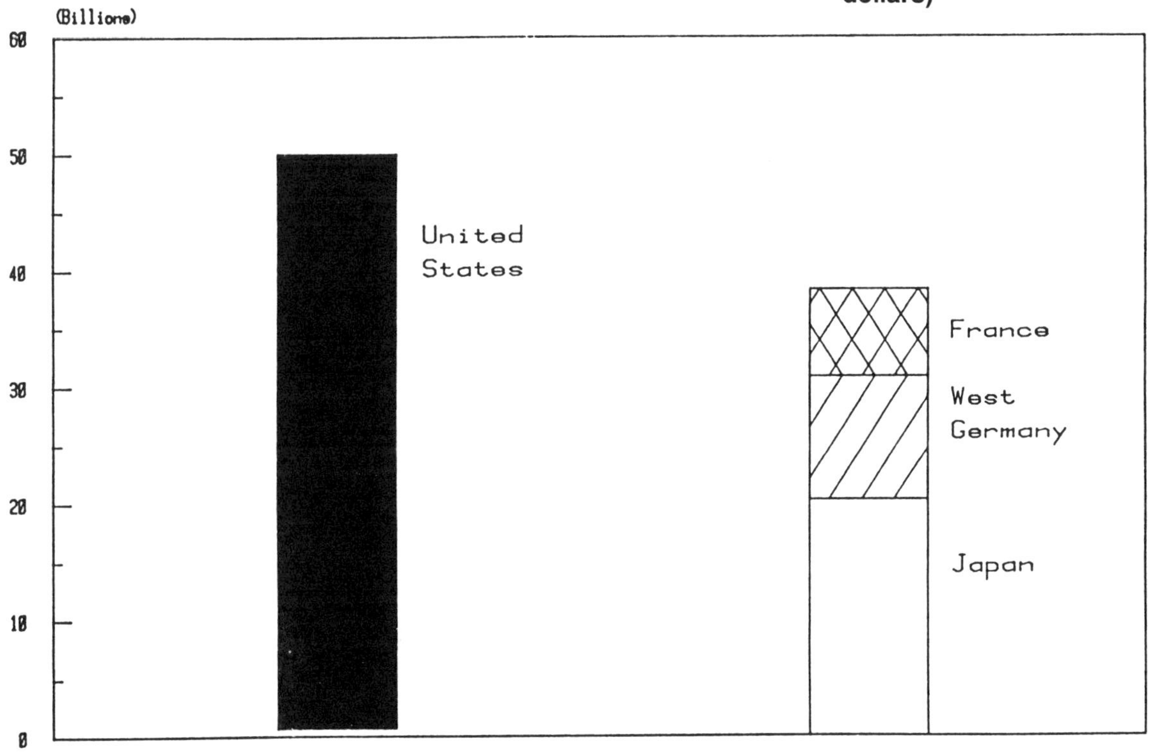

1/ U.S. dollar conversion based on purchasing power parity exchange rates and deflators.

Source: National Science Foundation

Comparing the different R&D GNP ratios of these countries (Figure 12-4), there has been a converging of such ratios since the early '60s to the point where now the ratio for the United States is roughly equal to that of West Germany and only slightly above the ratio for the other industrialized Western nations. Pulling out defense-related R&D expenditures, however, puts the U.S. ratio at 1.8 percent, below the 2.5 percent for West Germany and 2.3 for Japan.

Before leaving the subject of international comparisons, there is another group of countries that I think we should

FIGURE 12-4
International R&D Expenditures as Percent of GNP

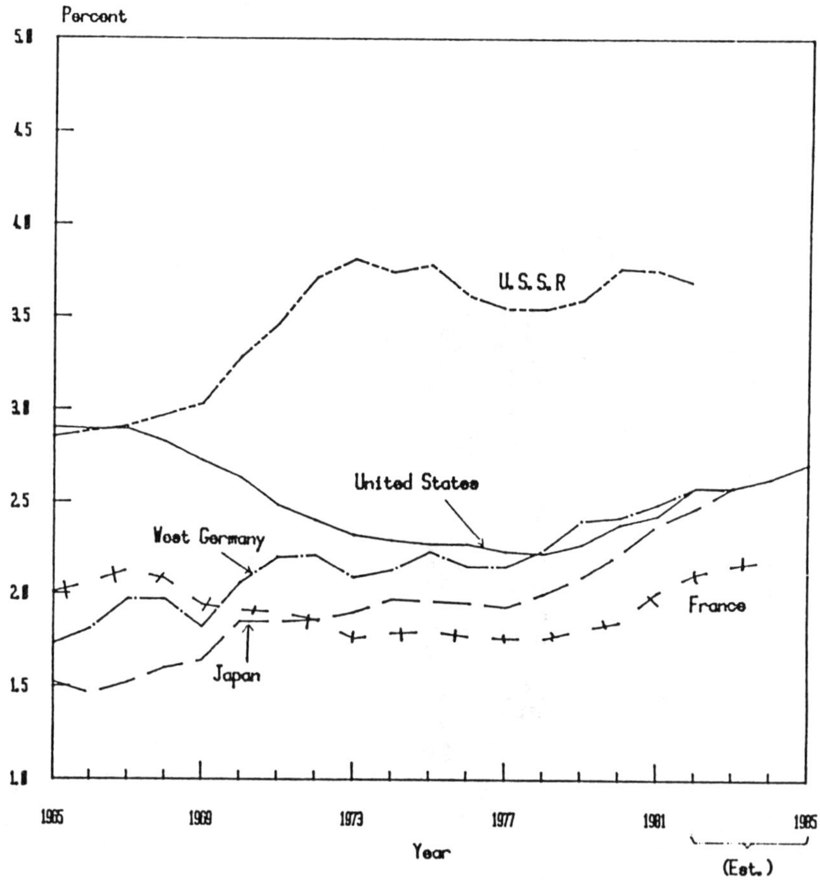

(NOTE: The U.S. devotes a higher proportion of its economy to R&D than most countries except the U.S.S.R. and West Germany.)

Source: National Science Foundation

take note of, countries which will soon, if they haven't already, move into head-to-head competition with the countries mentioned earlier, namely: South Korea, Taiwan, Hong Kong, and Singapore. While we have been concentrating on Japan and the European developed countries, these newly industrializing countries have become increasingly important in international trade. If you compare the combined deficit figure for these countries with the United States, you will find that it increased fivefold between 1980 and 1984. That is, from about $4 billion to $21 billion. And Taiwan alone accounts for about half of that deficit.

Unfortunately, we haven't developed the science and technology indicators for these countries to demonstrate their level of investment in R&D and science and technical personnel. That is on our agenda. But, clearly, this is a group which must be considered major competitors to the United States and to other participants in international markets.

One reason the federal government is so important in the nation's R&D picture is that it provides nearly one-half of all R&D funds, and another reason is that it distributes most of these funds to the other sectors. During the past few years, there have been only minor changes in the relative distribution of federal R&D moneys among the sectors (Figure 12-5). Industry remains the largest recipient of federal R&D funds and has gained slightly, over the years, reaching 50 percent in 1985. Federal intramural dropped slightly to 27 percent in 1985, which is a move consistent with the Administration's design to shift more intramural work to the private sector. Universities have received about 12 percent to 14 percent over the years, and another 5 percent or so for university-administered FFRDCs. The nonprofit sector, which is not shown, usually accounts for around 5 percent of the federal total.

Earlier, I spoke of the increase in federal funding for defense-related R&D programs. Figure 12-6 shows rather dramatically just how much the defense programs have grown. Ten years ago, defense accounted for one-half of the total R&D funding. In the 1986 budget, defense R&D funding was close to three-fourths of the federal R&D total. Funds for space research dropped slightly, but have returned at least to the 1976 level in the 1986 budget.

FIGURE 12-5
Federal Research and Development Expenditures by Performer (1972 constant dollars)

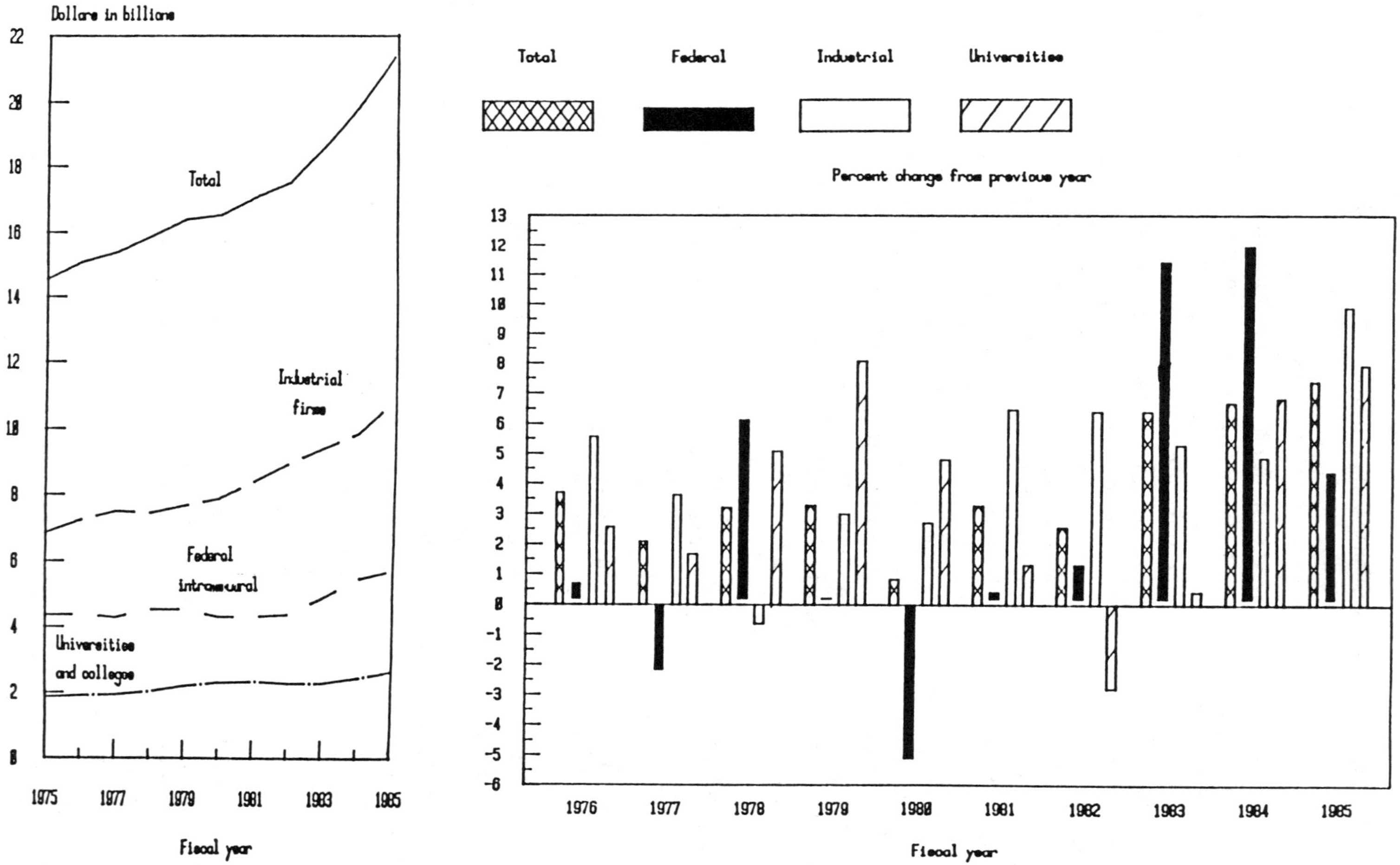

Source: National Science Foundation

Applied research and development provide the basis for the technological growth needed to keep the U.S. competitive in the world markets. Basic research provides the basis for advancement through applied research and development. You can say that is a gross over-simplification of the innovation to market process, but I think you will agree that there is more than just a grain of truth in that statement.

In 1985, the nation is expected to invest $70 billion in development, an increase of 8 percent over 1984 in constant dollars (Figure 12-7). The federal proposals for defense spending, as reflected in the DOD and DOE budgets, and U.S. industry's historically heavy emphasis on development activities primarily account for the increase. This compares with an average annual 4.7 percent increase between 1975 and 1982, and nearly an 8.5 percent average annual growth between 1982 and 1985.

About two-thirds of the nation's total R&D expenditures are generally devoted to development. Industry has been the major source of development funds, about 55 percent; and also the major performer of development activities, about 85 percent.

The nation is expected to increase applied research by some 4 percent over 1984 in constant dollars (Figure 12-8). Industry, again, surpassed the federal government as a major source of applied research funds in 1980, and, in 1985, it will account for 55 percent of the national applied research total. In real terms, industry support in this area has increased about 50 percent since 1980. Federal applied research support over the same period has remained fairly level. The only decrease in constant dollars in the federal side occurred in 1982.

Funding of basic research in 1985 (Figure 12-9) is expected to increase nearly 6 percent, in real terms, over 1984, about the same as the average annual increase since 1982, but well over the average growth during the 1975-1982 period. The current dollar level for basic research in 1985, estimated at $13 billion, is three times the current dollar level recorded a decade earlier. About one-half of the basic research money is spent in university laboratories, of which the government provides about seven out of every ten dollars. In constant dollars, national basic research support has grown about 50 percent since 1975, with growth in federal funds falling slightly behind that in the private

FIGURE 12-6
Federal Funding by Selected Function
(1972 constant dollars)

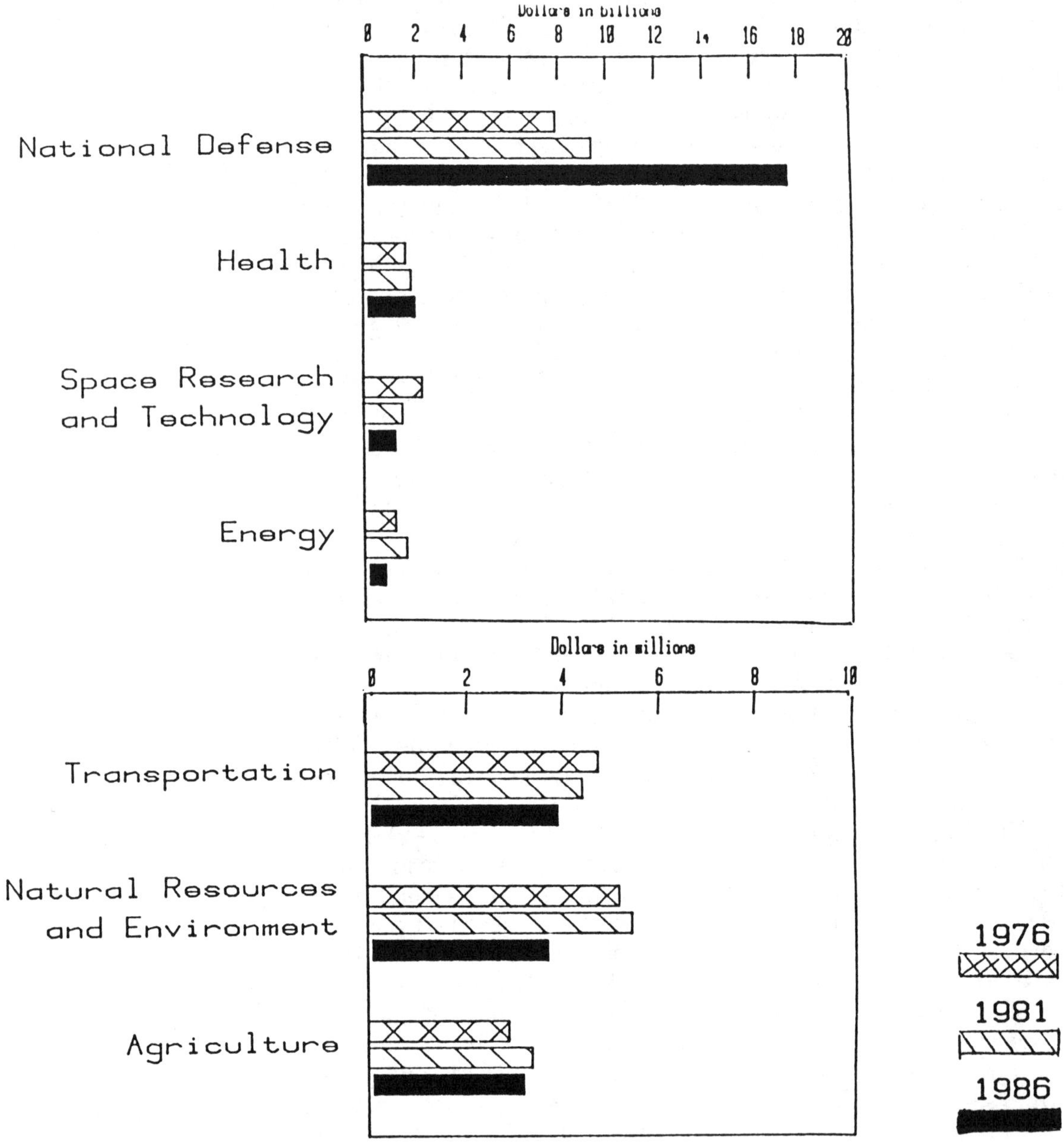

Source: National Science Foundation

FIGURE 12-7
National Expenditures for Development by Source

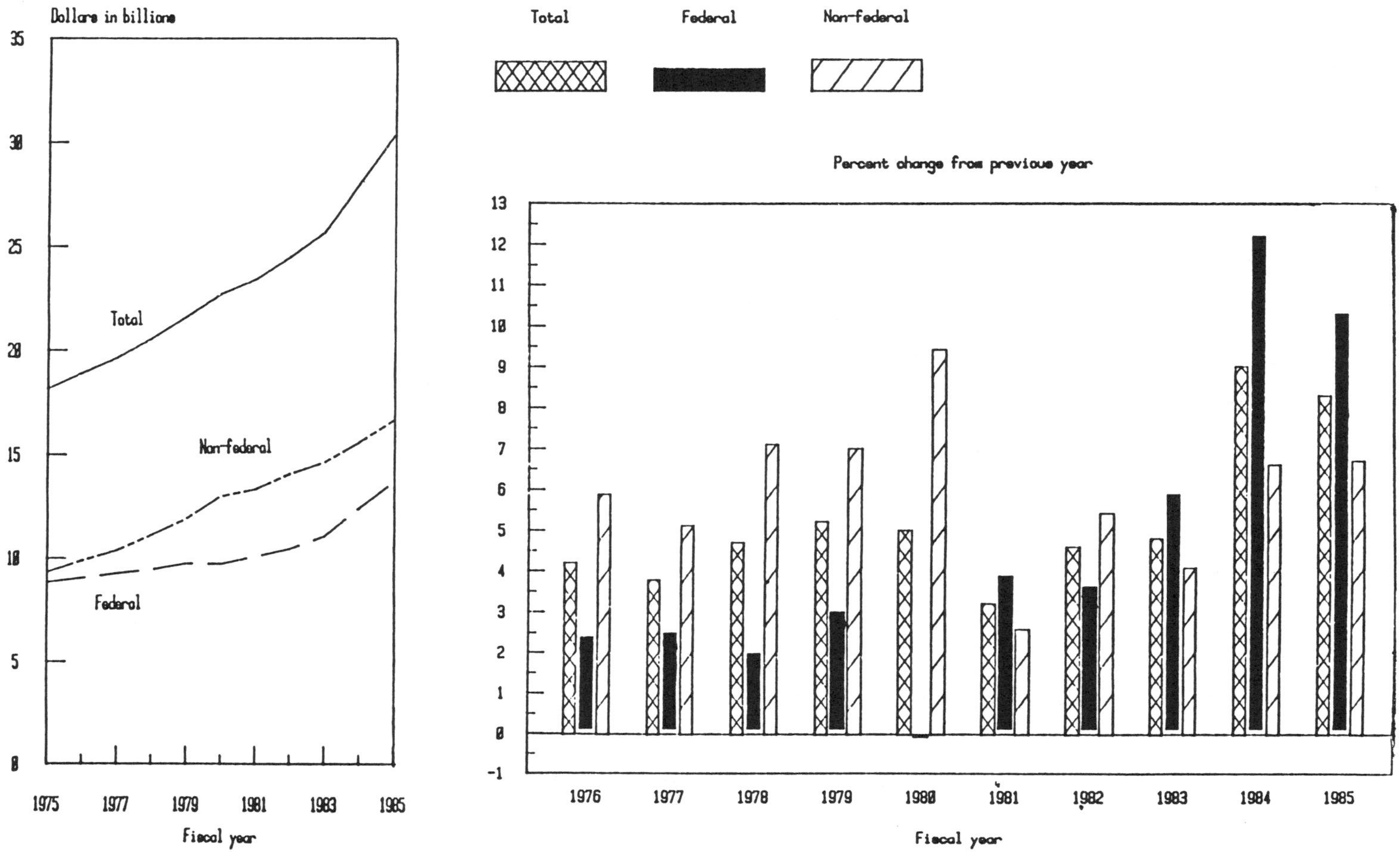

Source: National Science Foundation

FIGURE 12-8
National Expenditures for Applied Research by Source (1972 constant dollars)

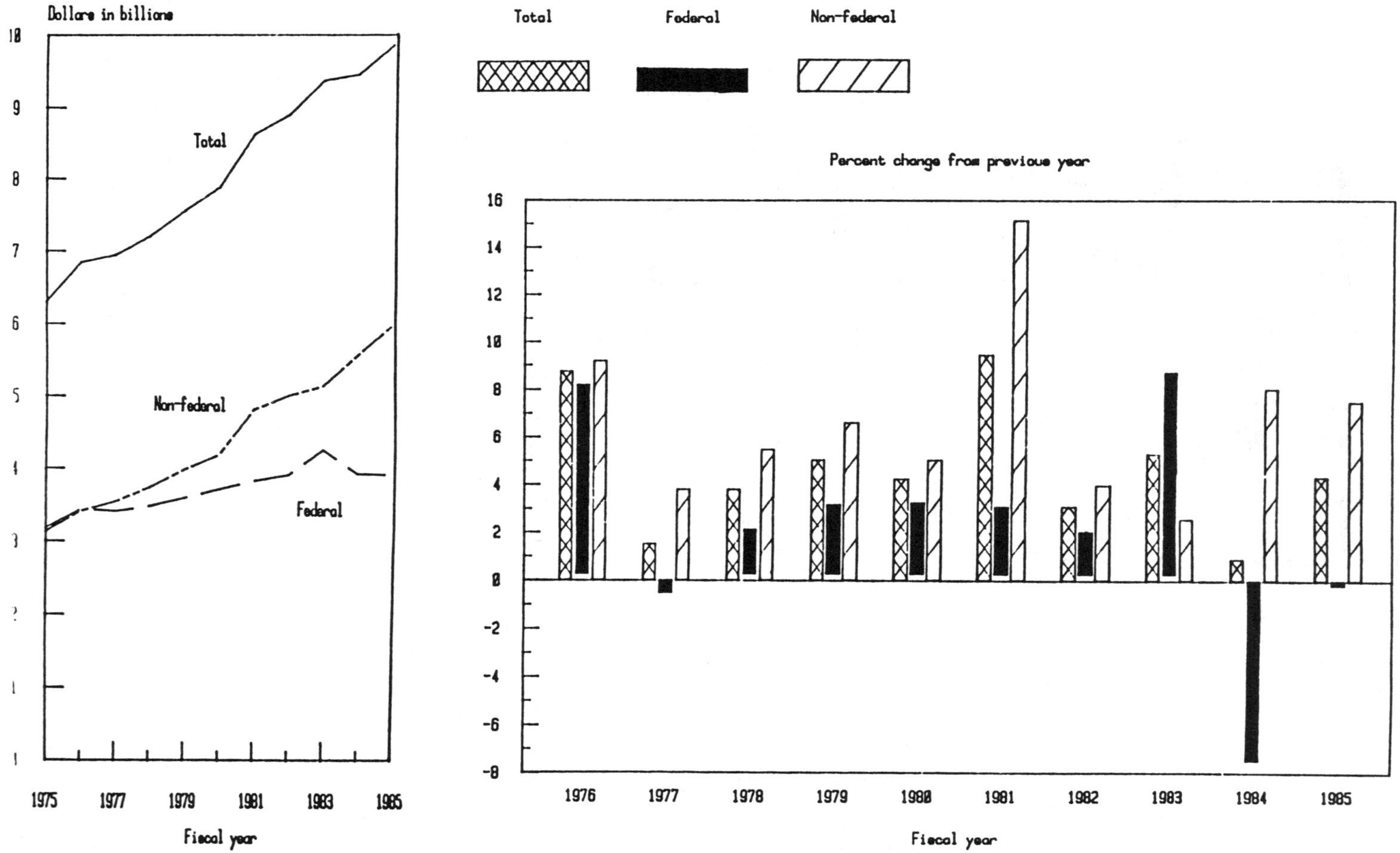

Source: National Science Foundation

FIGURE 12-9
National Expenditures for Basic Research by Source (1972 constant dollars)

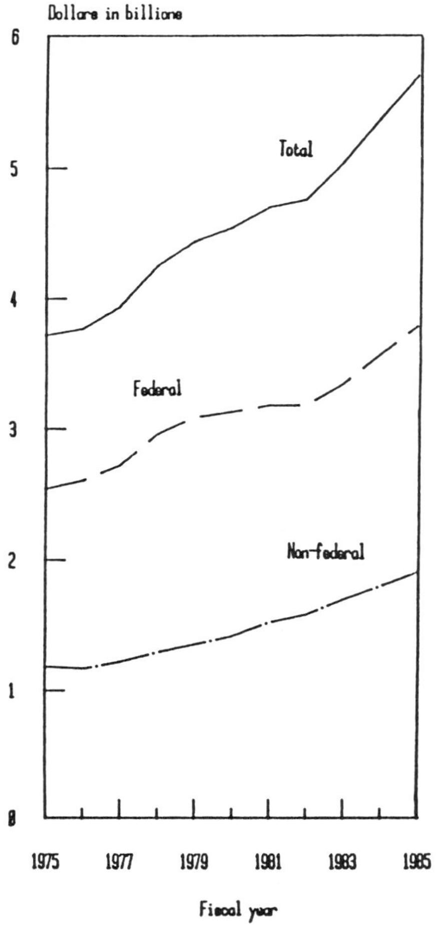

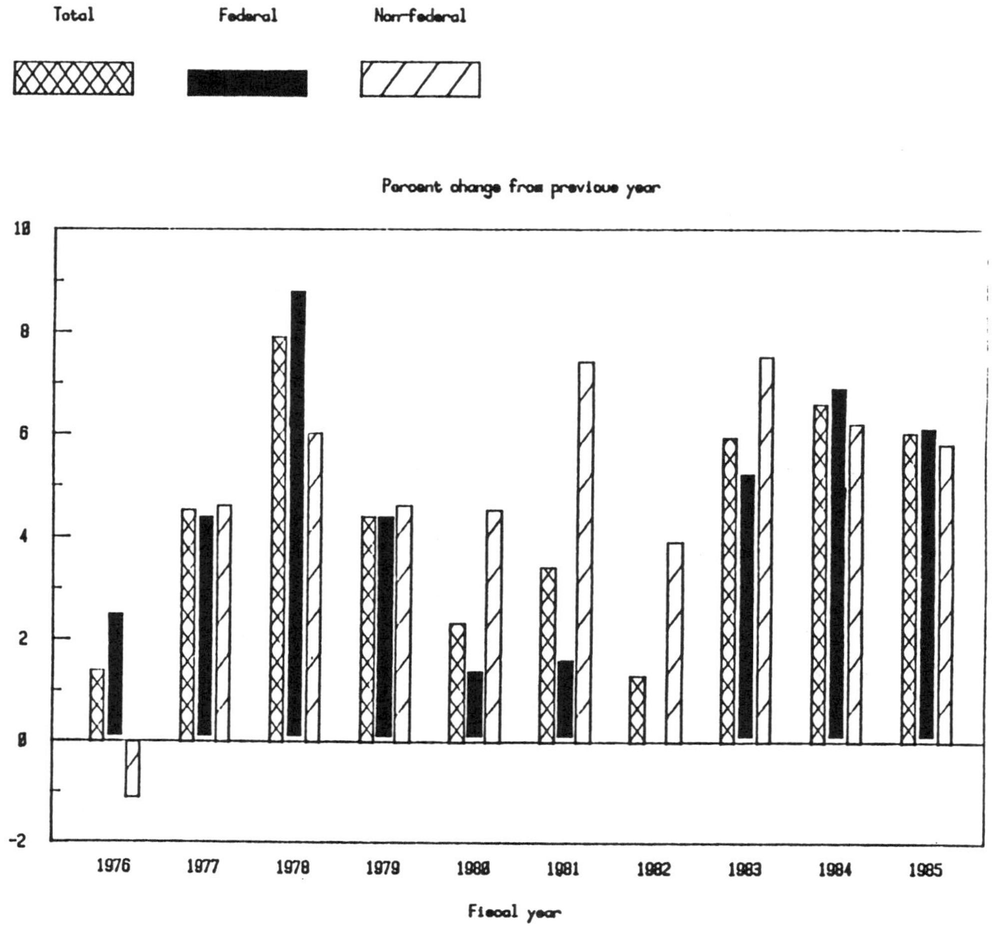

Source: National Science Foundation

sector. Still, you should note that, in constant dollars, federal basic research funds have increased 13 percent since 1983, compared to an 8 percent decrease in federal applied research funds.

As I pointed out earlier, U.S. industry has become the leading sector in support of research and development. Indeed, as Dr. Keyworth stated in a recent issue of Technology Review, "The commercial market for technology has expanded tremendously in the past decade or so. Today, industry, not government, is pushing hardest at technological frontiers in many areas."

What I said earlier about the increasing gap between the federal and industrial R&D funds applies equally to the funding of research and development performed within the industry sector, at least during the first half of the 1975-1985 decade (Figure 12-10).

In 1975, federal funds accounted for 36 percent of the R&D expenditures in industry. By 1980, federal funds were down to 32 percent, where they have remained until the present time. Real growth in company's own R&D funds rose an average of 6.2 percent per year over the 1975-1985 period. Federal R&D funds to industry rose 4.7 percent per year. The combined average of the federal and the company funds come to an average annual increase over this period of 5.7 percent.

Looking at the development activities of industry, they typically account for slightly over three-fourths of industrial R&D expenditures (Figure 12-11). Here, again, we see the gap between federal and industry R&D funding widening at about the same rate as for the total. Federal money picks up sharply, as you will notice, in the 1984 and the 1985 period, again, associated with the defense buildup. The real growth record, over the 1975-1985 period, is only slightly below that for total R&D expenditures, except that the federal development funds grew at the same rate as shown on the previous chart, which was about 4.7 percent.

Roughly one-fifth of industrial R&D expenditures (Figure 12-12) finances applied research. The increase over the 1975-1985 decade, 6.1 percent a year, is due mainly to growth in industry's own funds for this activity, which averaged 6.8 percent per year. Company funds typically account for about three-fourths of the industry applied research total. The sharp drop you see in federal funding in

FIGURE 12-10
Industrial Research and Development
Expenditures by Source (1972 constant dollars)

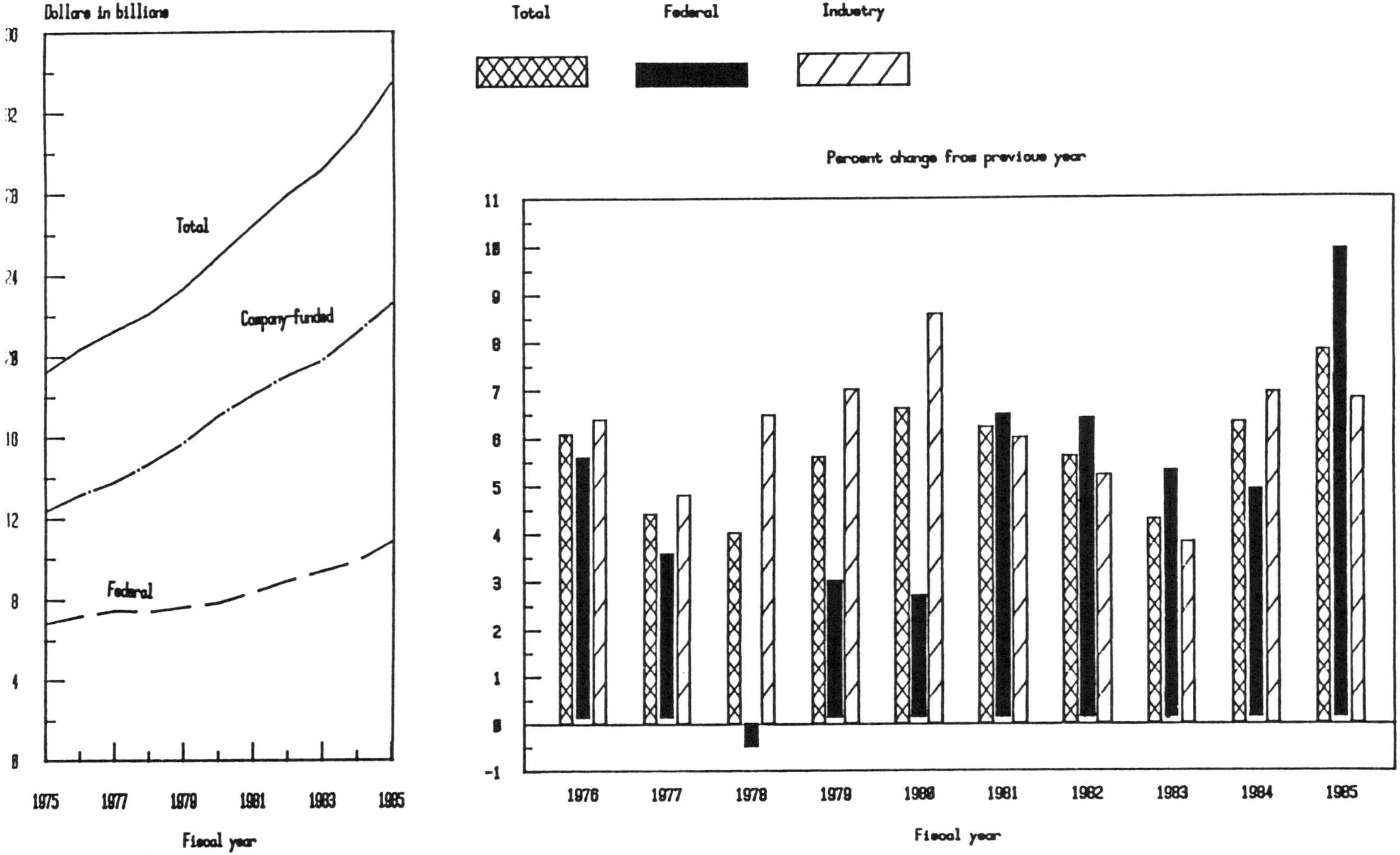

Source: National Science Foundation

FIGURE 12-11
Industrial Development Expenditures by Source (1972 constant dollars)

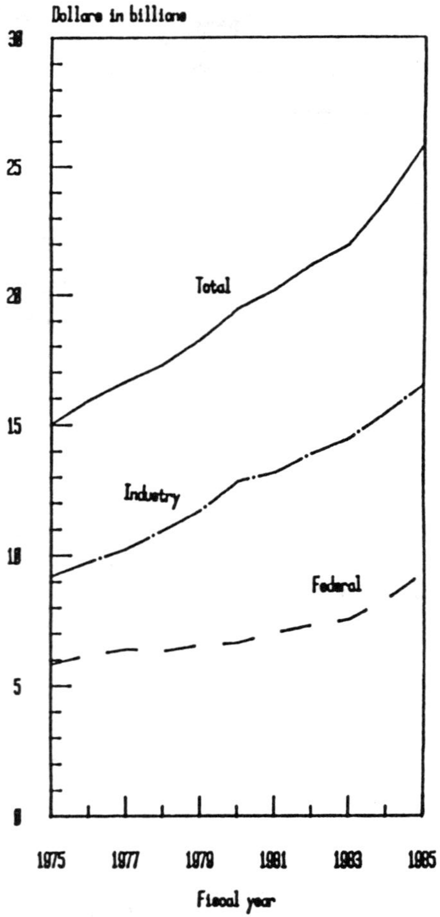

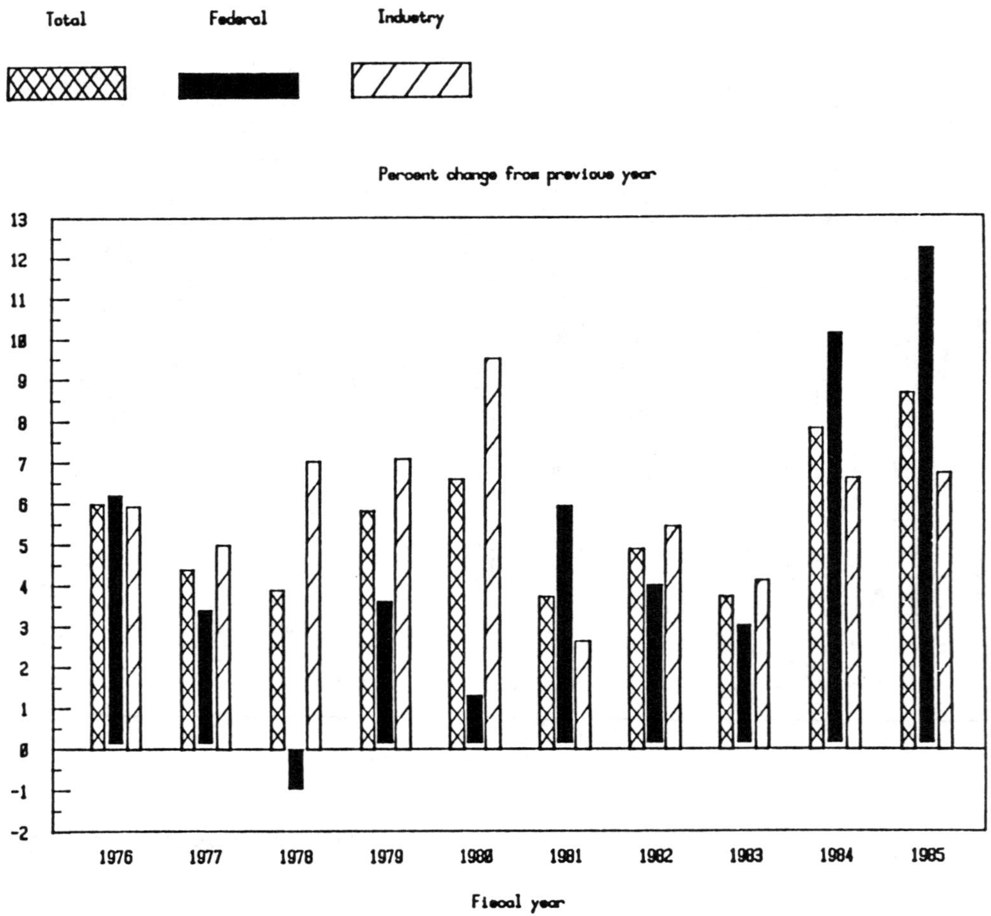

Source: National Science Foundation

FIGURE 12-12
Industrial Applied Research Expenditures by Source (1972 constant dollars)

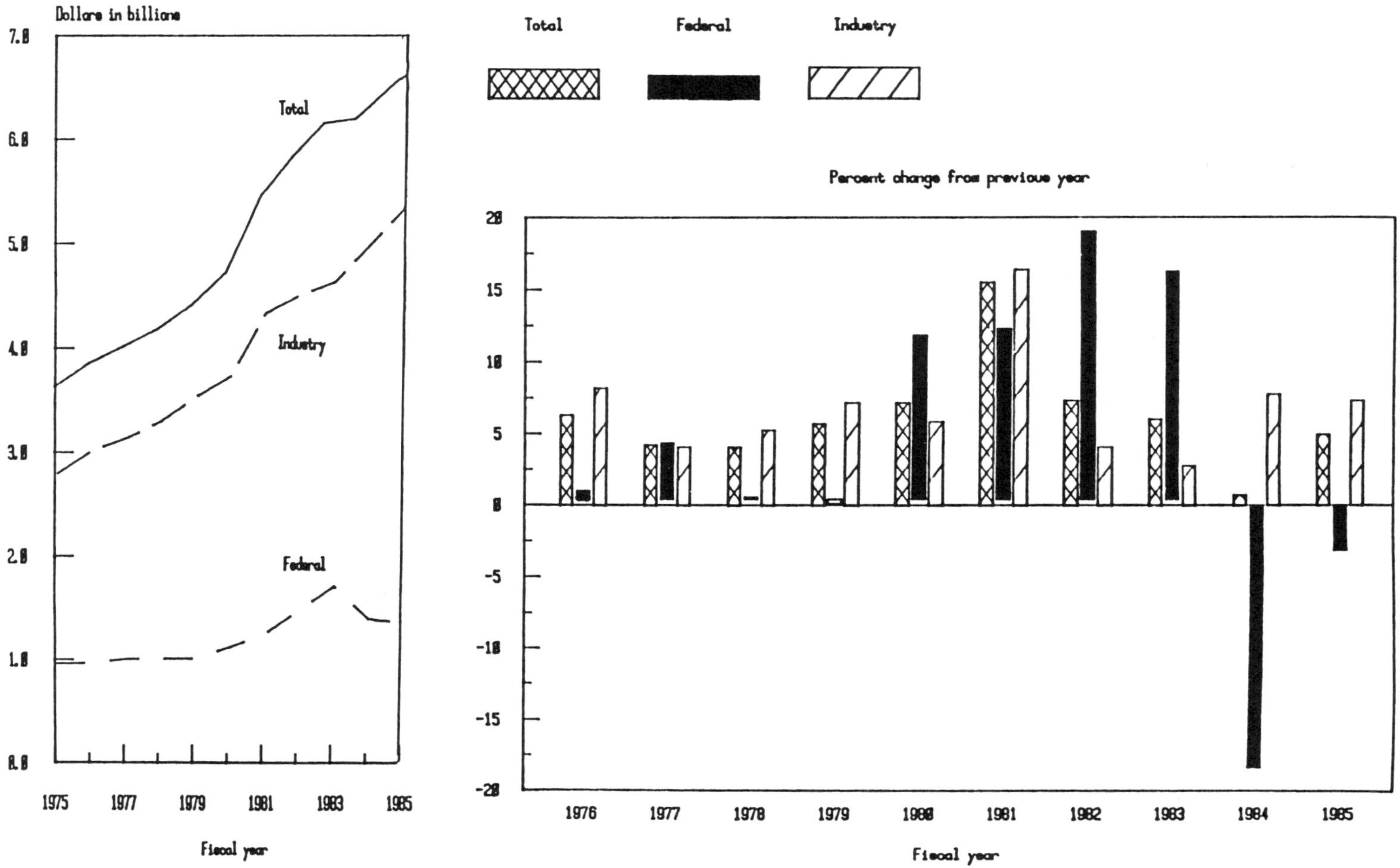

Source: National Science Foundation

1984 is the result of a very large shift in funds from a few firms, reflecting movement of projects from the applied research stage into development. It is more evident on Figure 12-12 than on Figure 12-11 because of the relative size of the applied research funding versus development funding.

I would like to spend a few minutes to pass on the results of a survey we conducted of our industrial panel on science and technology in the late spring and early summer. The results are significant, not only because they helped us develop our R&D estimates for 1985 and 1986, but also because they provided some insight into industry's outlook for the economy and for market conditions over the next year, at least from the perspective of the officials responsible for companies' R&D activities.

After reviewing the comments of the 1986 planning figures, and the 1986 planning figures provided by company officials, we estimate that company-funded R&D expenditures will reach $58 billion in 1986, an increase of 9 percent in constant dollars over the 1985 level. Between 1976 and 1984, for perspective, the average annual percent increase in R&D expenditures in constant dollars was 6.3 percent.

Company R&D officials predict a slightly lower average annual constant dollar growth rate of 5.8 percent for the period 1984 to 1986, with greater growth coming in 1985 than in 1986. R&D spokesmen attributed the 1985 increase in R&D spending to a strong commitment in research and development as a means to protect profits and market shares, and an increased focus on process-oriented research and development to improve productivity and competitiveness. There is also increased emphasis on rapid transfer of new technology from the labs to the operating units, and on effective research project management.

In this year's survey, unlike other years, company officials expressed a great deal of uncertainty about the economic and marketing conditions for the following years, and frankly, they were even hesitant to give us final 1985 R&D growth figures, based on their current estimated budget plans. R&D officials noted that, while the economy seems relatively healthy, they see a lack of consistency in the direction of various economic measures, and it makes it difficult for them to finalize their plans for next year's R&D activities.

a. Machinery

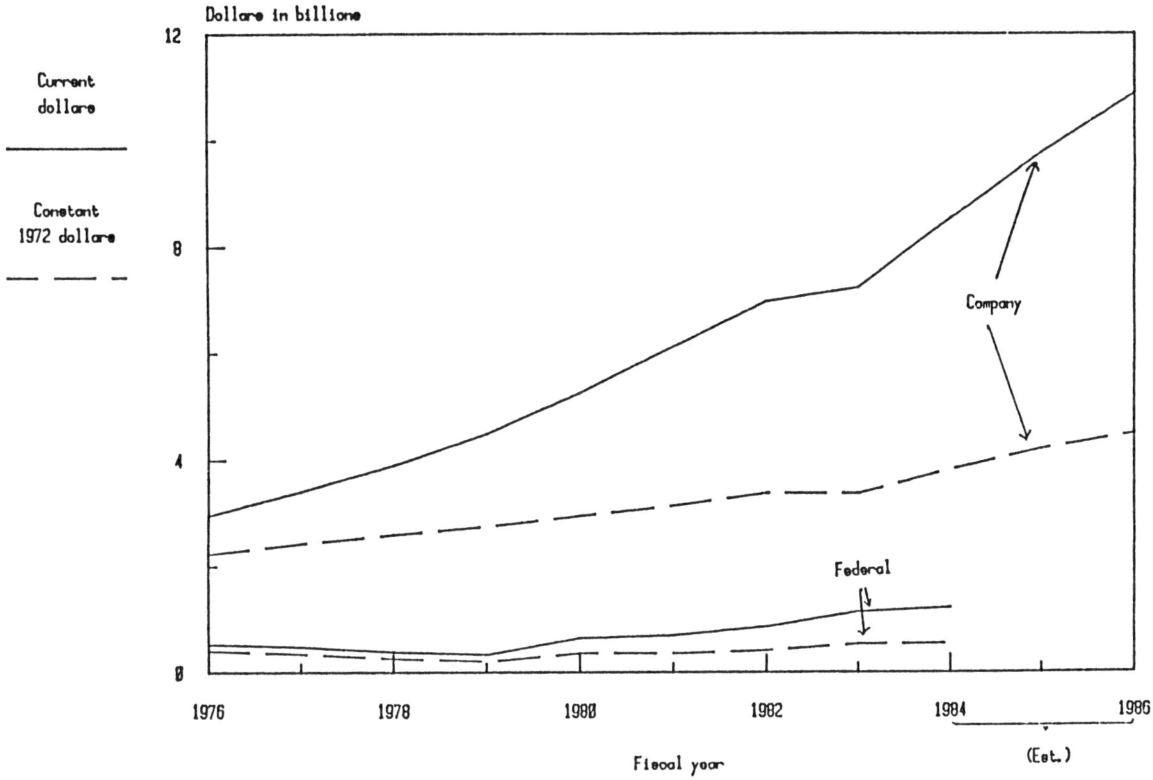

Source: National Science Foundation

FIGURE 12-13 Company Research and Development Funds

Let's look briefly at each of these six R&D industries (Figures 12-13a through f), looking first at machinery (Figure 12-13a). This time we are covering the years through 1986. The machinery industry projecting the largest increase in R&D spending from 1984 to 1986 is fueled by a 14 percent average annual increase in the computer segment. This reflects increases in computer R&D of almost 16 percent in 1985, 12 percent in 1986. The industry as a whole is dominated by the computer segment's 70 percent share of the machinery research and development total and expects increases of almost 15 percent in 1985 and 11 percent in 1986. Officials in the computer segment emphasize that their budget estimates for R&D growth in 1985

b. Aircraft

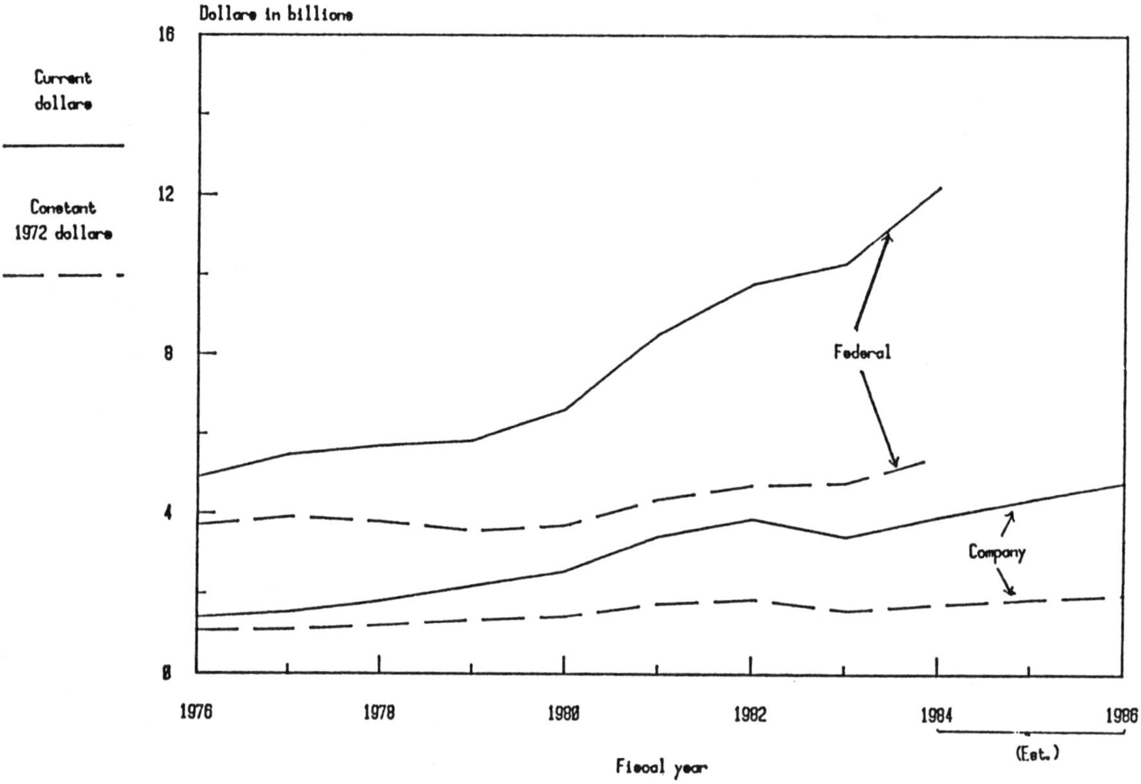

Source: National Science Foundation

FIGURE 12-13
Company Research and Development Funds (continued)

and 1986 are tentative because of possible fluctuations in market conditions, especially in the international sphere. Sales overseas improved for many companies, but R&D officials were reluctant to forecast levels of international sales and/or revenues, since they are dependent both upon the value of the dollar relative to other currencies and upon conditions of other countries' economies.

In the United States, lagging computer sales reflected cutbacks in capital spending by some industries and a postponement of purchases in anticipation of more advanced and powerful products. R&D directors, again, would not speculate on the level of sales to industrial customers in 1986. Officials in machine tool, farm equipment, and

c. Electrical Equipment

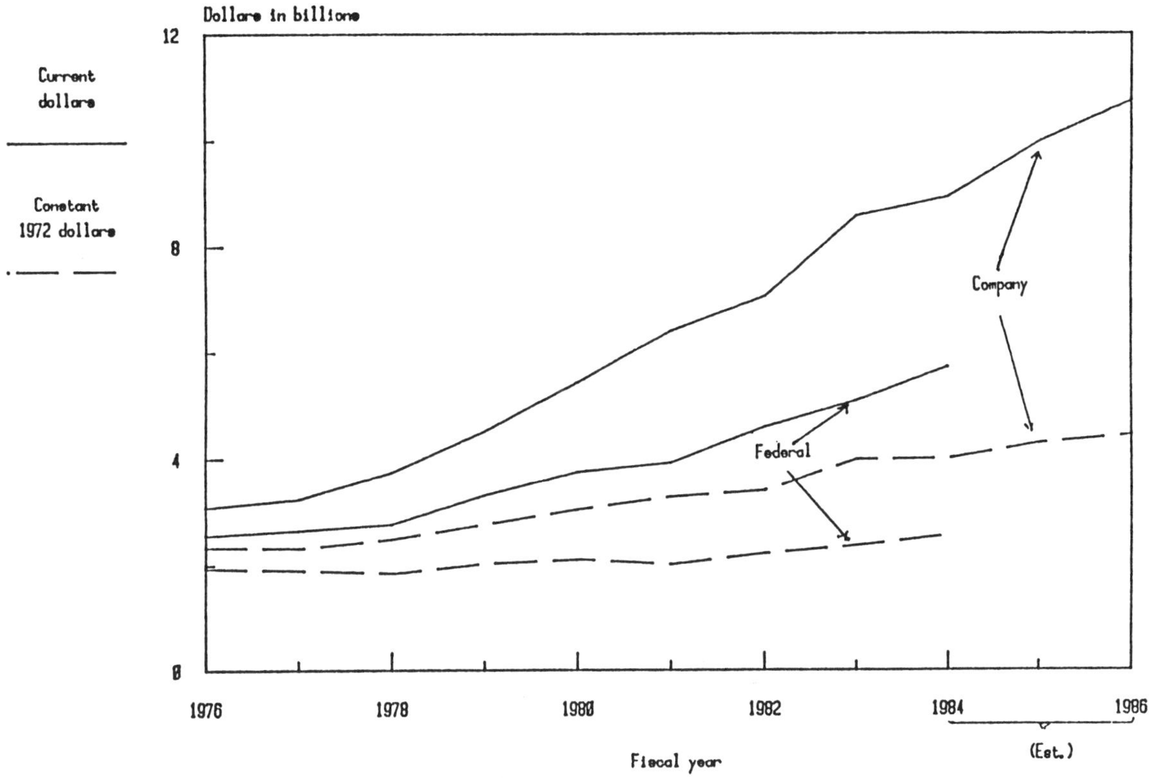

Source: National Science Foundation

FIGURE 12-13
Company Research and Development Funds (continued)

robotic companies blamed poor sales, especially overseas where the dollar remains strong, for the limits on their R&D budgets. A typical explanation was that, due to the pressure of imports on both sales and margins on product lines, companies were forced to make budget cuts. The opportunity for new and improved manufacturing equipment is great, but they feel the risks today are even greater because of the strong dollar.

The aircraft industry projects increases in company-funded spending of 11 percent in 1985 and 9 percent in 1986 (Figure 12-13b). These increases are in sharp contrast to the 11 percent decrease in company-funded research and development in 1983 which followed 1982 decline in sales

d. Chemicals

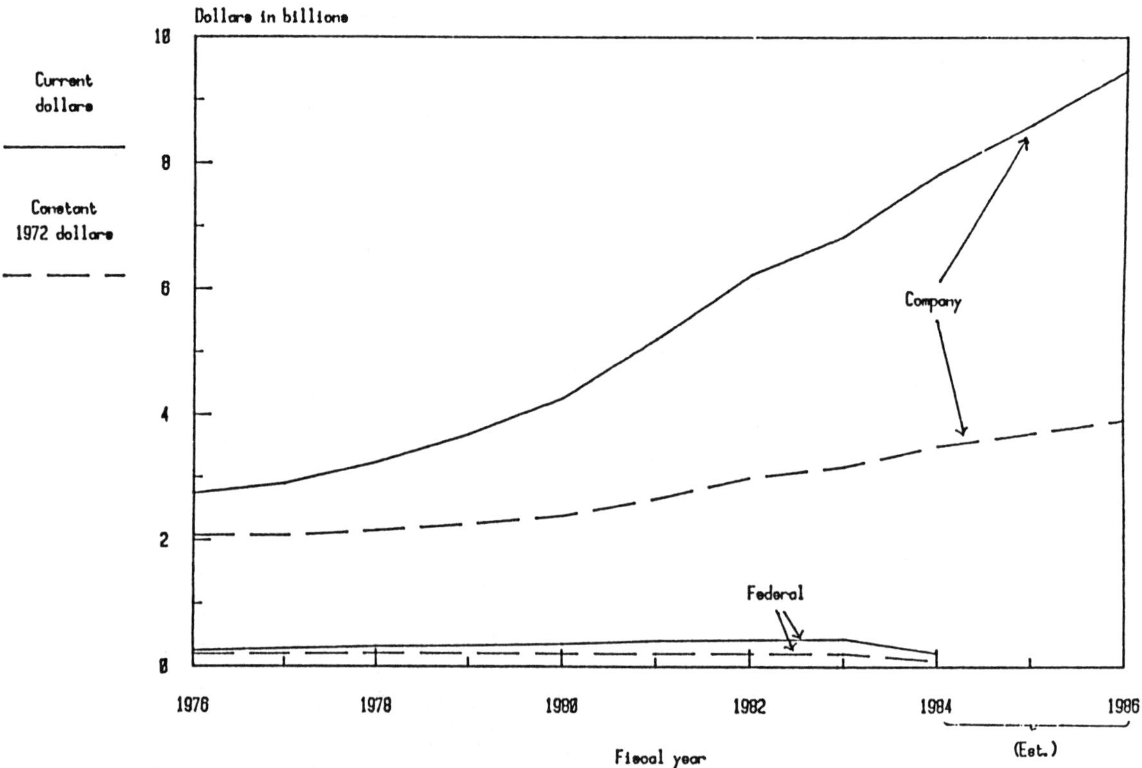

Source: National Science Foundation

FIGURE 12-13
Company Research and Development Funds (continued)

and profits in the aircraft industry. R&D directors anticipate a continuation of the current recovery in worldwide sales through 1986.

The electrical equipment industry projects R&D expenditures increasing almost 12 percent in 1985, 8 percent in 1986 (Figure 12-13c). The communications sector is leading the industry with a 14 percent increase in 1985, 8 percent in 1986. You see this pattern of where the 1986 percentage is less than that of 1985 in almost every industry.

The segments of the electrical equipment industry making consumer goods expect more modest increases in R&D spending because of lower sales and markets saturated with Japanese products. Semiconductor companies also expect

e. Motor Vehicles

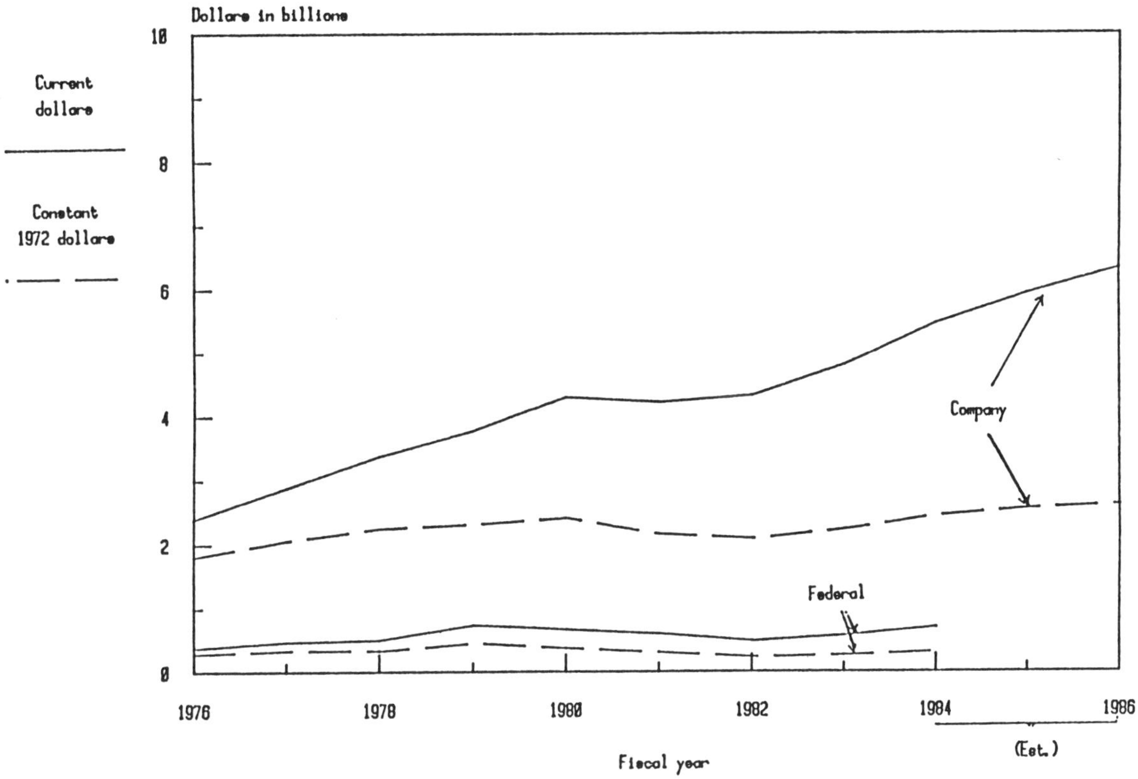

Source: National Science Foundation

FIGURE 12-13
Company Research and Development Funds (continued)

tightened R&D budgets in 1986 because their sales and profits have been limited by Japanese competition.

The chemicals industry predicts R&D spending increases of 10 percent in both 1985 and 1986 (Figure 12-13d). Drugs and medicine companies lead the increases, predicting 12 percent growth in 1985 and 13 percent in 1986, as they work on virus treatments and biotechnology, including DNA and genetic engineering. Industrial chemical companies forecast increases of about 9 percent in both 1985 and 1986.

The motor vehicles industry expects to increase R&D outlays almost 9 percent in 1985, 7 percent in 1986 (Figure 12-13e). R&D officials in this industry believe that

f. Instruments

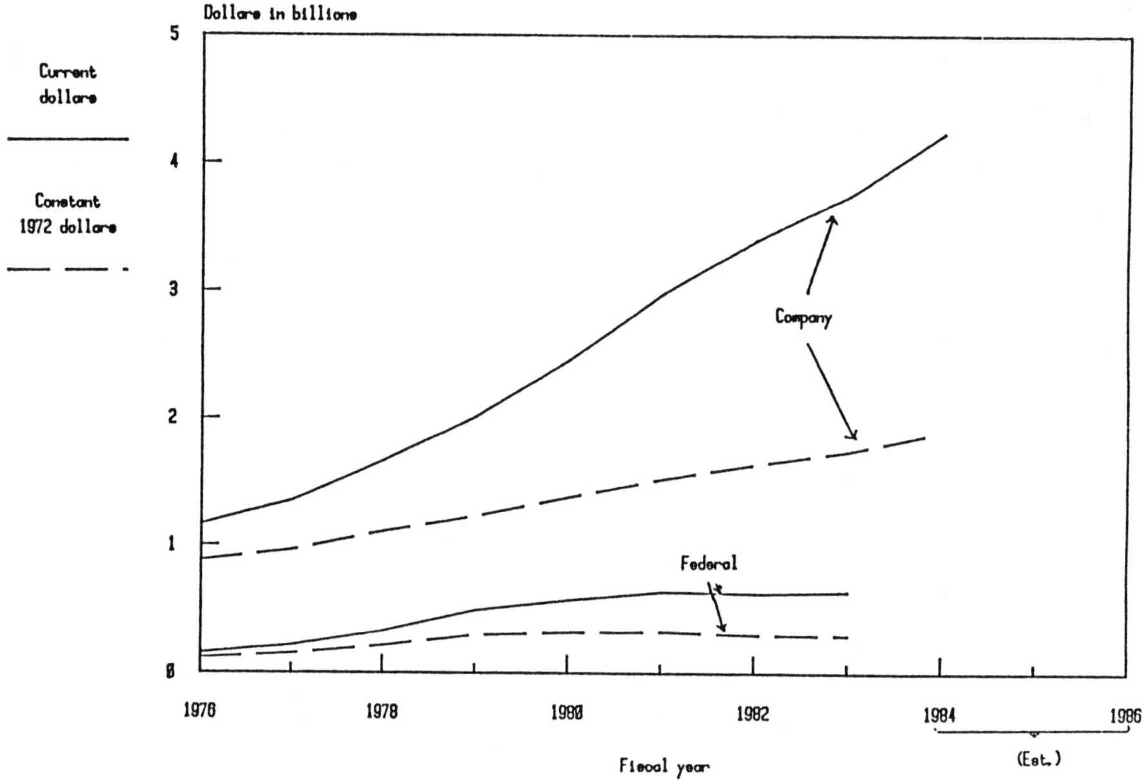

Source: National Science Foundation

FIGURE 12-13
Company Research and Development Funds (continued)

economic times are better for them now, vis-à-vis two or three years ago, and that the removal of import restraints is the driving force for the research and development to meet quality and product goals.

The professional and scientific instruments industry projects an increase in R&D spending of 8 percent in both 1985 and 1986 (Figure 12-13f). These estimates of growth are well below the 17 percent predicted a year ago, as companies now have lower expectations of domestic and foreign sales. Medical equipment companies have modified their R&D spending plans since hospital administrators have cut back their budgets for major equipment purchases.

Let me look briefly at the academic R&D expenditures (Figure 12-14), just to give you a quick rundown of where they are. Here it is not the volume of R&D performance that is so important as is the fact that they account for one-half of the basic research performed in this country. Except for a brief period during the early 1980s, academic R&D expenditures have seen uninterrupted growth, even after adjusting for inflation. The interruption occurred when several federal agencies, particularly NIH, had very little growth in their current R&D, and NSF even declined in current dollars. Inflation, at that time, was still in double digits. Therefore, the decrease that you see reflected in Figure 9-14 in 1982 stands out as the single year of decline for universities' R&D programs.

The basic research picture (Figure 12-15) is pretty much a reflection of the total R&D.

Some are surprised to learn that universities do as much applied research and development (Figure 12-16) as they really do. Actually, universities' applied research and development make up about one-third of their total R&D expenditures. Furthermore, while federal funds account for well over half of what universities spend on applied research and development, over the past two or three years non-federal funds have grown much more rapidly than federal to the point where they are now almost equal.

In summary, let me just point out that, after some slow years in the last 1970s and early 1980s, research and development activity has enjoyed a fairly strong and healthy condition. There appears to be some concern among industry officials of what is going to happen to their markets and to the economy in 1986. But, by and large, we see that there is no reason to expect research and development activity of this country to fall behind. The factors that will be affecting economic growth, I think, are outside of the science and technology area, and have more to do with international politics, trade agreements, general wage levels, inflation, and the value of the dollar in foreign countries.

FIGURE 12-14
Academic Research and Development Expenditures by Source (1972 constant dollars)

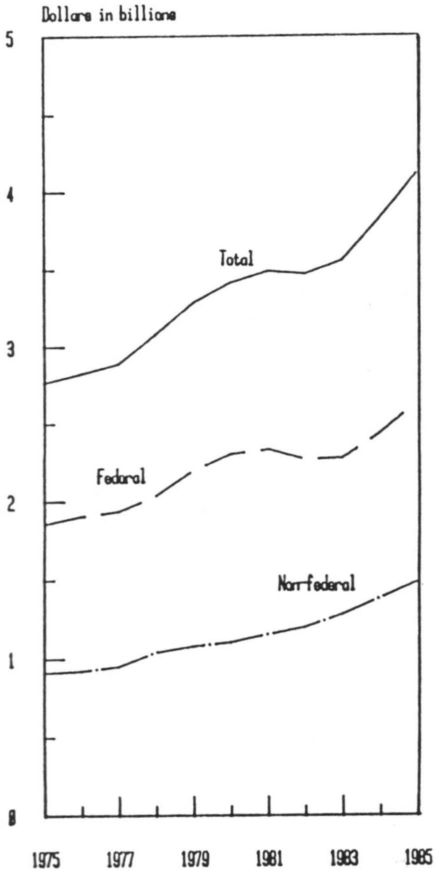

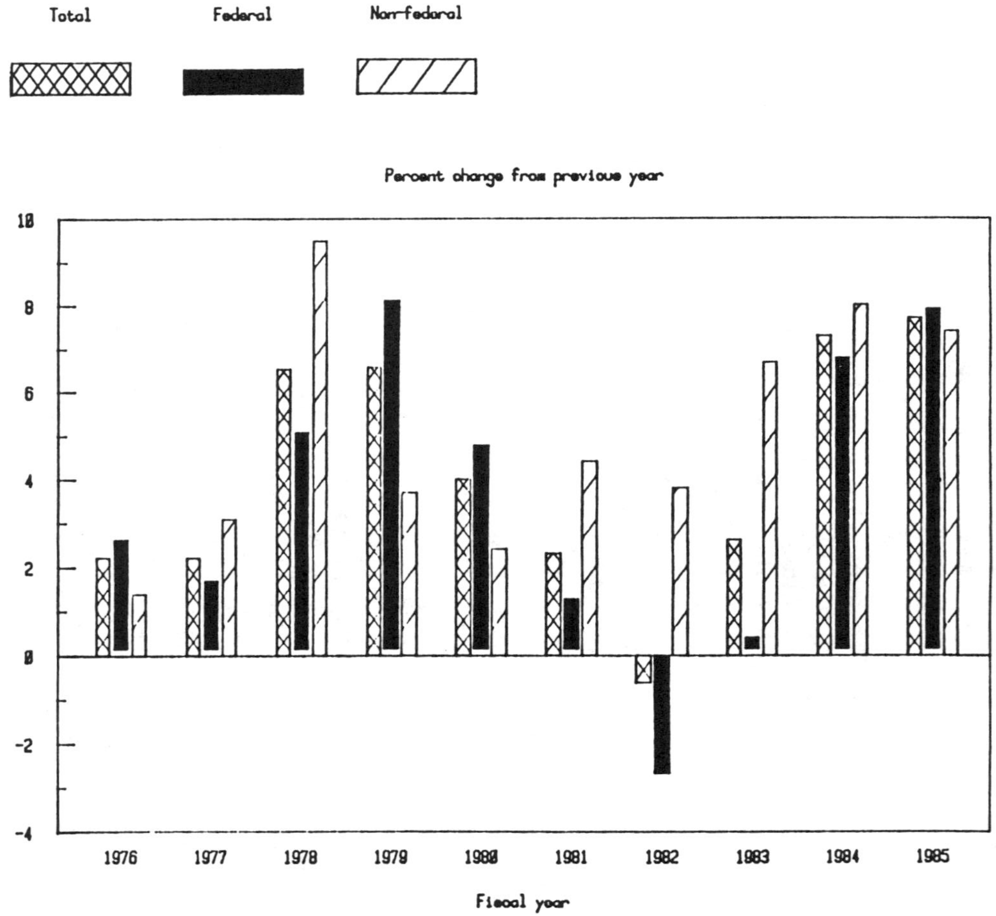

Source: National Science Foundation

FIGURE 12-15
Academic Basic Research Expenditures by Source (1972 constant dollars)

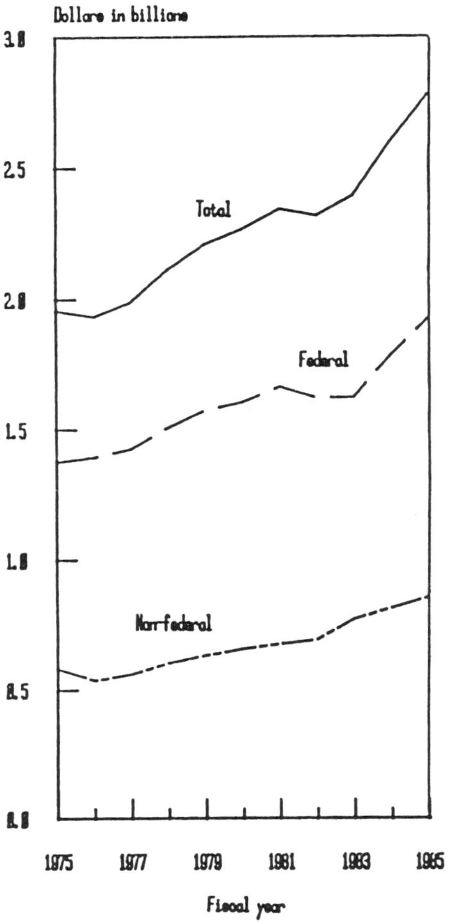

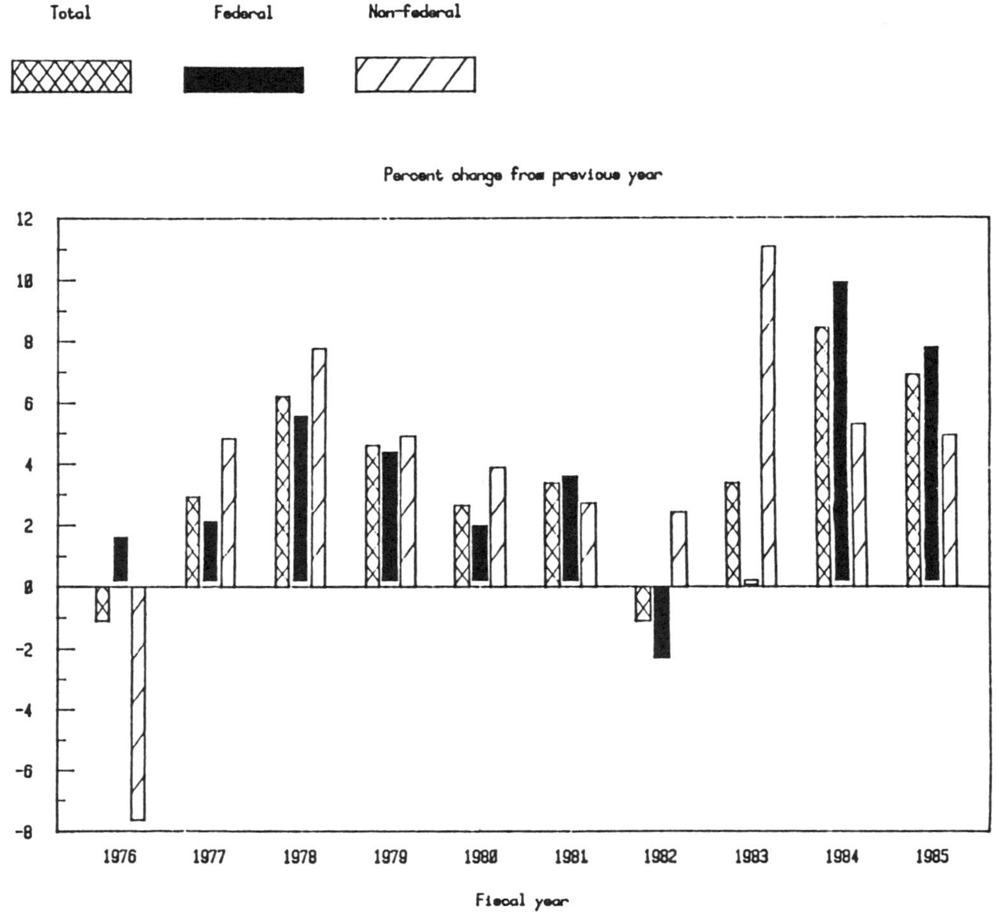

Source: National Science Foundation

FIGURE 12-16
Academic Applied Research and Development Expenditures by Source (1972 constant dollars)

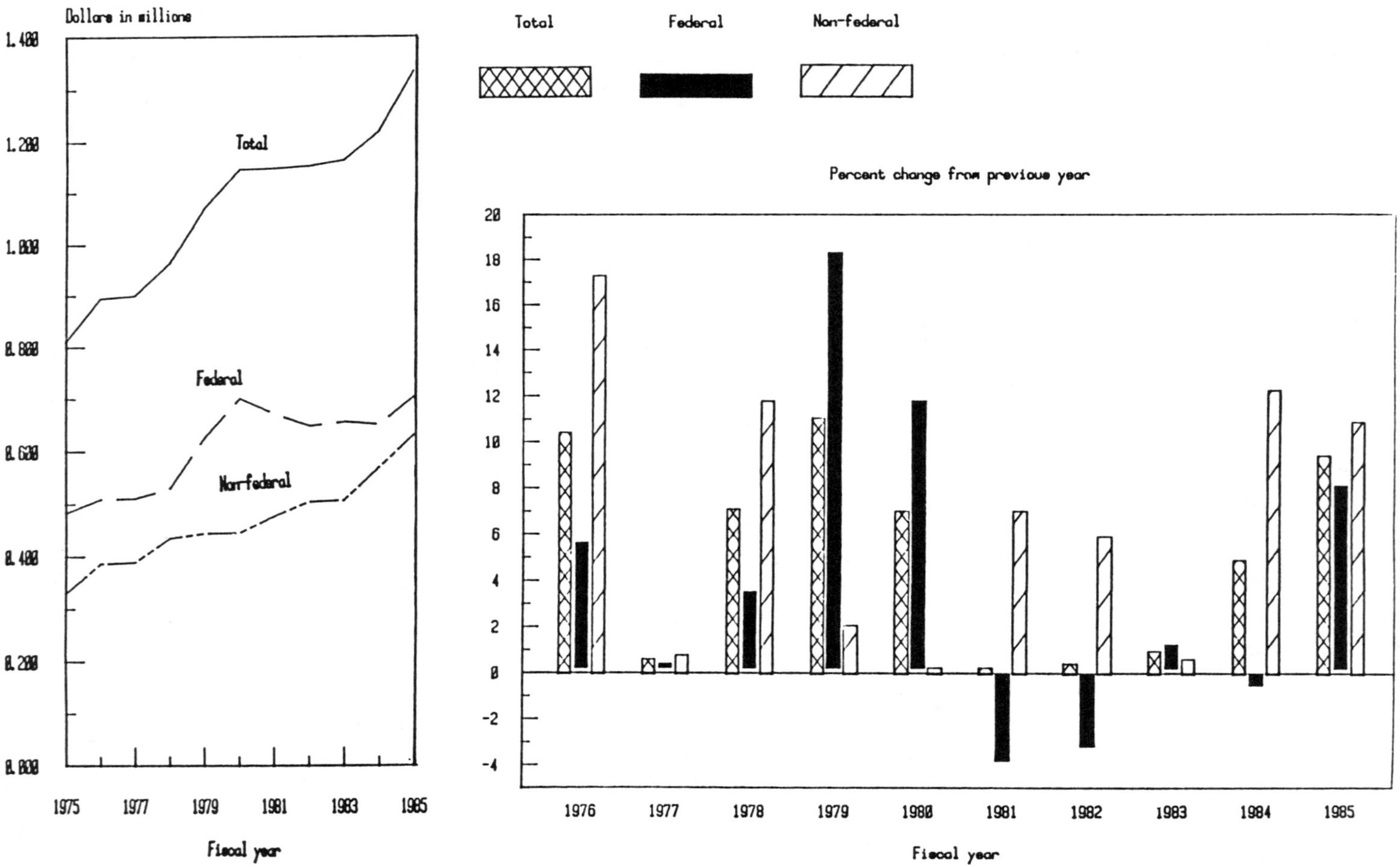

Source: National Science Foundation

13

THE IMPORTANCE OF ENTREPRENEURS AND SMALL BUSINESS IN THE ECONOMY OF THE UNITED STATES

Joseph W. Duncan
The Dun & Bradstreet Corporation

What I will talk about today is at the opposite end of the spectrum from very large industrial enterprise. I'm going to talk about entrepreneurs.

When you think about the structure of firms by size, there are not many billion dollar-plus corporations. Out of the total population of firms, as measured by the Internal Revenue Service, by the Census Bureau, or by the Dun & Bradstreet file, 90 percent have fewer than 20 employees. Therefore, when we talk about the birth and death of business, in terms of the large numbers, we are really talking about what is happening to very small and medium-sized enterprises.

We are going to examine the dynamics of that change; that is, the growth of small business, which is the result of new business formation and older business closings. I also will address the role of smaller businesses in the employment patterns of the past several years because it is a very significant feature of our current economic system.

Next I'd like to talk about changes in business attitudes that are reflected by the various businesses that Dun & Bradstreet deals with through its surveys and its data collection.

Finally, I'd like to end on a philosophical note that relates some recent changes in entrepreneurship to some of the fundamental changes in the American political system and the American business scene for the balance of this decade.

Most of you are familiar with Dun & Bradstreet. You probably think of it in terms of a business that rates your credit. And that is, indeed, one of our important businesses.

Dun & Bradstreet also has recently acquired A.C. Nielsen, so we conduct consumer product research. We have long owned Donnelley Marketing and Donnelley Directory, the Yellow Pages people, so, if you roll through the 25-plus major divisions of Dun & Bradstreet, the common element is: We collect an enormous amount of information about U.S. businesses and households. In fact, our household file, which has information on 87 percent of the households in America, permits us to track some of the important changes in income, employment, and spending patterns as a backdrop for economic analysis.

One of our divisions, Technical Economic Services, focuses on collecting data on manufacturing facilities in the country. It has data on 20,000 facilities, including the type of energy used and alternative sources of supply. It also has data on process technology, operating rates, and so forth.

An illustrative point: The 20,000 plants in their file generate about 80 percent of total U.S. manufacturing production. We're now expanding that coverage to about 100,000 plants, which will move us up to a little more than 90 percent. So, my small-entity feature really draws on that information not normally identified in other surveys and studies like the Conference Board or the Fortune 500.

Thus, the growth of small business is an important dimension of our economic system.

One way to think about new business is to look at new incorporations. Dun & Bradstreet collects data from across the country, using field reporters who go to state houses and record the number of new incorporations. An examination of the general pattern over the past two and a half years shows that there has been a significant growth in the number of new incorporations. This probably overstates new business formations because incorporating is a legal activity; it doesn't necessarily mean that there is business activity involved with a new incorporation. For example, somebody takes an existing firm and breaks it into parts, in terms of incorporations, in order to limit exposure to liability.

But it is significant that last year new incorporations, as a whole, grew about 7 percent over the previous year, and the same amount of growth is expected this year. On an annual basis of more than 600,000 new incorporations per year, that is a very high level of entrepreneurial development.

A study of new incorporations, for example, in the first few months of this year (data assembled through May because some of the state houses take a long time to release the information), shows that the rising trend of last year has been basically repeated so far this year. And that rise in 1985 is happening during a period of general economic slowdown.

Dun & Bradstreet also measures new business activity by counting business starts. Business starts are new businesses which are added to our information file each year. A business start is defined as a new activity added to our file whose birth date is the current year. In order to qualify as a business start, by our definition, an activity has to have been formed during the current calendar year; therefore, first-quarter data are generally soft. Many companies, for example, that start at the end of the previous year may not enter our file until the next year and hence don't count as business starts by our definition. That data base is being changed. Starting next January, Dun & Bradstreet will have a broader definition and a more realistic set of numbers about new businesses.

Annual business starts number about 100,000, a great deal less than new incorporations. The reason is twofold:

the dating of the birth period and the fact that a lot of new incorporations are nonproductive activities from an economist's point of view.

So, on the birth side, it is difficult to assess exactly how many companies are being born each year. However, if we look at the relative levels of activity, we will see a little bit later that the dynamics of this are quite impressive.

The other side of the coin from starts, of course, is failures. Many businesses don't make it. Something goes wrong. The competition gets heavy. Plans are not realized. Business failures shot up dramatically during the last recession, increasing 35 to 50 percent per year over the 3 years of the recession.

Shortly after joining Dun & Bradstreet, I revised Dun & Bradstreet's data series and we increased our coverage. Consequently, I don't have a historical data chart. But if you look at the monthly data, which start in January 1983, you will see that the number of business failures on a relative basis has plateaued. Despite the fact that there has been a strong economic recovery, many businesses continue to fail.

Now, why is that? There are three reasons: First, interest rates have remained relatively high for the small businessman who pays prime plus two or three percentage points. With the continuing stress on the balance sheet, companies have continued to fail, despite the fact the overall economy is robust.

Second, business bankruptcy filing has become a legitimate business strategy. Big companies are doing it and little companies are doing it. A decade ago, companies that would not have filed for bankruptcy are today filing bankruptcy as a method of reorganization and realigning their businesses. That keeps the number of failures high.

Finally, and the good side of the coin from my point of view, particularly the spurt that we see from early 1984 on is a by-product of the entrepreneurial activity of new businesses being formed. Companies are starting in high-risk, high-tech areas; some of them don't make it. The result of that high level of business formation with a number of them failing, is that the failure rate remains high. What we're seeing, then, is that stress remains in business as far as people failing. In other words, filing bankruptcy. It is also there in terms of people just giving up and closing their doors, that we call business discontinuances—a company that leaves no outstanding debts but says, "I have had it. I want to retire. I'm going out of business. I'm going to give up the entrepreneurial activity that I had in the past."

The present economy is softening, reflecting a slight decline in the number of people going out of business. This could be a product of people worrying about the future and not closing their doors. They want to have some alternative income source in years ahead. But, on a relative basis, we're still seeing a very high level of discontinuances.

Add them together and we have what is called business closings. Simultaneously, we have both a growing number of new incorporations or births and a stable level of business closings.

Now, what is the impact of this on the total economy? One way to look at it is to look at the employment associated with these activities. Every year, Dun & Bradstreet conducts a survey of companies of all sizes, large and small, and asks them to forecast their current year gain in employment. The total employment of the smaller companies, those with fewer than 20 employees as a total universe, is expected to grow by more than 5 percent this year, based on what companies tell us. We have a statistically representative sample that permits us to estimate total labor force impacts.

Very large companies with 25,000 or more employees, have had a very modest growth in terms of the total amount of employment associated with big firms in our economic system. This can perhaps be shown more dramatically if we look at how the employment growth—which we estimate to be two and a quarter million new jobs for 1985—is distributed.

Twenty-eight percent of that employment growth is in companies with fewer than 20 employees and another 24 percent is with companies from 20 to 50 employees. In fact, 75 percent of employment growth this year will be in companies with fewer than 500 employees, the Small Business Administration's definition of "small." I happen to think "small" is less than 100, but regardless of how you define "small," it is quite clear that in our current economic recovery and expansion, most of the employment growth has come outside of the big firms.

In fact, there is an interesting phenomenon which I call

the "outsourcing of labor." Basically, what is happening is that big companies are running tight ships. They cut their overheads during the recession and they are not reinstating it now. They are buying outside services from small companies. In fact, temporary help is the most rapidly growing four-digit standard industrial classification right now, growing nine times the rate of growth of total employment, because people are buying temporary help rather than adding them to the permanent payrolls. This is a fundamental change in our total economic structure, something I want to come back to in a moment.

Studying business attitudes is one of the ways we monitor the health of business. Every quarter, Dun & Bradstreet conducts a survey which asks companies: Do you expect sales in the coming quarter to be more or less than they were in the comparable quarter one year ago? We calculate an optimism index. It is a very simple concept. If 80 percent of the people say they expect an increase, 10 percent expect a decrease in sales, the optimism index is 70, that is, 80 percent increase minus the 10 percent decrease. The 10 percent that remain unchanged do not impact the index.

Businesses, large and small, fundamentally still see a lot of strength in the economy, despite the economist's analysis of GNP which shows a very weak growth. That is a product of the fact that domestic sales in the United States are growing at more than 5 percent per year. Domestic final sales include imports, and although there are many markets and many smaller firms thriving very well in the import business, the bigger companies are the ones that are having trouble with import substitution.

However, the basic optimism of business remains quite strong, despite the fact that we have come out of a period of very weak total national economic growth.

Another way to look at it is to look at employees. While it is true workers peaked in 1984 when there was an enormous growth in employment, there is still a lot of optimism in the sense that people plan to continue to add workers, albeit at a somewhat slower rate than last year.

Selling prices, of course, are the good news. In this case, we want to have a low index because if prices are not increasing that means that inflation is under control. And that is the case. Fundamentally, consumer prices for this year will be under 4 percent for the year as a whole. Inflation, people are finally beginning to believe, is somewhat under control and even the pessimistic forecast for the next couple of years is at a low level of total inflation, and that is because people do not plan to increase their prices.

The good news, of course, from a businessman's point of view is bottom line: profits. Despite the stories of the slowdown in profit growth, recognition must be paid to the fact that many of the published numbers are for very big companies which deal with the extraordinary gain of last year's profit base. It follows then that many companies are not seeing further growth in their total profits. However, smaller companies see higher levels of profits, which kept the index at a relatively high level for this period in the business cycle.

Let's now go from the fact that small business is dynamic and growing and relatively optimistic to the political and economic context for the next several years.

Around the beginning of this year, The Economist magazine published an article which provided a very interesting perspective on the future of the entrepreneurial spirit in the United States.

A group called Oxford Analytica in Cambridge, England, developed an interesting perspective of what has been happening in the United States. Their view, of course, is from a European prism. The following quotes from The Economist are illustrative.

> Reaganism has added political ideas to the cultural ones with which America influences the world. American politics is thought by everybody, especially Americans, to be a non-ideological contest between parties that pragmatically represent interest groups.
>
> But Ronald Reagan is one of the most ideological presidents America has ever had. The fact that his few clear ideas are uncomplicated ones adds to their electoral stength. One reason why the democrats have failed so badly against Mr. Reagan is they have not developed a new set of simple ideas of their own.

Now, when we think of ideological conflicts, we frequently think of ideology in the nineteenth-century concept of class and class conflict. But The Economist correctly points out that class concepts do not really mean much in the United States in the 1980s.

"Today's important political thinking in the United States," The Economist says, "is about the freeing of markets and individuals from the authority of the state in an age of technological change."

I'd like to explore that philosophical framework by looking at several features that help explain the ideological shift toward entrepreneurship that is already evident.

First is the population shift, the Sunbelt phenomenon. Three states, California, Texas, and Florida, accounted for more than 40 percent of population growth between 1970 and 1980, according to the U.S. Census. Those three states also accounted for 25 percent of the nation's Gross National Product, and the sum of their electoral votes represents a third of the number a candidate needs to win the presidency.

It is interesting, when we think about this shift to the South and to the West, to think of the ideological characterization which The Economist made of those regions. It said, "These regions are more distant from Washington, more resentful of government controls, closer to the frontier, and less familiar with the old world across the Atlantic." Obviously a British perspective.

But The Economist points out, correctly so, that these states are more inclined to embrace individualism as the basis of political philosophy, and all you need to do to verify that is to look at some of the legislation that has characterized those states in the last decade.

The second phenomenon is immigration. The U.S., with 5 percent of the world's population, takes in 50 percent of the world's nonrefugee immigrants each year. It takes in 450,000 immigrants legally, and certainly an equivalent number illegally. The flow of immigration is different today than it used to be. Eighty percent of the immigrants to the United States during the past decade came from the Third World, changing U.S. ethnic composition. The rush of immigrants from Mexico and Central America, The Economist believes, is likely to affect traditional views of America's links with Europe. Once again, a British point of view.

Finally, there is the role of technology. Technological change is my favorite topic, starting when I was at Battelle in Ohio working on the impact of science on society.

Returning to The Economist's article which commented on the role of technology in U.S. society today, they estimate that there is now more capital stock for each information worker than there is for each manufacturing worker in the United States. In other words, the return to capital is very great in the intellectual component of activity, ranging from the marketing department through research, advertising and other activities, as well as banking, finance, insurance, or the information industry that Dun & Bradstreet represents.

The significance of that, however, goes a little deeper. The Economist notes that "America's race toward new technology is accelerating the pace of economic change. Products, even whole industries, come and go much more quickly than they used to."

The significance of this is that Americans are more prepared than ever before to let economic development be decided by what a free marketplace bursting with entrepreneurs happens to turn up. The birth and death of business reflects that there is an enormous entrepreneurial spirit in this country.

I recently returned from Europe where I talked with economists from a number of countries. They are uniformly distressed that they cannot figure out how to develop a policy that will create the entrepreneurial spirit that we have in the United States. That is because it is part of the social ethic which The Economist has so correctly pointed out.

In its final paragraph, The Economist captured the tone of its article titled "The Rediscovery of American Resilience." It concludes:

> There are reasons for thinking that the change in American opinions will last. One is the permanent shift of American gravity to the south and west regions where the old liberal faith and the beneficence of government has scant appeal. A second is those technological wonders which create conditions that shall reward small entrepreneurs in free markets. Above all, America's new mood brings it back to one of its main intellectual roots: Will a people whose inalienable rights include life, liberty, and the pursuit of happiness let a government get in its way?

The dynamics of the birth and growth of small business is the strength of the U.S. economy today. The diversification that flows from the outsourcing of labor puts us in a better position for a slowdown or a recession in the future. We've created 8 million new jobs in the past 3 years and, for the balance of this decade, we are well on our way to a period of economic growth without recession. But first, we must solve the federal deficit problem.

14

TRENDS IN THE U.S. AUTOMOTIVE INDUSTRY

Albert J. Sobey
General Motors Corporation

From an automotive industry standpoint we are interested in three primary impacts of energy. First, is the impact on the total economy—the health of the nation to which we want to sell products; second, the impact on the decision to purchase a vehicle—the cost of energy, fuel at the pump, and what that does to buying big cars or small; and finally, and not necessarily least, is the impact on our manufacturing cost. I will touch on each of these.

To digress briefly to the use of scenarios, the GM economics staff uses two different kinds of scenarios for assessments. GM has a product planning cycle scenario that spans ten years because that is when we start our product planning, and it gets more rigid as you get closer to the present date. We prepare three internally consistent forecasts of economic growth, productivity, energy prices, etc. that are basically the variations on a theme. They take the present kinds of things we know and extend them toward the future.

We use other scenarios when we assess the future technical or new product development activities and contingencies such as energy supply interruptions. What are the possible scenarios, what could happen, and what should we be prepared to respond to if it does are questions I want to address.

It is hard to put labels on those scenarios, but for my own internal descriptive use, I refer to them as short-term and long-term oriented. The first is typified by the decisions of the politician that wants to get reelected and the manager who wants a short-term return on his profits; and the second, long-term oriented, is something that we sometimes refer to as a "statesmanlike approach." But, of course, you have to get through today to get to tomorrow, so I tend to give greater probability to short-term policy decisions.

In these scenarios we look for the significant changes in social attitudes, resource availability and cost, technology, jobs, productivity, what is happening to our cities, innovations in vehicle technology, the economics of the world at the national level, and finally, but not least of all, regulations, the things I would call "trip wires."

First, in the social area—both of our industries (automobile and utility) have been buffeted over the past 15 years by changes in people's attitudes. We are gradually learning how to live with it, although not without grave concern for the future. Similar changes in attitude also impact why people buy our product and we have to understand what is happening in people's product decisions.

One group that is important in both contexts are those who, because of their knowledge and lifestyles, are identified by the SRI as the experiential people; people who buy Fieros and diesel Chevettes. SRI's studies indicate that there are about 35 million of those people in the United States. If the U.S. gross national product grows 2.5 or

3 percent per year there should be 55 million of those people of like mind by 1993. And these are the people who tend to take on causes, such as the environmentalists.

The attitude of many of these people about the automobile a decade ago was that the automobile was unsafe, antisocial, and the cause of lowered air quality. Many felt that public transport was the answer. Clearly, that has turned out to be a nonanswer since it provides unacceptable service and unaffordable prices. Today these same people are forcing a change in the role of utilities through a transition from the social contract which until recently was recognized by most people, your requirement to provide power independent of who wants it and for what purpose, to great concern about rates and environmental issues.

We found that part of the automobile problem was the way we looked at it, not just the way that the environmentalists and the activists looked at it. We did not think that "irrational, noneconomic" requirements for safety or performance standards could be mandated. We were concerned that some of the people were using those issues as a means to get across hidden agendas. I suspect some were, but most were not. And we have to understand the way they think about things. We haven't found all of the answers, but we do find that many things we say are just not understood. For example, if the press reports that profits increased 50 percent in one year following a 50 percent drop the previous year, the general public thinks we would come out even. As you know, we don't. We need to put things in a context that is more meaningful to the public—the benefit of jobs, quality of life for the future.

Let me move on to one of our major problems: the price of our most convenient energy source, petroleum. A decade ago, people thought we were running out of oil. We know now that there is at least 70 years of supply out there but the U.S. supplies are limited. We may have from 6 to maybe 18 or 20 years at our present rate of production, perhaps more if the price rises to $40 or $50 a barrel.

The question is: How high will oil prices get? One way of getting a feel is to evaluate the factors which could limit the long-run prices. For example, the low long-run trends are limited by investments. If you are going to have much product outside the Middle East, the marginal barrel has got to earn at least the market price. Some recent projections indicate that at the turn of the century, in 2000, the cost of discovering and producing each incremental barrel is going to be somewhere between $25 and $35 a barrel. The minimum price of oil has got to be in that range.

The maximum price trends assume that OPEC or more specifically, the Persian Gulf nations, regain the ability to control the capacity of oil in the world and thus act to maximize their revenues. This will occur somewhere around the mid-'90s. Most oil companies are now telling us that they expect at least half of the U.S. oil to be imported by the end of the century. Some think as much as 90 percent could be imported if prices remain near or below present levels.

But, as we consider the conservation and substitution effect (elasticity), OPEC maximizes its revenue at somewhere between $35 and $50 a barrel, it would be self defeating to run prices up much higher because even below these prices we see economic alternatives coming in as substitutes for oil. A major factor in reduced demand for oil, and its price, of course, has been the use of coal, gas, and nuclear power by the utilities.

One of the major alternatives for propulsion is natural gas, which has limited capabilities. It will be the lowest cost fuel for use in the future, about half that of gasoline. We expect a growing but still small market for compressed natural gas vehicles. The other major alternative is methanol. Methanol can be made from many resources: natural gas, coal, biomass. Even nuclear reactors may be, in the long run, an economic source of coproduced methanol fuel.

We have been studying these alternatives very carefully. We have obtained estimates of the technical feasibility and cost of methanol facilities from the companies who have designed most of these facilities in the world. There are few real unknowns. Two billion gallons a year of methanol are now produced in the United States. Our arithmetic, supported in some cases by price proposals, shows that, in the United States, you could build a "Greenfield plant," that is, a new factory on a new site, to produce methanol using U.S. natural gas and have it provide an attractive return on investment when the average retail pump price is $1.30. Not bad potential. We are very close to that point.

There are three reasons it won't happen soon. One is

that we haven't said we will produce cars and trucks which can use it. Second, the Persian Gulf nations can still "whip saw" the prices. The final reason is that you can produce methanol cheaper overseas in countries that have flared natural gas or where they have no local demand for natural gas. We expect to see methanol first in those nations that are oil-poor and are spending a large share of their foreign currency on OPEC oil. That could start in 5 to 10 years. It could start here in 10 to 20 years. There will be a gradual transition because the alternatives will put a cap on oil price. The result is that many of those barrels of oil that we know we have in the ground may never actually be required, at least not in my lifetime and presumably in most of yours.

Methanol has many technical problems, all of which appear to be resolvable, but we will not be ready for the market until we have all the problems solved. We don't want to have product problems come back to haunt us. There are positive environmental factors that are in its favor; methanol is probably less toxic than gasoline. The only reason I can't give you a number for that is that we don't really know how toxic gasoline is.

Now, let me move to the future technical developments which will impact the design and manufacture of our automobiles and trucks. The major change in our approach now is that we are emphasizing the manufacture of the vehicle in its design as well as safety, styling, comfort, and performance—the almost infamous Saturn approach. We start with how you build it. We will be able to provide a better product for the customer incorporating the desired performance at a lower price.

The Saturn concept will permit us to reduce the size of our plants. This is consistent with other trends which indicate that industrial plants may have exceeded the economies of scale. There are unlikely to be any new integrated steel plants, for example. The new car plants may be much smaller than we would have expected five or ten years ago.

From the standpoint of the product itself, there are two major areas of progress: The first is materials and the second is power plants. The battle over what we build cars out of is not over. Given a straightforward, simple, economic decision today, high-strength, low-alloy steel might be our choice. But plastics have many attractive characteristics. They require a substantial improvement in tooling and reductions in cycle time that offset many of the economics of the product itself. The problem is in predicting technical progress; the two materials are playing leap frog; when one improves, the other counters.

Automobile materials technology may best be described as "mid" technology, increasingly sophisticated but with cost being a major factor which precludes use of many exotics being developed for aerospace. The more exotic composites will be used where necessary for performance. Clearly, we will use ceramics in engines; we will have plastics in places we never would have thought of a few years ago, but those are not going to be in big quantities. The use of exotic composites in large quantities to get the weight out of the vehicles, will only come if they are, in effect, mandated as a means of achieving corporate average fuel economy standards. But we also have another concern about plastics that is not significantly different from our concerns on petroleum or metals. The center of the production of the plastics may be moving to the Persian Gulf. The Saudis have invested much of their earnings in petrochemicals. We are very concerned about that.

To give you an idea of our progress in energy use in our manufacturing, we consume somewhere between 1 and 2 percent of most energy types in the country, but we are not energy-intensive in the sense of other industries. Each automobile contains the equivalent of about 180 gallons of oil; about 15 of that is actually oil or gasoline. We have increased our efficiency of energy use in our plants around 1.5 percent per year in the past decade. On the other hand, our use of electrical power has been going up. On a per car basis, we use 19 percent more electrical power now than we did in 1973—roughly a 1.5 percent per year increase—and we see no reason for that to change. The factory of the future will, if anything, increase our electrical power consumption relative to other power sources.

From the standpoint of electrical power and its technology, I want to raise a couple of issues. One is availability of supply. I'm extremely concerned when the chief financial officer of a major bank, in this case, Chase Manhattan, says in a public talk that they will invest in no more large utility power plants—coal or nuclear—and when others say, at the same time, that a company or operation that is electric power dependent must have its own secure supply. It ought to be even more disturbing to you. Whether it is rhetorical or not is a question; they do put in one parenthetical phrase, "until the regulatory system is changed." I think that this is a little like asking for the sun to come up in the west. But what this situation does is put us in a position of having to look at our requirements in a new light. We are concerned about that in two contexts; one is the cost of our manufacturing and the other is its impact on the economy.

We have taken some Electric Power Research Institute projections, adjusted them to GM's expectations for economic growth and potential additions and cancellations of new nuclear power plants. We have compared what we call free-expression electrical power availability with the constrained power availability, and find that there is a risk of significant loss of GM car sales.

What are the options? Well, one of them is to provide our own electric power. I'm not advocating this as a national policy, I'm saying that under the circumstances, this may be the kind of thing that we will be required to do. We are looking at cogeneration quite extensively. We are not a good cogeneration candidate. We are not energy-intensive in the sense of other companies. But if we can buy a plant for less than the capital cost of an equivalent large utility plant, and come up with a per kilowatt cost similar to what a utility would charge, it begins to be attractive, and we see the technology as going in the direction of reducing these costs in time. Parenthetically, we have a division that provides a substantial portion of the world's cogeneration equipment.

Similar decisions are going to be facing the residual market and some of the other commercial markets, as well. One of the things which I discovered a year or so ago was that about the time Edison was building the first central power stations, some other people were supplying self-contained technologies, incidentally, including a coal fuel cell. Many thousands were built and worked, and the ownership costs were attractive and competitive with the central power station, but superior marketing and reliability, etc., gave the central station an advantage, and the self-contained systems became unpopular.

There has been some progress in the technologies of

small stationary plants that may make it attractive to own your own, particularly if you are in a position like I am. Two or three times a winter I have electrical power interruptions at my home and, in some cases, they have been as long as a week. Most of my neighbors have bought "Little Joes" to run sump pumps and furnaces. If a little more money would buy something that would serve all my electrical needs, I begin to wonder if the security premium isn't sufficient to offset some cost penalties.

I'm raising questions. I'm not trying to say that these are the right answers, but there are some parallels. We have one in the transportation field. Years ago the way you got to wherever you wanted to go was to walk or ride a horse. In the early 1800s the horse car came along leading to the public transit utility. Then, as we increased our income and our ability to build good, economic vehicles, personal mobility came back in the form of the private automobile. A similar development has occurred in engineering from the slide rule to the mainframe computer, and now to the PC. These technical developments are things we have to think about.

Moving on to jobs and employment—the automobile business is a mature industry. And all mature industries are facing increasing competition, as was said yesterday, from overseas. We have several choices; build a better product, cut costs, be innovative, but primarily, improve productivity. If we do that, industry can avoid moving overseas. Productivity increases imply better management and more effective use of labor.

If we paid for no labor whatsoever for our present cars, our costs would still not be as low as the Japanese, so labor is not our only problem. Other mature industries have the same kinds of problems. These industries can almost be considered as suppliers of commodities. When anyone with enough money can go to a consulting firm and buy the design for a product, the factory, etc., it will operate almost like a commodity in the marketplace. That is where we are. On the other hand, the United States has an entrepreneur climate creating new jobs in new industries and small companies more rapidly than the increased work force. Most of them are the corner hardware store equivalent—they are not high tech—they are using what is available in the way of technology, sometimes with new marketing concepts like

Federal Express. We think those should be encouraged and are optimistic. We do share the opinion of some demographers that within the next decade there could be a labor shortage in the United States.

Cities—where are they going? It is important to GM because our cars are designed to operate in cities and on interstates. It's important to you because it is the people you serve. We see signs of some significant changes in urban development. Historically, cities grew to support industry or commerce. They expanded in size as transportation permitted. When walking was the only mode, cities averaged about a mile in diameter. The street car expanded it, and then the automobile, and now with the expressway, we may have reached the limit of transportation distance. We are encountering congestion and delays. At the same time, we are also being impacted by a second set of technologies; communications technology—the change in economy of scales of industries, which as I indicated earlier, will probably be significant in industries other than automobiles. There is no longer going to be a need for megatowns to support megaindustries. We wouldn't build the same cities we have today. If we were to start over again it would lead to a different urban environment.

Some people thought that the energy crisis would force return to older cities (redensification). Those of you who have tried to recondition older houses know that it is not an economic approach. And it is clear that cars sitting in a traffic jam are not being used effectively. The growth centers, as we see them, will be in the suburban developments and in the smaller cities that have good transportation access.

We expect the economy to continue to grow, somewhere between 2.5 to 3 percent per year, adequate to meet the expectations of our population, assuming we don't do something foolish. We think electrical consumption will increase about as fast as GNP. Using our own example and our studies as evidence in specific industries, the multiple may be anywhere from maybe 1.2 to perhaps 1.4 times product output. This growth, as I implied before, may not necessarily benefit utilities. The total growth may be reduced if high-energy-using industries move offshore or turn to internally generated electric power.

Moving a moment to the regulatory environment; this is

potentially the most damaging from our standpoint. For example, in the early '70s there were a lot of myths, there was no elasticity to energy use and many rules and laws were created to artificially reduce energy use—most have been counterproductive.

You have had trouble with the Fuel Use Act. We have had our troubles with fuel use, too. The responsible members of the automobile industry are very worried about the corporate average fuel economy standards. Oil prices went down—our sales went up substantially. Our sales were something like 17 percent higher in 1984 than they were projected to be in 1980 when we thought the gasoline prices were going to be as high as $2.00 per gallon.

If we are required by law to meet CAFE, the Department of Commerce has estimated that it could cost more than 100,000 jobs. The benefit presumably is extending the life of the U.S. auto reserves, but if you look at that objectively, you will find that a one mile per gallon improvement in the total automobile fleet, not just the new cars, for all time, will extend the life of our proven oil reserves by one month. A very low benefit for a very high cost, which could divert intellectual and financial resources from finding long-term solutions.

I'd like to make one other comment. Government involvement takes many forms, part of which is regulatory, part of which is investment in research and development. I recognize that the utility industry is different from the automobile industry, but it is not clear to me that government research, in the transportation area, has been productive.

We delayed the production of our new bus by nearly ten years because of regulatory involvement. If I am an entrepreneur and the government is going to conduct research in my field of proprietary interest, do you think I'm going to put my money into it? Actually, government R&D may delay progress. When we start talking about R&D investments, we need to think a little bit about how meaningful it will be to the economy.

To summarize, the automobile industry will decrease its energy use in manufacturing except, probably, in electricity. We see methanol as the fuel choice for transportation which will help place a cap on the price of transportation fuels. We see some changes in our cities, in our cars, and in our products. We see new power systems, new services for the automobile.

We expect further changes in the electrical industry. We are in a window in time, and many innovations in both technology and business approaches can be initiated. They couldn't have a generation ago, and probably not a decade ago, and they may not be possible a decade from now. It is an opportunity for the utilities. I believe that, in the future of the United States, energy will not be a problem. I have faith in our future growth and our quality of life again, with one caveat, "unless we make it so."

QUESTION-AND-ANSWER SESSION FOLLOWING PRESENTATION

MR. SAM SCHURR, EPRI: Can you tell us what the elements are in increasing the growing intensity of electricity used per automobile, and the factors involved in that?

MR. SOBEY: I don't have all of the details. I do have a report published by our plant energy people that gets into that in more detail and I would be glad to send you a copy. It fundamentally is in changing from natural gas, for example, to induction heating, and moves of a similar nature, but I'm not that close to the process to answer your question.

I should make one other observation. I know this group is interested in the battery electric car, and I should say, that in our judgment, that has essentially no chance of being a significant factor in our economy. I will give you two reasons. One is that, if you study battery cars and ask "What do you want them to do?"; and then compare battery cars with cars running on compressed natural gas, you will find that the compressed natural gas car meets all of the requirements with perhaps one exception, and that is environmental, in a closed space. It does it at a much cheaper cost; it is here today. Yet, there are only 20,000 to 30,000 natural gas cars in the United States.

To anyone who says you can sell hundreds of thousands of battery cars, I will ask "Why aren't we already selling equivalent numbers of a car which is better?" There is a fundamental reason, the customers' desire for range—cost and service and the difficulty of doing something new. We do see fuel-cell technology coming along quite rapidly, and could have commercial fuel cells in commercial use before the end of the century. Ultimately they could be used in electric cars, but they are not battery electric cars.

MR. BOB CILIANO, Dun and Bradstreet: Two questions. First of all, in terms of new concepts like flexible manufacturing, the much publicized use of robotics by the auto industry—I know that there has been a controversy over what it will do in terms of net increases to load and to the load shapes. Do you have any comment or insights on this to share with us?

MR. SOBEY: I wish I did. I have been going through the same kind of arithmetic. You can make a case both ways. You use more electricity in the process, but you use less in the environmental-control side. It looks, right now, like the environmental-control side may win in the amount of electricity used, assuming it doesn't change in other areas.

MR. CILIANO: Secondly, in terms of, I guess what you were alluding to as security of the supply issue in the minds of the Chase Manhattan and others, instead of the concept relating to cogeneration or on-site generation, might it not make more sense, in an area like Detroit, for GM to actually consider taking over some base-load generation facilities, and even part of the T&D system if—real or not real in the minds of bankers—the security issue is so important?

MR. SOBEY: It might, but we, as a matter of policy, have decided we would address our cogeneration on the basis of our own needs and not on the sell-back. It is a matter of policy at the moment. We would have to look at it differently.

MR. CILIANO: So it is more of a policy than an economic issue?

MR. SOBEY: It was a policy decision made after a discussion with some of the utility people.

MR. CILIANO: Thank you.

MR. BLAIR SWEZEY, EPRI: You project that electricity use will grow 1.2 to 1.4 percent times faster than GNP. Can you tell use where that growth comes from? Is it primarily industrial sales, residential, or commercial, or just what?

MR. SOBEY: I guess the answer is yes. I think that the biggest growth will be on the industrial side. The one big caveat on that, that I know that we'll be talking about here this afternoon, is industry-specific. If you move some of the big energy-consuming industries offshore like aluminum, the total may decrease. The growth may not add up to a total increase.

Residential use will go up, but not that dramatically. One place we do see it going up faster is in the government sector. It hasn't leveled off at all.

15

TRENDS IN THE U.S. STEEL INDUSTRY

*James F. Collins**
American Iron and Steel Institute

I first came in contact with the steel industry when I was in the government and some leaders of the steel companies in the United States came in to see us to tell us they had a trade problem, and, of course, like all good government officials we said, "We are free traders; go out and solve the problem yourself."

---

*This paper represents a verbatim transcript of the presentation given by the author at the EPRI seminar.

Neither we nor they fully understood the problem at that time, and, as that problem developed, steel probably became the most studied industry sector in the United States, and it still continues to be an industry of major interest and concern. What kind of an industry is this? The Director of Purchases of General Motors, at a seminar in which we both participated with some government officials, said: "Steel is a fascinating industrial material. It is the most pervasive industrial material. It is versatile. There are plenty of raw materials worldwide to make it. We will never run out of raw materials to make steel." In contrast to an industry like aluminum, which had only 6 or 7 million tons of shipments last year, steel had about 74 million tons of shipment. So, the scale is very large compared to many other industries.

Yet it is an industry beset with problems. The American steel industry has been in a state of decline since 1981, but that decline must be put in some kind of world perspective. The largest steel producer in the world today is the Soviet Union. Next is Japan, with about 105 metric tons of crude steel production last year; and then the United States with approximately 92.5 million net tons of crude steel production and about 74 million tons of steel shipments last year.

So the American steel industry is still one of the largest steel industries in the world—it is third in the world—and we expect that, while it will continue to decline somewhat, there will also be commensurate declines in other major world steel industries. For example, the European Community Commission announced about a month ago that it had set a target for a reduction in European steel capacity of about 27 million tons by 1990.

The American steel industry has raw steel capacity now consisting of about 134.3 million net tons, and most expert observers would conclude that this industry will further decline, to a level of capacity between 117 and 123 million net tons by 1990, but that its shipping capability—and I believe this is important to the electric utility industry—will not decline that much because it will be installing continuous casters and achieving a higher yield from the raw steel produced by the industry.

With respect to its shipping capability we estimate today's shipping capability of the industry to be 95 to 100 million tons. There are approximately a hundred million tons—98, we believe, this year—of consumption of steel in the American economy. Arthur D. Little just did a study of the American steel industry and concluded that the American industry's capacity is not overbuilt for the American economy. Obsolete capacity in the industry exists, but, in terms of the capacity to consumption ratio in the U.S. economy, unlike these same ratios in many other countries, ours is pretty much in balance.

The EEC in contrast, has a serious problem in terms of its capacity versus consumption. Japan also has a very serious problem. With upward of 145 million tons of raw steel capacity and with at least a 90 percent yield on 145 million tons, Japan's nominal shipping capability of 130 million tons for short periods is far in excess of domestic requirements plus export demand. Japan, for example, had about 105 million tons of crude capacity utilization last year and shipped approximately 95 million tons of steel. Japan is still the largest steel exporter in the world, exporting routinely about 30 million tons of finished steel products. Last year Japan sold 9 million tons of that 30 million tons to mainland China at prices substantially below cost of production. But this is not an irrational economic act on the part of Japan or on the part of those other countries who ship into export markets at prices below their cost of production, because they segment markets and price-average, resulting in an average price either at a break-even level or one which permits them to incur lower losses than they would if they had not engaged in these kinds of sales.

The steel industry is one of the largest industrial consumers of electricity in the United States. In 1984, the industry consumed more than 56 million kilowatt hours of electricity, of which 49 billion kilowatt hours were purchased from utilities and approximately 7 billion kilowatt hours were generated by steel companies themselves, primarily from by-product fuels.

Table 15-1 shows the consumption of energy in the United States steel industry from 1972 through 1984. The primary uses for electricity in steel mills are as a primary heat source in electric arc furnaces. It is also used to drive rolling mills, pumps, fans, and other electrically powered equipment, and, to a lesser degree but growing extent, induction heating of materials in preparation for hot rolling.

Electric arc melting is the principal method for producing steel from recycled scrap and this method of

TABLE 15-1
Energy Consumption in the United States Steel Industry

Energy Source	1972	1975	1980	1981	1982	1983	1984
			Total Consumption Btu (trillions)				
Coal	2059	1780	1463	1395	847	874	990
Natural gas	667	503	562	585	402	362	391
Petroleum	204	230	115	101	58	62	56
Purchased electricity	125	117	141	143	98	105	111
TOTAL	3055	2633	2283	2226	1407	1403	1550
			Energy Efficiency Million Btu/Shipped Ton				
Coal	22.33	22.71	18.96	16.92	18.04	15.40	15.87
Natural gas	7.23	6.42	7.28	7.09	8.56	6.38	6.27
Petroleum	2.21	2.93	1.49	1.21	1.24	1.08	0.89
Purchased electricity	1.35	1.49	1.83	1.74	2.08	1.84	1.77
TOTAL	33.14	33.58	29.58	26.99	29.94	24.73	24.84
kWh per ton of shipped product (purchased electricity only)	395	436	536	510	609	538	518
			Percent of Energy Use by Source				
Coal	67.4	67.6	64.1	62.7	60.2	62.3	63.9
Natural gas	21.8	19.1	24.6	26.3	28.6	5.8	25.2
Petroleum	6.7	8.7	5.0	4.5	4.2	4.3	3.6
Purchased electricity	4.1	4.5	6.2	6.4	6.9	7.5	7.1
TOTAL	100.0	100.0	100.0	100.0	100.0	100.0	100.0
Percent electric melting	17.8	19.4	27.9	28.3	31.1	31.5	33.2
Percent continuous casting	5.0	9.1	20.3	21.6	29.0	32.1	39.0
Percent of industry represented in survey	98.1	94.0	94.9	92.5	87.4	84.5	84.9

steel-making has steadily increased in application; more than 33 percent of steel was produced by electric arc melting in 1984, compared with less than 20 percent ten years ago. The electric furnace melting operation alone utilizes 700 to 800 kilowatt hours per ton of finished steel products.

Table 15-1 shows that purchased electricity as a percentage of energy use by source has increased from a little more than 4 percent of the total in 1975 to 7.1 percent in 1984. However, it is clear that the total number of Btus obtained from purchased electricity is affected by the level of steel industry operations and shipments.

In 1975, for example, the industry used 125 trillion Btus. In 1981, the last relatively good year for steel, the industry used 143 trillion Btus and in 1984, a bad year, only 111 trillion Btus. So the level of steel production and shipments directly affects the amount of purchased electricity that the industry will use.

I'd like to briefly give you some facts about the industry before we return to electricity. Sales for the industry in 1984 were $53.6 billion. Approximately $30 billion of that total were steel segment sales. There are also $10 billion of oil and gas sales in that $53.6 billion number, which essentially are those of U.S. Steel's Marathon Oil operation. Steel is still the fourth or fifth largest industry in this country. It had 236,000 employees last year and 170,000 hourly employees. It paid $7.5 billion in wages and salaries last year. Raw steel production, however, had declined from 151 million tons in 1973 to approximately 92.5 million net tons in 1984. As I said, domestic shipments have dropped to 74 million tons in 1984 from 111 million tons in 1973.

We have lost 217,000 jobs in steel since 1979, which represents a 50 percent drop in total employment in just five years. In the economy, it is estimated that 1.7 million manufacturing jobs have been lost in the past four or five years, and steel has lost about 13 percent of those jobs. Steel has been singularly impacted by a series of events to which I will refer later.

Our industry lost about $6 billion between 1982 and 1984. It is the only major industry sector in the United States to have lost money in each of the past three years. The industry has lost approximately 50 percent of its equity in the past three years, and the debt/equity ratio of the industry has doubled from approximately 40 percent to 80 percent in the past three years. It is tough to get capital to invest in steel these days, not only in the American industry but worldwide. Yet, as I said, there is barely enough capacity in the United States to satisfy U.S. demand, if one doesn't include imports.

The current western world raw steel capacity of 600 million tons is approximately 140 million tons over free world demand. Most of this excess capacity was built or maintained with foreign government subsidies. When the boom in steel consumption occurred in 1973 and 1974, that triggered an enormous round of expansion in western Europe; steel producers substantially overbuilt their capacity. Particularly in Japan, which overbuilt its capacity by at least 40 million tons, Japanese steel producers are in the very embarrassing position of taking down relatively new capacity on which they spent billions of dollars. None of their growth trend analyses showed that they would be able to use their excess capacity in the coming decade.

In the past ten years, the U.S. industry made capital expenditures of $32 billion and achieved net income of less than $3 billion. Steel companies have been inverting more money into their steel segments than they have been receiving cash flow from their steel segments, for at least 20 years. Table 15-2 shows the steel segment cash flow and reinvestment back into steel, and you will see that the way our industry has accomplished greater investment in steel than the cash flow derived from steel operations has been to borrow money, increase the debt, and attempt to raise the level of investment through a combination of cash flow and debt. Nevertheless, the investments in steel have been, by all expert accounts, inadequate.

Our industry, however, has been making great progress on several fronts in the past three years. We have dropped our production costs in the steel industry between 15 and 20 percent, depending upon the company. Product quality has improved exponentially. Ford Motor Company has indicated to steel producers that its reject rate for domestic steel has dropped from close to 10 percent to less that 2 percent, and they say they are highly satisfied with the quality of American steel. Roger Smith from General Motors has also made similar remarks, though I would hasten to add that

TABLE 15-2
United States Industry Cash Flow Statement for the Steel Segment (millions of dollars)

	1984	1983	1982	1981	1980	1979
Cash Provided from operations						
Net income	$ (248.3)	$(2,235.0)	$(3,384.4)	$1,653.8	$ 681.4	$ 806.6
Depreciation	1,388.9	1,292.1	1,481.2	1,636.1	1,547.2	1,502.7
Deferred income taxes	54.2	(722.9)	(854.4)	762.9	255.8	(144.6)
Unusual noncash P&L items	113.6	1,366.2	2,036.6	(39.8)	8.2	677.5
Total from earnings	1,308.4	(299.6)	(721.0)	4,013.0	2,492.6	2,842.2
Changes in working capital	(567.8)	(94.1)	1,677.3	(463.6)	50.0	(297.2)
Net cash (used) provided from operations	740.6	(393.7)	956.3	3,549.4	2,542.6	2,545.0
Cash provided (used) for nonoperating units	(143.5)	(160.6)	63.3	(161.2)	(83.6)	(45.1)
Cash (used for dividends)	(215.3)	(203.3)	(413.8)	(774.9)	(714.6)	(521.3)
Long-term investments:						
Additions to plant & equipment	(1,188.9)	(1,850.2)	(2,258.2)	(2,370.6)	(2,650.6)	(2,469.4)
Disposals of plant & equipment	293.4	48.5	602.8	166.2	227.1	51.7
Other provided (used)	97.6	701.0	(940.6)	279.6	(69.1)	2.7
Net cash provided (used) for long-term investment	(797.9)	(1,100.7)	(2,596.0)	(1,924.8)	(2,492.6)	(2,415.0)
Cash provided (used) by operations	(416.1)	(1,858.3)	(1,990.2)	688.5	(748.2)	(436.4)
Cash provided from (used for) financing net additions/ reduction in:						
Transfers (to) from corporate	(49.0)	46.5	(242.6)	(186.4)	(87.5)	94.5
Long-term debt	(135.3)	524.1	1,814.7	(638.4)	805.8	244.1
Short-term debt	283.4	51.3	106.7	121.3	(179.5)	51.8
Stock	273.5	934.1	87.2	11.5	63.3	19.1
Net cash provided from (used for) financing	372.6	1,556.0	1,766.0	(692.0)	602.1	220.5
Increase (decrease in cash and marketable securites	$ (43.5)	$ (302.3)	$ (224.2)	$ (3.5)	$ (146.1)	$ (215.9)

General Motors is indicating that, rather than purchasing what it calls the average of the worst for its stamping presses, which had been its purchasing policy with respect to steel in past years, it is now engaging in a policy to purchase the average of the best. This has had a profound effect on the quality efforts of the American industry, because it is forcing the industry to do all it can to improve its quality. The other American automotive producers are likewise concerned, and quality is improving exponentially.

Half of the steel produced in the United States is in the form of sheet products, and most of that goes to the automotive industry, so the automotive industry continues to be steel's biggest customer. Steel is paying very careful and close attention to its major customer. One of the principal objectives—in competition with plastics, for example—is to reduce tooling costs of the automobile producers on short runs. Individual steel companies selling to Detroit are working with the automotive producers attempting to find ways to improve the quality of their steel in order to reduce automotive tooling costs, a significant and important component of automobile production costs.

Another series of projects is aimed at extending the life of steel in the automotive body. Significant flows of capital have gone into two-sided galvanizing plants which will permit the automobile companies to guarantee the auto body sheet for at least five years and, when perfected, probably ten years. This will eliminate rust-through in auto body sheet.

The point is that the industry has become exceedingly concerned about the quality of steel, and it recognizes that its available capital must go into quality improvements and process control improvements.

The average price of a ton of domestically produced steel sells for less today than it did in 1981. Between 1981 and 1984, there was a 38 percent increase in imported steel mill products into the United States. The average price of these imports dropped from $513 per ton to $389 per ton, a 24 percent decrease.

We believe the basic problem is that subsidized worldwide steelmaking capacity resulting in unfairly traded imported steel has combined with the strong dollar to produce the current problems of the domestic industry.

Our industry has only invested about a billion and a half dollars annually over the past two years in steel operations, and all our analysis indicates that we should be investing about $5 billion a year to modernize this industry adequately. We are not investing at a level to do that. Under the proper public policy circumstances our industry could achieve a level of investment which would still leave us about a billion and a half to $2 billion short of that $5 billion; or about $3.5 billion of investment, given reasonable demand for steel and proper public policies applying to steel.

We are not going to reach the total of $5 billion annually for the next several years, but we could get close enough to it to achieve a level of modernization which will permit the industry to retain its competitive position into the next century.

There is some progress today in the steel sector. Our productivity has increased exponentially, and the reasons are obvious. We have lost half our work force in the past five years. The Japanese work force, for example, has remained static. They haven't had much of a decline, if any.

Our productivity is now about 5.4 man hours per ton for carbon-integrated steel production, versus 6.7 man hours in Japan, 8.4 man hours in West Germany, 9 man hours in the United Kingdom, and 9.7 man hours in France. We have increased our productivity about 20 percent in two or three years. On the other hand, our labor problem is not solved because, when you multiply 5.4 man hours per ton times $22 per hour you get an entirely different result than you do when you multiply 6.7 man hours in Japan times $12.50 or $13 a ton, etc.

Clearly, the 6.5 to 7 man hours per ton required to produce a ton of steel in Korea, times $3 a man hour, doesn't put us in the game with Korea at all in terms of comparative employment costs.

In comparison to costs in other major industrial countries, we still have a unit labor cost problem that we must solve. You have been reading the papers, I'm sure, about the strike at Wheeling-Pittsburgh Steel and the desire of that company to try to move out of Chapter 11 by reducing its employment costs per ton. Other integrated steel companies will be making similar efforts as they move into the labor negotiations in 1986 with the United Steelworkers of America.

There are other potentials for production cost savings in our industry. We cast a little more than 39 percent of steel on continuous casters of our total steel production last year and we achieved a yield of 80 percent. The EEC countries have continuous cast about 60 percent of their total output and have achieved a yield only a little more than 78 percent. So our industry is doing something right. Of course, Japanese output is almost totally continuous cast. Japan's yield is up around 90 percent. Our industry expects to improve its yields, which will lower cost of production, as we continue to install continuous casters. We are now installing more continuous casters than any other country. Of course, we have further to go, but as those casters go on line, our potential for further cost reduction will improve.

Our industry is also working with the Federal Labs and privately among companies on what we would hope would be some technical breakthroughs including direct steelmaking and strip casting. These have great potential for the American steel industry and for productivity increases and cost savings.

Although there are large opportunities in steel, you will note that steel is not a very profitable business in the world. From Table 15-3 you will note that losses in steel companies have been very large, and most of the European industry would have been wiped out over the past six or eight years had not the governments of the European Community countries injected about $30 billion of government capital into rationalizing and modernizing their industries.

There are still superlatively efficient steel industries in the world. Japan has probably the most efficient steel industry in the world today, without doubt. You will note, however, from Figures 15-1 and 15-2—this is Morgan Guaranty Bank's data—that the U.S. industry's cost position would improve dramatically against those of the other major industrial countries if the dollar came back down to a more rational level against other currencies.

Comparing the cost differences associated with 1979 and 1984 exchange rates, the picture is entirely different in each case. Our $510 of cost in 1984, according to Morgan Bank, versus the comparable cost for most of the western European countries and Japan shows that, inclusive of their

TABLE 15-3
Fifteen Companies with the Largest Pretax Losses in 1983

Company	Consolidated Pretax Income (millions)
Italsider	$ -834.2
Usinor	-632.2
Sacilor	-621.4
Republic	-393.5
Armco	-360.0
National	-310.4
Cockerill	-289.5
BSC	-259.3
LTV	-251.8
Iscor	-219.3
Ensidesa	-186.9
Inland	-182.8
Algoma	-151.5
Voest-Alpine	-137.4
Krupp	-135.8
TOTAL	$-4,966.0

transportation and duty costs to enter the U.S. market, they would not be competitive in the American market if the dollar exchange rate came down.

The electric utility industry of this country thus has a large stake in exchange rates levels and in the future of the international monetary policy. The Bretton Woods agreement was terminated about ten years ago and the floating exchange rate approach is not working well. It is certainly not working well for the United States, and many other governments are complaining bitterly about the over-valued dollar.

Although the steel industry does not have a formal position on this issue, some of us believe that a totally free float is not the way to manage the U.S. exchange rate versus other major rates, and possibly a managed or targeted range is the proper approach. It is really an

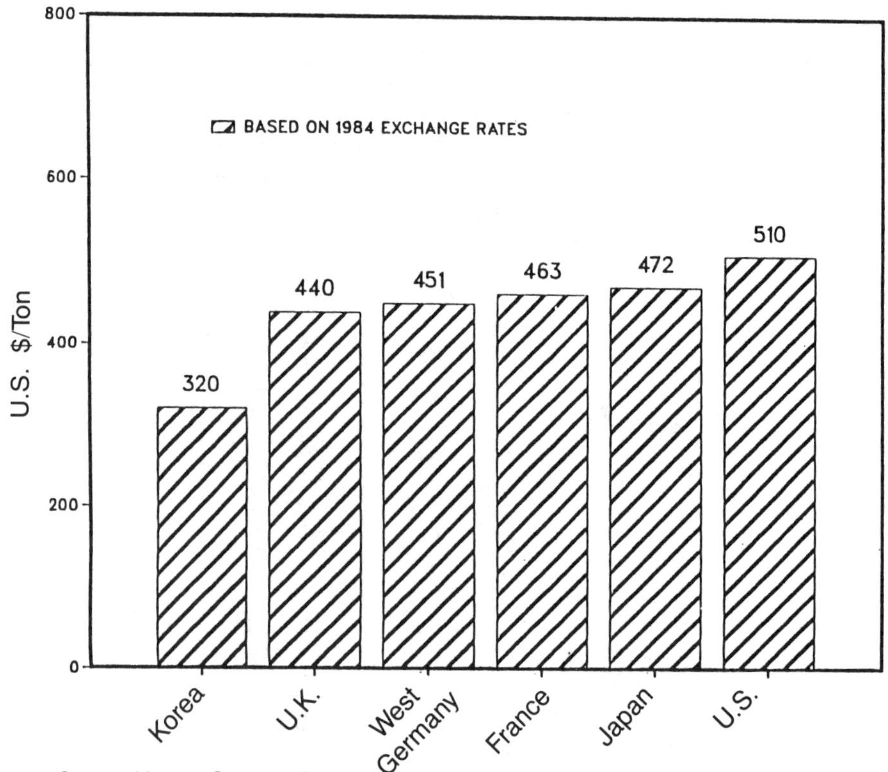

FIGURE 15-1
1984 Pretax Costs Foreign Exchange Rates (1979 vs. 1984)

international disgrace to have a $30 billion merchandise trade deficit with Japan last year, which may go up to $50 billion this year, and have the yen at an extremely low level of 236 to 240 to the dollar. Something is wrong with the exchange rate relationship. Exchange rates are supposed to reflect national inflation rates and national trade balances. One might also make the same comment with respect to Canada. The Canadian dollar is only 70 percent of the U.S. dollar, yet Canada had an $18 billion merchandise trade surplus with the United States.

But when the United States moves into current account deficit and cumulatively from 1981 to 1984, it will have a deficit in a current account of around $120 billion. As the dollar does not come down as it should as the country moves into massive current account deficits, something is wrong with the exchange rate mechanism. It has very direct relevance for steel because it is driving large quantities of

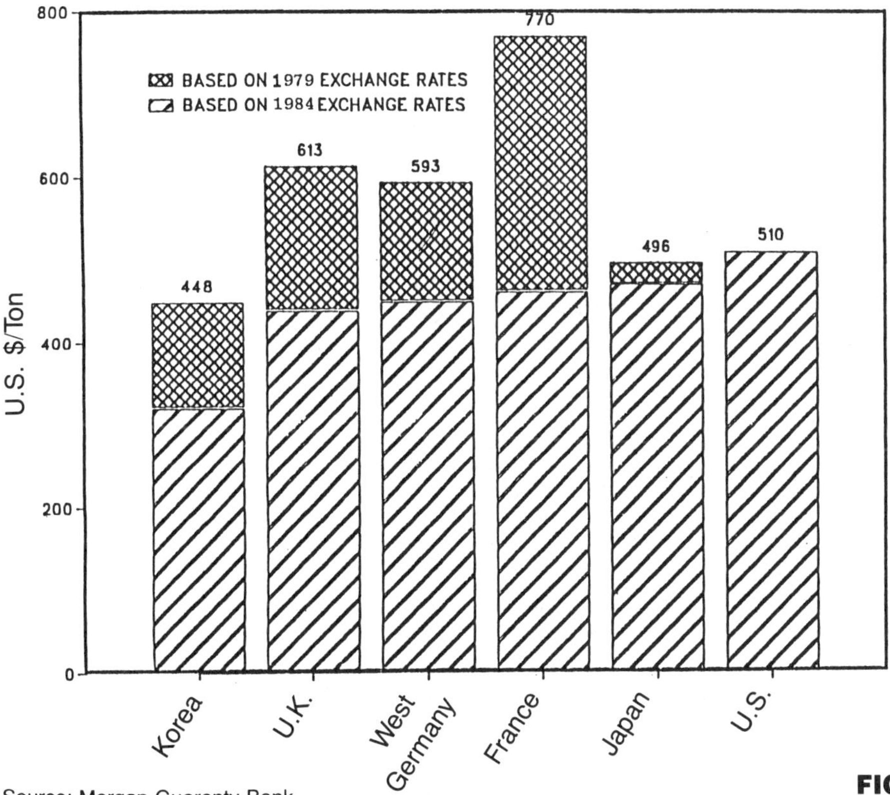

Source: Morgan Guaranty Bank

FIGURE 15-2
1984 Pretax Costs Foreign Exchange Rates (1979 vs. 1984)

steel imports into the United States at prices at which those producers could not otherwise compete, except for the exchange rate differentials.

What is the future for our industry? We have a consumption in this country of about 100 million tons. You will note comments and data to the effect that the steel intensity has gone down in various applications: down-sizing in automobiles, for example; less use in fasteners; less use in semi-durables like refrigerators. And that is true.

But if you look at steel consumption in the American economy, you will see fairly constant steel consumption, in terms of tons. As a percentage of GNP, of course, it has gone down, but in terms of tons, it is still about a 100-million ton market. The question is: Who gets the market? The level of U.S. steel imports has about doubled since 1974-1975, from about 13 percent to 26 percent of consumption last year, and that is the big issue. The indirect trade in steel is also hurting us badly. The balance in U.S. trade in manufactures has moved from a surplus of $11 billion in 1981 to a deficit of $81 billion in 1984, a swing of $92 billion due to the exchange rate distortion.

The distorted exchange rates are hurting manufacturing across the board in the United States, which is obviously of direct relevance to both our industries. Our industry has lost several million tons of steel shipments to industries that otherwise would have manufactured those products in the United States, to be consumed in the United States. Foreign manufactured imports have cost your industry substantial losses in electricity sales.

Down-sizing has had an effect, but so have imports, and so has the indirect trade in steel. The imports and the indirect trade in steel are a problem exceeding the problem of down-sizing that has occurred in the United States, although we have lost tonnage from all three. But it is not down-sizing alone; it is down-sizing plus imports, plus the loss of sales from the indirect trade due to the shift in the balance of trade in manufactures.

The current outloook is for a fairly static market in steel consumption. We don't see any large growth coming in steel consumption. The question is this: How much of that static market will our industry get? We think that the exchange rate eventually must come down; if it doesn't, then we are in deep trouble, and so are most other U.S. manufacturers, as are the electric utilities in this country. As the dollar comes down and we continue to work with our government on the question of subsidized and dumped steel imports, we think our industry will have a reasonably good future.

We don't project any large growth. If it hits 1 percent annually, we will be very happy. But, as I said, if we get up to anywhere from 80 to 90 million tons of shipments—and we are at an annual rate of 73 at the moment—we will be a highly profitable industry in this country. We will be getting the volume of production and shipments we need. The unit cost reductions from the volume combined with the other cost reductions that have already been effected in the industry will improve our cash flow and, therefore, permit us to expand our investment in modernization of the industry. As I said, we look to see further capacity reductions, but we think that will level off, and that the steel consumption level in American industry will stay around 95 to 100 million tons through the remainder of this decade.

QUESTION-AND-ANSWER SESSION FOLLOWING PRESENTATION

MR. RAY SQUITIERI, EPRI: I wanted to ask you about your statement that we should be investing $5 billion a year in steel modernization. As far as I can see, the banks don't believe it; they won't lend the money. Wall Street doesn't believe it. A lot of steel companies don't believe it—for instance, U.S. Steel. And the companies who do believe it have been taking a beating: Republic invested a lot; they are now broke. Kaiser invested in the '70s; they are out of the steel business. Bethlehem has made a lot of investments; look at their stock prices, and they have not been doing well in the eyes of investors. And Inland has invested perhaps the most as a percentage, and they are certainly the best managed of the integrated companies by reputation; they are looking for a Japanese buyer.

MR. COLLINS: I don't think that is correct, but go ahead.

MR. SQUITIERI: It seems to me that if we have 30 percent world-over capacity, as you have suggested, and the United States doesn't have any comparative advantage that is visible in steelmaking, why should we be investing $5 billion?

MR. COLLINS: First of all, it is not a question of whether they believe the $5 billion. We have had no serious challenges to the estimate that we should be spending about $5 billion a year to do a full-scale modernization of this industry. No one really argues with that number. Some might say it is $4 or $4.5 billion but, by and large, most experts who have looked at it have agreed with the number.

The issue you are raising is: Where does the capital come from? You are right. Capital is scarce for steel in the United States. But the point I'm also attempting to make is that capital is scarce for steel, worldwide. Steel is not a favorable business, in terms of the efficiency of capital utilization, comparing the potential returns in steel to the potential returns in capital invested in other industries. Once you have said that, then take another look at steel. Why is $30 billion of capital from EEC governments going into their steel industries? Why did the Japanese government use the Bank of Japan to funnel capital from the Tokyo banks into the steel industry to make sure that steel, a target industry, was expanded in the 1960s and 1970s? Why is the Korean government using Korean government funds to inject money into the Korean steel industry? Why is the Brazilian government routinely injecting government capital into the steel industry? Because, as the General Motors Director of Purchases said: It is an exceedingly pervasive and necessary industrial material for any advanced industrial country. You have got to have it.

The EEC is not going to let its steel industries go down. Japan isn't. These other newly industrializing countries want to expand their steel industries. From a public policy perspective the question we have is this: Can you imagine a country the size of the United States, with a GNP like that of the United States—still the greatest manufacturing country in the world and we hope it stays that way—operating without a steel industry? The answer of course is: No.

For the long term, those steel companies still in the business know that the only way they can stay in the business is to get the best technology and latest equipment they can. That's what Inland did and Inland spent a billion dollars on new technology. And I don't think anybody argues that if you are going to stay in the steel business, you have got to find the money somehow and invest that money in your steel operations, or else you are merely prolonging the agony and eventually you will phase out of the business. The question is this: Is the capital available? Not in those increments I described today, surely. But it is also not available in comparable increments versus the investment needs in any country in the world, unless it comes from governments. That is our problem—government subsidies.

MR. SQUITIERI: Don't capital markets put capital where it gives the most return?

MR. COLLINS: Precisely.

MR. SQUITIERI: They give it to the most efficient uses of capital, and they are obviously not doing that in the world. The reason other countries are investing is because their steel industries and labor unions are even better organized and have even better lobbies than they do in the United States, which is not necessarily in their national interest. Why should we repeat the same mistakes that these countries are repeating?

MR. COLLINS: Some feel, particularly in the newly industrializing countries, they have certain factors of comparative advantage. Clearly, Korea has a labor cost comparative advantage. Brazil has indigenous raw materials.

Last year in Santiago I addressed the Latin American steel industries. After my presentation, the Brazilians came to me and said, "We have comparative advantage for your market. We have low-cost raw materials."

I responded, "Correct."

They said, "We have low-cost labor costs."

I said, "Correct."

They said, "Why should we have a quota on our steel entering the United States? We should be selling in your market."

I said, "Have you looked at your financial costs per ton of steel? They run $120 a ton. When you add those to your raw materials cost—ignoring those subsidies for the moment—and your labor costs, your total cost of production does not permit you to compete in our market. Our own government found that to be the case. That is why Brazil said to the U.S. government 'We will take a quota instead of your countervailing duty anti-subsidies determination.'"

That is the kind of a game it is worldwide in steel. The Brazilians said, "Oh, we don't count our financial costs per ton. That is money from our government given to us for social purposes to provide employment."

So, of course, steel has been a very tough industry in which to make a reasonable return, but some people are making a return in steel. If the American steel industry continues as it has done for the past two or three years, you are quite right, sir, we will have no steel industry in this country, because this is a private economy. Capital gravitates toward the most efficient utilization. No capital will gravitate to the industry. It will phase out over the next 10 or 15 years.

But I think something else will happen. Exchange rates are going to change. Please refer to Figures 15-1 and 15-2. You will find that compared to other industrial countries, the American steel industry has relative comparative advantage for the American market—at more balanced exchange rate levels. We have excellent labor productivity. With some reasonable public policies, our industry will survive. I'm convinced it is going to survive. There is an enormous upside potential in the American steel industry, and when those exchange rate shifts come, as inevitably they will, this industry is going to make considerable money. It is going to be smaller—no doubt about it. It is going to be leaner; it is going to be tougher and smarter, with a much smaller work force, but it is going to survive.

MR. BOB CILIANO, Dun and Bradstreet: In terms of the figures that you just alluded to, your comment seems to suggest that these are prices for the U.S. market alone.

MR. COLLINS: Prices for the U.S. market; correct! That includes roughly $70 worth of importation charges. But, remember, we're approaching this in the context of 26 percent of our market having gone to imports.

MR. CILIANO: But if you were to subtract the labor costs at the rates you quoted for us—the Japanese, and the Koreans—it would suggest—which seems to be incorrect—that we have very, very high levels of capital productivity in our steel industry versus these other industries, and that would, as I say, seem to be contrary to what we are led to believe about . . .

MR. COLLINS: No. That is not true. The American steel industry was the most profitable industry in the world for two decades, up until 1981. Our capital productivity was substantially higher than that of the other major steel industries of the world. There is no question about that. Even today, in some of these newly industrializing countries, they are rapidly expanding their steel capacity. But they are not competitive with us in terms of cost, if you include their financial costs per ton of steel produced. Brazil is a prime case in point.

MR. CILIANO: This is confusing to me. Why should that be so? If we look at your rationalized exchange rates and subtract the direct labor costs at the rates you have quoted, why is it that we should have such capital productivity advantages over the Japanese?

MR. COLLINS: Well, one reason is that traditionally, the Japanese have had a debt-to-equity structure that was about 80 percent debt, 20 percent equity. Japanese producer interest costs on that debt have traditionally—aside from the increasing debt level in the U.S. industry in the past two or three years—been substantially higher than U.S. interest costs per ton.

The newly industrializing countries obtain funds from their governments. The governments give them the funds. They are never going to pay back the capital. Thus it is a pure subsidy. But the governments charge them interest on the capital invested in the steel companies and those interest costs per ton in Brazil and in other LDCs are quite high. Thus they have high financial costs per ton.

MR. ROBERT CAMFIELD, Geogia Power Company: I have enjoyed your talk today, and I think it is focused on the competitiveness of the United States with other nations.

For us here, of course, we are concerned about the competitiveness of various regions of the United States, and, considering that the industry you represent uses energy and electricity increasingly intensively, do the industries, or should I say the manufacturing firms in your industry, look at regional differences in electric cost in terms of location of output?

MR. COLLINS: Certainly the steel industry is not one behemoth. As Mr. Sheppard indicated, there are many different divisions in the chemical industry. The steel industry is a composite of many different product-line industries. It covers the large integrated steel producers producing sheet in the heartland of sheet consumption,

Detroit. It is little mini-mills producing rebars for construction applications at 2.8 man hours per ton.

The mini-mill companies have tended to cluster in the Sun Belt areas, because they can produce there at the location of large construction activity. They can operate without union labor. The major integrated producers have remained relatively close to the major manufacturing centers of the United States—the midwestern and north-midwest quadrants, the east-northeast quadrants—and their plants are generally located in those areas.

Also, there are centers of steel production in the South. U.S. Steel recently established a half-billion-dollar oil country tubular mill in Alabama. Investments in various parts of the country depend upon the portion of the industry involved. The specialty tubular producers sell to the high-tech industries, to the aircraft industry, and some of them have production centers close to those areas—a regional warehousing operation—to better service their customer groups.

The West Coast has been a disaster for the American industry, and that market has largely been ceded to imports except for some mini-mill production and some integrated production serving the West Coast.

MR. CAMFIELD: Another concern, of course, for us, is the prospect for deregulation on wholesale capacity and energy. Do you anticipate that steel producers will be concerned or be interested in not necessarily relying upon the local electric utility for their source of supply, but would be interested in importing it from other producers of energy?

MR. COLLINS: Energy constitutes about 20 percent of the cost of production of a ton of steel today. In our industry, the companies have been exceedingly concerned about reducing that cost. As you can see, they have substantially reduced their oil consumption and have even reduced gas consumption to some extent. Electricity consumption has risen. Much of that is due to the increase of the electric furnace operations in steel companies.

I would hasten to add that about 55 percent of the electric furnance capacity in the United States is owned by the integrated producers. They know how to operate electric furnaces. Thus it is not solely the mini-mills that are consuming electricity. They are interested in lowering their Btu costs, as is every other manufacturing industry in the United States. The importation of energy units certainly might be a consideration with respect to lowering such costs. I don't know whether that is feasible with respect to electricity, however.

16

TRENDS IN THE U.S. CHEMICAL INDUSTRY

William J. Sheppard
Battelle Columbus Division

In the 1967 movie, The Graduate, our hero, played by Dustin Hoffman, is told the secret of making a fortune in business—"plastics!" From about 1950 through about 1970, the chemical industry was regarded as a prime example of a growth industry. However, since then, it has become a mature industry—sometimes called a cyclic industry, other times called a smoke-stack industry. This paper will review the structure of the chemical industry as it exists today, the trends for the next 15 years, and finally, energy use in the chemical industry.

INDUSTRY STRUCTURE

The chemical industry is formally known as Standard Industrial Classification (SIC) 28, Chemical and Allied Products. Categories within SIC 28 are shown in Table 16-1. The industry includes large-volume industrial chemicals and small-volume, high-value products such as drugs and perfumes. Prices range anywhere from 20 cents per pound to more than $1,000 per pound. The names of the categories into which the industry is classified reflect the diversity of the industry. Some are named for the raw material that is used, for example, wood and gum chemicals. Others are named from the process by which products are made, for

TABLE 16-1
Standard Industrial Classification 28: Chemicals and Allied Products

281	Industrial Inorganic Chemicals
2812	Alkalies and chlorine
2813	Industrial gases
2816	Inorganic pigments
2819	Industrial inorganic chemicals, nec[a]
282	Plastics Materials and Synthetics
2821	Plastics materials and resins
2822	Synthetic rubber
2823	Cellulosic man-made fibers
2824	Organic fibers, noncellulosic
283	Drugs
2831	Biological products
2833	Medicinals and botanicals
2834	Pharmaceutical preparations
284	Soap, Cleaners, and Toilet Goods
2841	Soap and other detergents
2842	Polishes and sanitation goods
2843	Surface active agents
2844	Toilet preparations
285	Paints and Allied Products
2851	Paints and allied products
286	Industrial Organic Chemicals
2861	Gum and wood chemicals
2865	Cyclic crudes/intermediates
2869	Industrial organic chemicals, nec[a]
287	Agricultural Chemicals
2873	Nitrogenous fertilizers
2874	Phosphatic fertilizers
2875	Fertilizers, mixing only
2879	Agricultural chemicals, nec[a]
289	Miscellaneous Chemical Products
2891	Adhesives and sealants
2892	Explosives
2893	Printing ink
2895	Carbon black
2899	Chemical preparations, nec[a]

[a] nec = not elsewhere classified.

example, fertilizer mixing. Others are named by their actual composition, for example, carbon black. And, finally, some are named from the uses to which they are put, for example, printing ink and drugs. This diversity has made it difficult for economists to generalize about the industry as a whole.

Firms in the chemical industry span a range from billion dollar companies to small garage operations, where detergent powder is added to water and sold as janitorial supplies. Table 16-2 lists the top 15 companies in 1984, based on chemical sales. The top two companies are traditional chemical companies, but six of the top 15 are oil companies. Exxon's chemical sales make it number three in the chemical industry. Of the 15 companies, only 7 are primarily chemical companies.

The largest-volume chemicals are primarily industrially used inorganic chemicals; the top ten are shown in Table 16-3. Of these, only one is an organic chemical, ethylene, which is used for making polyethylene plastic, ethylene glycol for antifreeze, and many other organic compounds. The largest-selling chemical is sulfuric acid, which is used as the workhorse acid throughout all manufacturing industries, but is mostly used in converting phosphate rock into phosphate fertilizer. Nitrogen, in liquid form, is used in refrigeration, particularly freezing foods. In gaseous form, it is used to provide an inert atmosphere. Ammonia is used primarily as a fertilizer.

Lime is the cheapest base and is the workhorse in many industries where base must be employed, for example, the pulp and paper industry. The largest single use of oxygen is in the steel industry, where the basic oxygen furnace has almost completely replaced older methods of converting pig iron to steel. Sodium hydroxide is a base widely used where water solubility is important. The fertilizer industry is the chief consumer of phosphoric acid, where it is used in acidulating phosphate rock to produce triple superphosphate fertilizer. Sodium carbonate is a cheap base widely used in the glass industry and elsewhere. It is an interesting example of a chemical that used to be made synthetically by the Solvay process, but now is entirely produced in the United States by mining of mineral deposits in Wyoming. One of the chief reasons for the switch was that the Solvay plants were unable to meet pollution control regulations economically.

TABLE 16-2
Chemical Sales for the Top 15 Companies (1984)

Basic Chemical	Petroleum	Specialty Chemical	Other	Billion Dollars
Dupont				12.0
Dow				9.6
	Exxon			6.9
Monsanto				6.0
Union Carbide				4.2
	Shell			3.4
Celanese				3.3
	Arco			3.1
		Grace		2.8
	Amoco			2.8
	Mobil			2.6
			Kodak	2.4
Allied				2.4
	Phillips			2.4
			General Electric	2.1

TABLE 16-3
Top Ten Chemicals (1984)

	Billion LB
Sulfuric Acid	80
Nitrogen	43
Ammonia	32
Lime	32
Ethylene	31
Oxygen	31
Sodium Hydroxide	22
Phosphoric Acid	22
Chlorine	21
Sodium Carbonate	17

The distribution of the value of shipments and the value added for the various subcategories of SIC 28 are shown in Figure 16-1. The largest single category is SIC 286, organic chemicals, and the second largest, SIC 281, inorganic chemicals, are characterized by a low ratio of value added to value of shipments. On the other hand, SIC 283, drugs, and SIC 284, soaps and detergents, have a high ratio of value added to value shipments.

The chemical industry is its own best customer. Figure 16-2 shows that for the industrial organic and inorganic chemicals, which constitute the starting point for chemical manufacture, about half of the sales are to the chemical industry. A little more than a third goes to other manufacturing industries, and only about one-seventh is sold directly to the final consumer. Union Carbide Corporation, as an example, sells chemicals such as industrial gases to both the chemical industry and to the steel industry and others, but it also sells products such as Everready batteries, Glad Bags, and Prestone antifreeze directly to the consumer.

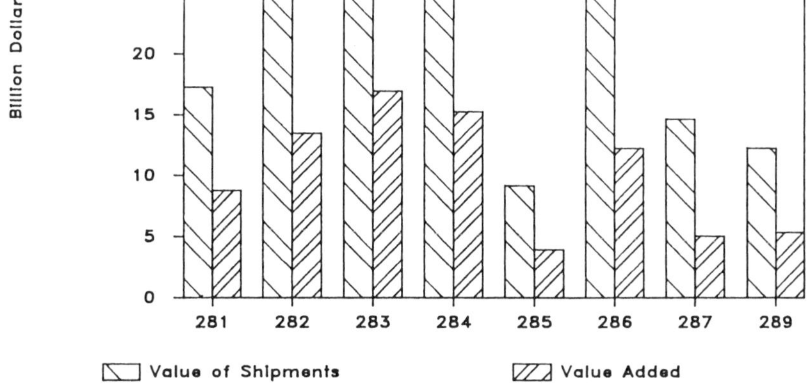

FIGURE 16-1 Value of Shipments and Value Added for SIC 28 Subcategories (1982)

FIGURE 16-2
Sales Distribution for Industrial Inorganic and Organic Chemicals

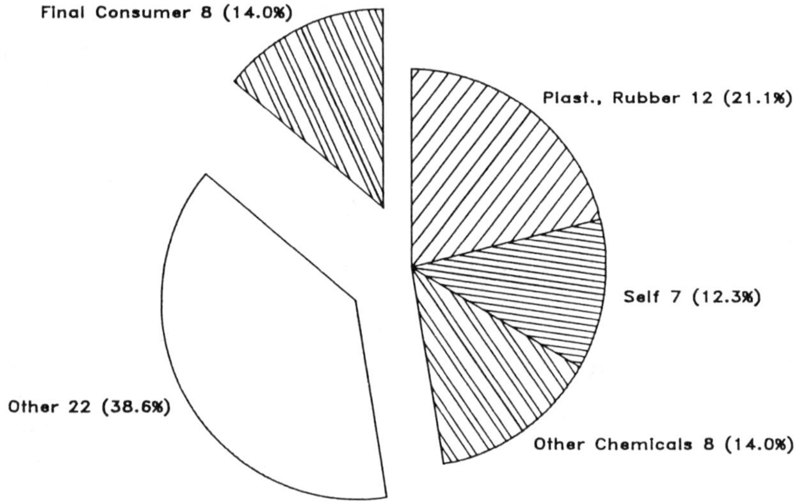

As a way to try to classify the nature of chemicals in the marketplace, it has been proposed that one separate chemicals depending on whether they are sold on the basis of composition or performance and that they be classified on the basis of whether they are large volume or small volume. The matrix structure is shown in Figure 16-3. Chemicals sold on the basis of specification of composition in large volume are termed commodities. If sold in small volume, but still based on specification, they can be termed fine

FIGURE 16-3
Classification Matrix

	Composition	Performance
Large Volume	Commodities	Pseudo-Commodities
Small Volume	Fine Chemicals	Specialty Chemicals

chemicals. There is no good term for the large-volume chemicals sold on the basis of performance. Some have referred to them as pseudo-commodities. The last category, small volume sold on the basis of performance, have been termed by some as specialty chemicals. Others prefer to call them chemical specialties.

Examples that will help to clarify the classifications are shown in Figure 16-4. An example of a large-volume chemical sold on the basis of composition is ethylene. A buyer is interested in purity, delivery, and price. An example of a fine chemical is aspirin, where the structure is known and purity is the criterion for purchase. The pseudo-commodities, on the other hand, are sold on the basis of what they can do. There are many different kinds of carbon black but only certain ones give the desired reinforcement and stability to a given grade of rubber. The specialty chemicals are materials often of proprietary composition and frequently are mixtures. An example is Excedrin®. One is buying headache relief, not just chemicals.

Commodity materials are characterized by importance of delivery and price. Often, they are sold at low margin and have little research and development support. On the other hand, they have long product life and low short-run

	Composition	Performance
Large Volume	Commodities Ethylene Sulfuric Acid	Pseudo-Commodities Carbon Black Epoxy Resins
Small Volume	Fine Chemicals Aspirin Dyes	Specialty Chemicals Excedrin® Clairol®

FIGURE 16-4
Examples of Chemical Classifications

risk (in the absence of imports). The other end of the scale, specialty chemicals, are characterized by the importance of sales and service. Frequently, advertising is important. R&D spending, as a fraction of sales, is quite high. The return is high but there is a higher degree of risk in specialty chemicals than in commodities.

It is interesting to compare the volume of chemical sales for a specific chemical with the price that chemical can command. Figure 16-5 shows data for 1974, in which the logarithm of production is plotted against the logarithm of price in dollars per pound. The data points form a broad band that is roughly a straight line. When one disaggregates

FIGURE 16-5
Price Versus Volume for Organic Chemicals by Type (1974)

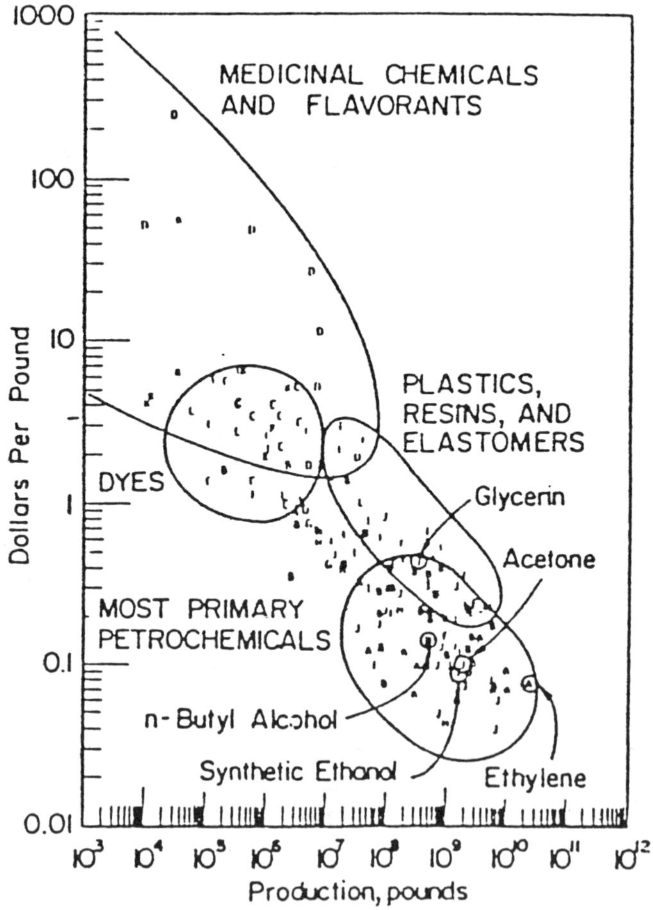

the data and groups them by the type of chemical, patches appear as shown by the closed curves in the figure. Most primary petrochemicals have a large volume and a low price. Those chemicals that have physiological activity, that is medicinal chemicals, dyes, and flavorants and fragrances, have a small volume and a high price. Thus, it would appear that a goal of a large-volume chemical with a high price is not likely to be achieved. (One problem in this analysis is that very large chemicals tend to be disaggregated in the data analysis. That is, low-density polyethylene is distinguished from high-density polyethylene. At the other end of the scale there is a tendency to aggregate, that is, Vitamin A alcohol, Vitamin A acetate, and Vitamin A palmitate all get grouped together and are listed as Vitamin A. This tends to make the slope steeper. Nevertheless, the concept is valid.)

TRENDS

The most important trend is that the chemical industry can no longer be considered a growth industry. Firms in the industry are reacting to this by trying to shift from commodity chemicals to specialty chemicals. What some people call rationalization is occurring in large-volume chemicals; plants with old technology and high costs are being shut down. Even some plants that have new technology are being shut down because of the poor market conditions. This is particularly true in chemicals that are only one step away from the original raw material and that can be imported easily. For example, methanol and ammonia. In contrast, there are certain areas of specialty chemicals that are considered "hot" and many companies are seeking entry by growth or acquisition. These include the use of biotechnology for the production of drugs and high-value chemicals and production of composites, for example, carbon fiber-reinforced epoxies for the aerospace industry. Another hot area is electronics chemicals, not only high-purity silicon and gallium arsenide used as substrate, but the special photoresist etchants, dopants, and even high-purity gases for blanketing. Advanced ceramics is another area in which there is great hope, for example, ceramics for coating turbine blades to allow them to be operated at higher temperatures and thus achieve greater energy efficiency.

There has been a trend in recent times for firms to spin off areas that are not of interest. For example, the newly created Allied-Signal Company is putting many of its commodity chemicals into a separate company. W. R. Grace has sold off its Herman Sporting Goods so it can concentrate on specialty chemicals. On the other hand, some of the companies desiring to get into specialty chemicals have made acquisitions. For example, the acquisition of G. D. Searle, maker of Nutri-Sweet and many pharmaceuticals, by Monsanto.

Chemical firms are trying to get more out of their investments. In just the last year there has been corporate restructuring by Du Pont, Union Carbide, Monsanto, Dow, Olin, Pennwalt, and many others. Frequently, this has led to reductions in staff through attrition, early retirement, and, in some cases, dismissal. In order to get more out of their investments some firms are going into consulting and contract services. Air Products, for example, has Stearns Catalytic as a major component. Companies like Du Pont that have invested heavily in improving energy efficiency now provide similar services to others for a fee. Many of the large chemical companies are involved in engineering and building plants. Finally, technology that at one time might have been kept secret is now available for license. For example, Union Carbide's Unipol technology for making linear low-density polyethylene. An interesting example of getting more out of an investment is Borden's "Chemical Condo" concept. At their plant in Geismar, Louisiana, a company cannot only get Borden to manufacture chemicals for them, but the other company can actually own a piece of the plant and do its manufacturing there.

Other trends include the shift in the chemical industry's balance of trade as shown in Figure 16-6. From 1973 to about 1979, the exports and imports of chemicals were both growing, but there was a healthy surplus of exports over imports. Starting when world oil prices rose rapidly in 1979, chemical exports jumped dramatically, while imports continued their historic growth pattern. This reflects the fact that in the United States oil prices were controlled by the government, making chemical feedstocks and energy relatively cheap. In 1981, one of the first things President Reagan did was to decontrol oil prices. Exports leveled off

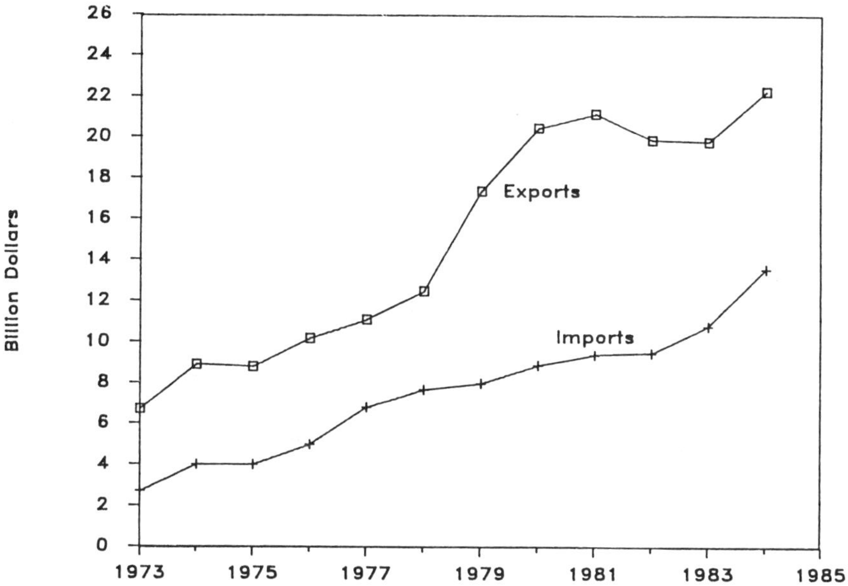

FIGURE 16-6
U.S. International Trade

at that time. Of course, at about this time, the value of the dollar increased relative to other currencies, which had an inhibiting effect on exports, and there was a worldwide recession. Today, there is still a surplus of exports over imports. This is in spite of the fact that energy-rich countries, such as Saudia Arabia, Canada, and Mexico, have targeted petrochemicals as a growth area and the United States as a market. Looking to the future, one can see the time coming when world oil is in short supply and a country like Saudia Arabia can require that petrochemicals be purchased from them as a condition for supplying needed oil. This has not occurred yet. However, the fact that many of the companies developing petrochemical industries are in the Middle East means that Europe will feel the impact first.

Another trend that can be seen is that many nonchemical firms with chemical subsidiaries are getting out of the chemical business. Examples are Beatrice, a food company; Tenneco, an energy company; and Armco, a steel company. European investment in the chemical industry has been increasing in recent years. For example, Imperial Chemical Industries (ICI), Bayer, BASF, and Schering have all invested heavily in the United States both by acquisition and by growing internally.

Government regulations are also impacting the chemical industry and will in the years to come. Even before the Bhopal tragedy, there was growing concern about workplace health and safety. Many states have regulations giving workers "the right to know" the identity and nature of the chemicals they use routinely. Transportation has become more closely regulated as a result of accidents in rail transportation. (In contrast to the oil industry, bulk chemicals are transported by rail tank car, barge, and truck, rather than by pipeline because of the smaller volumes involved and the need to assure purity.) Disposal of hazardous wastes has also become an increasingly expensive part of operations for several reasons. The Superfund Law put a tax on many chemicals to help the government pay for the clean up "orphan" disposal sites. In addition, operation of waste disposal facilities has become more expensive due to strict permit and manifest procedures, requirements for double lining of landfills and installation of monitoring wells, required monitoring of incinerator stacks, etc. Finally, the fact that a company placing a single drum in a landfill can be held liable for all clean-up costs if all other users go bankrupt (joint and several liability) has forced a reexamination of disposal policies at many companies.

Recent trends in profitability reflect the change in the chemical industry's status from growth industry to mature industry. During recent years, recessions have hurt profitability, return on sales has dropped, and return on equity has decreased. In spite of this, total R&D expenditures have grown in constant dollars, currently constituting about 4.5 percent of sales.

In coming years, it is expected that the fortunes of the chemical industry will reflect the state of the economy. Recessions will hurt the industry and growth will only be 1 to 2 percent faster than GNP growth in good years. Forecasts for individual sectors of the industry have been made by the Department of Commerce:

- Petrochemical markets will continue to grow but at a modest rate. Primary petrochemicals such as ammonia and methanol will be hit hard by imports until world demand catches up with capacity. A recent study by the American Gas Association predicts that when new capacity is needed, it will be cheaper to build and operate new plants in the United States than in developing countries. This assumes that gas prices continue their current stability.

- Chlor-alkali production will grow at 2 to 3 percent per year. There will be increased use of membrane cells, which will decrease electricity costs.
- Industrial gases will grow at about 3 percent annually over the next five years, while other inorganics will grow at about 8 percent.
- Fibers will continue to be hurt by imports as long as the government maintains a free trade posture.
- Industrial organic chemicals will grow slightly faster than the economy.
- Once stability returns to agriculture, the use of ammonia will grow at 2 percent per year and phosphate fertilizers will grow at 2.5 percent. Pesticide use will grow at 4 to 5 percent and much effort will be spent on pesticides produced by the use of biotechnology.
- The use of plastics will continue to grow. If the Pontiac Fiero proves to be a success, the automobile market will be an important one for plastics. (The use of plastics in automobile bodies is driven by the desire to reduce the number of parts and to increase the ease of fabrication and assembly. Weight reduction, which improves fuel efficiency, is also desired but is a secondary factor.)
- Rubber manufacture will maintain its historic 2.7 percent growth rate, while adhesives will grow at 6 percent annually.
- Soaps and detergents will grow at 3 percent per year, with low-temperature and liquid detergents growing more rapidly. Specialty surfactants should grow at 5 percent, while polishes and sanitation goods will have a 3.2 percent growth rate.
- Cosmetics and toiletries are expected to have a growth of 2.5 percent.
- Biologicals (blood products and vaccines) are expected to grow at 6 percent per year, with a large impact of biotechnology on manufacturing methods. Other drugs will grow at 3 percent annually.

ENERGY USE

The chemical industry (Figure 16-7) is the largest user of energy in the United States. As of the last year for which there is good data (1981), of the 11.6 quadrillion Btu used by all manufacturing, the chemical industry used 22.4 percent. (Note that in this and other comparisons, one kilowatthour is counted as 3413 Btu, the resistance heating value. Figures used in some publications use 10,000 Btu, the average energy content of the fuel required to produce the electricity. No feedstock use is included in this analysis.) When electricity alone is analyzed (Figure 16-8) the chemical industry ranks number two, exceeded only by the primary metals industry, which uses large amounts of electricity in aluminum smelters and electric arc steel furnaces.

The chemical industry uses natural gas as its primary fuel, 56 percent of all fuels purchased, as shown in Figure 16-9. Electricity is second, followed by coal. The electricity use is broken down by the industry sectors in Figure 16-10. Inorganic chemicals and organic chemicals, being large sectors, use large amounts. The figures inside the pie slices show the energy intensity, that is, electricity use per

FIGURE 16-7
Purchased Fuels and Energy (1981) (all manufacturing: 11.6 quadrillion Btu)

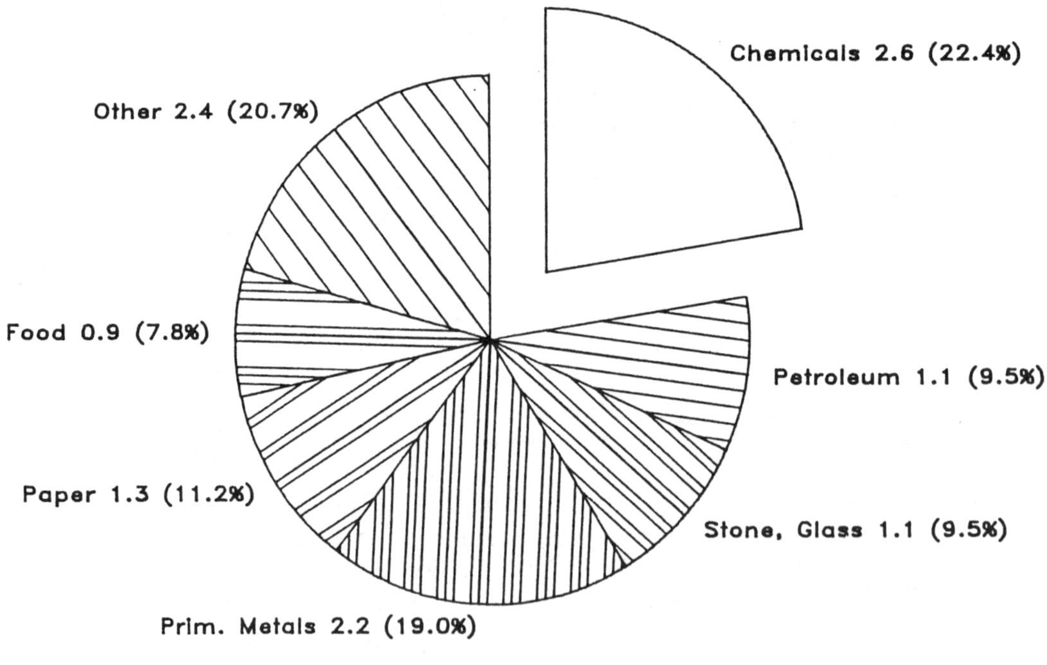

dollar of sales. Two subsegments of the inorganic chemicals, namely, chlor-alkali and industrial gases, have the highest electricity intensivity. In the chlor-alkali plant, electrolysis of brine is used to convert sodium chloride into sodium hydroxide and chlorine, requiring 2500 to 3500 kWh per ton of chlorine produced. In the industrial gases segment, electricity is used primarily for compression of gases, including compression done as part of the cryogenic process used in air separation. The Inorganic Chemicals Not Elsewhere Classified (NEC) subsector uses more electricity per unit of output than the remaining sectors, primarily because of the large amounts used in operating the uranium diffusion plants for the government. In other sectors of the chemical industry the use of electricity is primarily for pumping and compression.

Figure 16-11 shows one estimate of future use of energy and feedstock by the chemical industry prepared by the Office of Technology Assessment in 1983. The trend expected is a decrease in the use of natural gas fuel and an increase in the use of coal and electricity. These forecasts are based on the assumption that electricity will remain

FIGURE 16-8
Electricity Usage (1981) (all manufacturing—666 billion kWh)

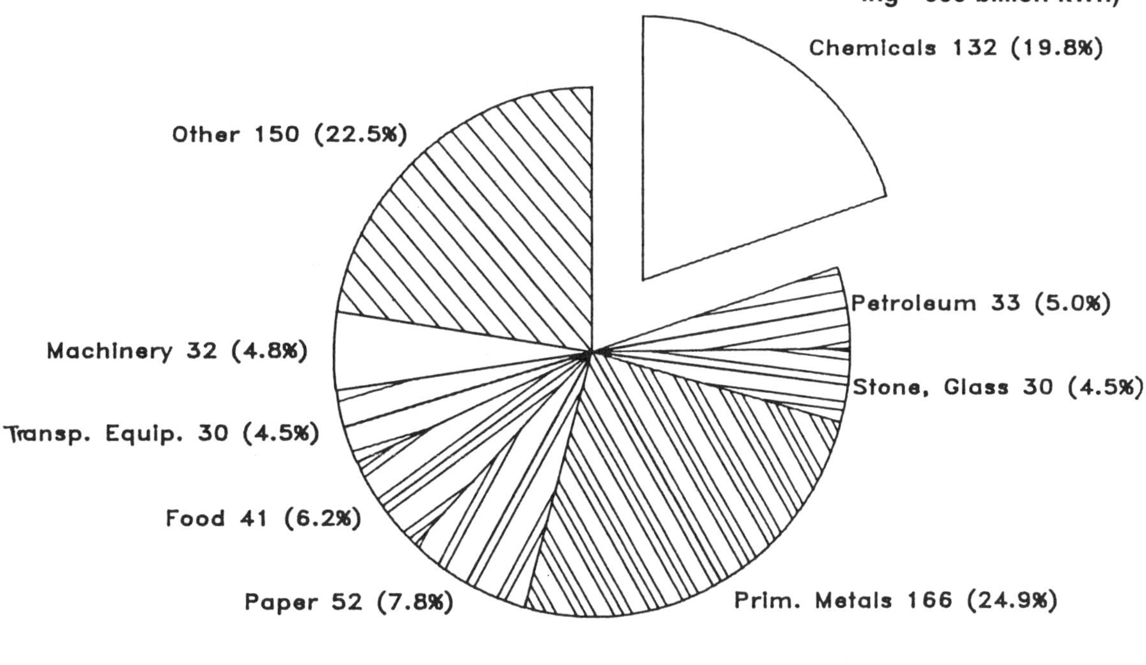

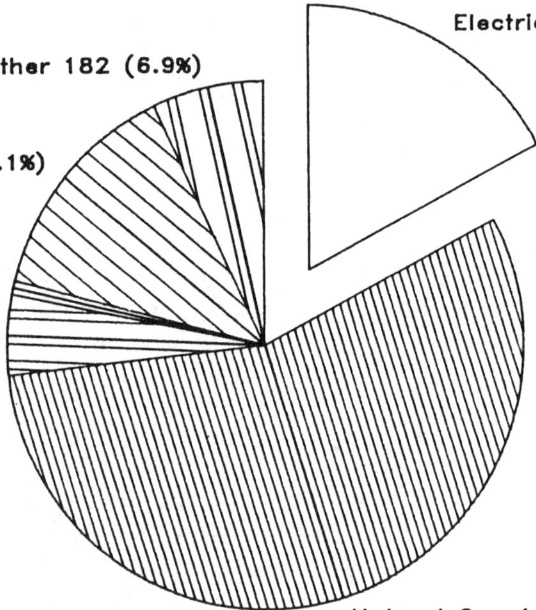

FIGURE 16-9
Chemical Industry Purchased Energy (1981) (2630 trillion Btu)

more or less constant in real terms and that the cost of oil and natural gas will go from $5 per million Btu to $9 per million Btu by the year 2000. Recent trends, of course, have been in the other direction, with oil and natural gas prices dropping and electricity in the Gulf states rising as nuclear-generated electricity is brought on line. At some time in the future, however, oil and gas will resume their rise, and it probably will be before the year 2000.

Other factors that will impact energy use are self-generation and cogeneration of electricity. By cogeneration is meant the production of steam and electricity, where the electricity is sold to the local utility at the cost the utility would have incurred to produce it (avoided cost). In practice, self-generation may not occur since a chlor-alkali plant, for example, can by law sell all of its electricity to the utility at the avoided cost and buy it all back at the average cost. Thus, although electricity use may well grow, net purchases from utilities by big users in the Gulf states are not likely to.

Technologies that may impact use of electricity are improved conservation, a resurgence of acetylene-based

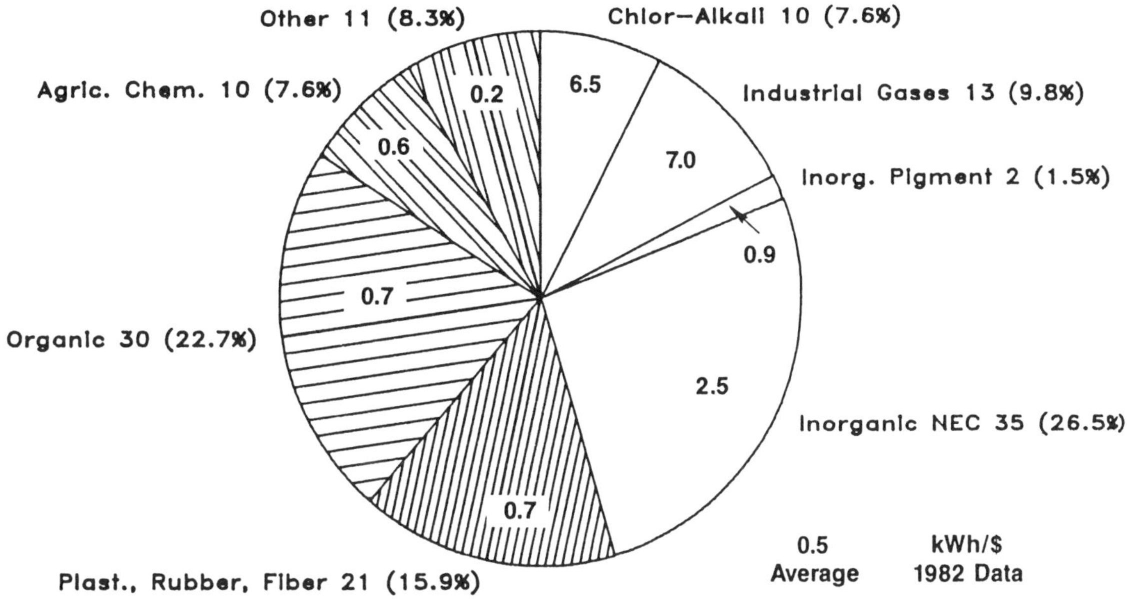

FIGURE 16-10
Chemical Industry by Sector Purchased Electricity (1981) (132 billion kWh)

chemicals, changes in uranium enrichment and chlor-alkali production, and electrosynthesis of organic chemicals. Although production increased more than 50 percent between 1972 and 1981, energy use declined by 12 percent. This conservation has been achieved primarily by improved housekeeping, that is, by better management of steam lines, improved control of furnace operation, and addition of insulation. Much of the potential for this kind of conservation has already been achieved. In the future reduced energy use will come from the design of new plants that have energy conservation planned into them and from the use of new energy-efficient processes, primarily the former.

A return to the use of acetylene, using coal as a raw material as a substitute for ethylene made from natural gas liquids and petroleum fractions, would increase electricity use if the acetylene were generated from calcium carbide made in an electric furnace. This is the pre-World War II technology. However, as of now, a shift back to coal as a feedstock by the year 2000 does not appear likely. If there were such a shift, the plasma process for generating acetylene from coal, which is much less electricity intensive, would probably be used.

FIGURE 16-11
Energy Use Projections

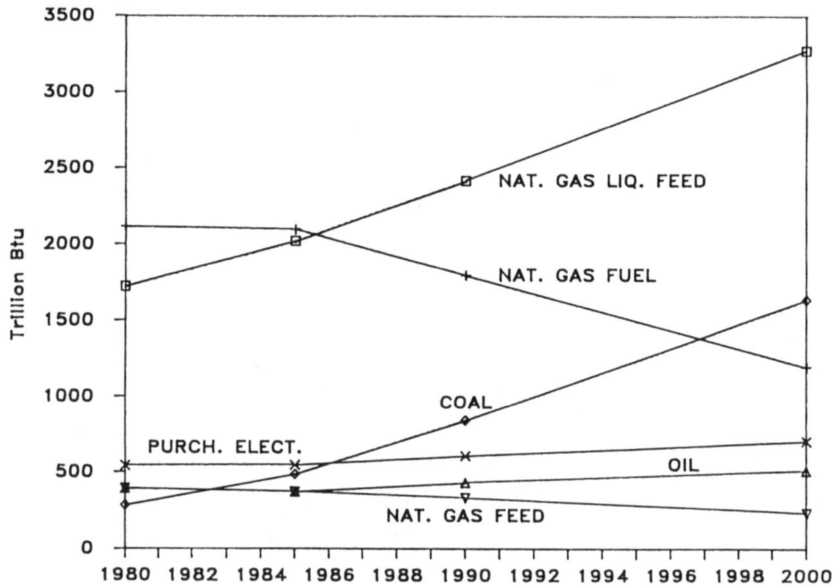

Two of the large current uses of electricity can be impacted by new technologies that will reduce unit electricity consumption. The use of the membrane cell for making chlorine, which is gaining widespread adoption currently, will reduce electricity consumption by 25 percent per unit. The use of centrifuge technology for uranium enrichment can reduce electricity use by 90 percent. However, because of the recent dramatic decline in the demand for enriched uranium the centrifuge enrichment plant now being built in southern Ohio may never be completed. In the more distant future, the use of laser-based enrichment technology would use even less electricity.

Electrolytic methods of the synthesis of organic compounds has been studied extensively. However economic analysis shows these methods to be too expensive to displace current technology. The one success story for electrolysis approach has been the production of adiponitrile, a nylon intermediate, from acrylonitrile. An alternative electricity-based technology is the use of a plasma to generate compounds such as ethylene from gaseous hydrocarbons. This is not yet a commercial technology. However, if half of all the ethylene made were to be made with plasma technology, total electricity use by the chemical industry would increase by about one percent.

This brief look at the chemical industry shows it to be the largest user of energy and the second largest user of electricity. Because the industry is mature, only a modest growth in production is foreseen. Unit use of electricity in some of the large uses will decrease and new, electricity-intensive technologies are not likely to have a large impact by the year 2000.

QUESTION-AND-ANSWER SESSION FOLLOWING PRESENTATION

MR. AHMAD FARUQUI, EPRI: Regarding the Saudis-are-coming situation, I have heard some people say that that is more a concern for the European chemical industry than for the U.S. chemical industry. Can you comment on that?

MR. SHEPPARD: Certainly it is a closer market and satisfying this market is the first domino in the effect of large Saudi production. Maybe 10 or 20 years from now, Iranian or Iraqi chemical production will compete in the European market. By taking a big share of that market, they will force U.S. exports out.

We're still exporting a fair amount to western Europe. The export and import situation is that we import just about as much from Europe as they import from us. A second factor, after closeness, is that a large number of American oil companies are involved in those petrochemical plants, and if oil ever gets to be short, the Saudis could force customers such as Exxon and Chevron to take petrochemicals along with receiving assured supplies of crude oil.

This did happen to Texaco and Amoco in Trinidad, where they had ammonia plants, and the Trinidad government encouraged them to take ammonia as part of deals on oil.

If you look just at transportation distances, it makes more sense to go into Europe and, yes, we will lose that European market first and that will back up exports into the U.S. and cause us problems.

But, then, there will be the second effect when the American companies that are dealing there have to take some chemicals in order to keep their stake in the oil over there. That is assuming that things don't change and that sooner or later oil becomes in short supply.

MR. ADAM KAHANE, Lawrence Berkeley Laboratory: I know it is very difficult to analyze energy or electricity use in the chemical industry because of the huge diversity of products and processes, but when you are using electricity, a lot of the end use seems to be in different kinds of motors—aside from the electrolysis to produce chlorine—a lot of motors for pumping and other parts of the processing.

Have you seen any information that would give you some idea of where motor efficiencies are going? There might be small changes but, with such a huge industry and so much motor use, I would think that a small change in the kind of motors being used could have a bigger effect than any of these other process changes.

MR. SHEPPARD: I would guess that, for noninorganic industries, where it was running on the order of about one or seven-tenths of a kilowatthour per dollar of sales, electric motors and pumps are very important. Improving the efficiency of pumps and compressors and electric motors, by maybe 20 percent, can be expected and electricity use there will be correspondingly reduced.

Every now and then, somebody says we have a breakthrough. Exxon bought Reliance Electric, thinking they had a breakthrough, and it didn't quite come about. But I have seen figures that on the order of 20 percent efficiency improvements can be expected. Maybe some of you have a better number.

MR. RALPH FERRARO, EPRI: I manage the industrial program where we are looking at electronic adjustable speed drives to push motors that are operating pumps and compressors, and we are seeing a 30 to 60 percent industrial process efficiency improvement by removing from the variable process the energy that is lost in the mechanical throttle, like the control valves or the inlet vanes. And, as far as motors themselves, maybe 1 or 2 percent efficiency improvements; the pumps, the same kind of number; but the real efficiency improvements are looking at the system and getting the losses out of the mechanical throttles.

MR. SHEPPARD: I might just make a comment about energy efficiency improvement. The chemical industry, since 1972 through 1983, improved its energy efficiency per pound significantly. The energy consumed per pound dropped by about 30 percent, and much of it was done through improved housekeeping and improved motors, improving ducts to reduce air losses, and by recovering heat. Electricity use went up on a total basis but not on a per pound basis, but use of almost all other energy forms dropped, on an absolute basis, between 1972 and 1983.

Much of the easy savings have been obtained, but now we have to go to the whole system and think about the whole process. If you lay out a new chemical plant, make sure that the place where the steam condensate collects is close to the boiler instead of being at the other end of the plant, so that you can afford to take that hot condensate and put it back in the boiler without losing a lot of heat. In 1972 and earlier, steam condensate was frequently discarded. Steam is generated at high-pressure, and where there is a need for low-pressure steam, it should be run through a turbine to generate some electricity. Planning for these things is what is meant by systems savings.

MR. BOB CILIANO, Dun and Bradstreet: If you consider the electrotechnology collectively, it requires more synthesis and laser technologies. As an educated guess, how much variance might that create over the next 15 years along your baseline projection for electricity consumption? Is it plus or minus 10 percent, 20 percent? I know you have expressed some pessimism regarding the rate of penetration, but how much variance is there in that baseline projection?

MR. SHEPPARD: I don't know about the uranium enrichment, because I think they are even talking about shutting down the centrifuge plant that is under construction, and how much uranium we really need is hard to say. So, let's ignore that one. But in the chlor-alkali industry—we are talking on the order of a billion kilowatthours just by getting a partial penetration of the chlor-alkali cells. If we go to plasma on a large scale for making something like ethylene, which is 30 billion pounds a year, this could have a much bigger effect—perhaps 20 billion kilowatthours per year, and this is out of a total that is on the order of 130 billion kilowatthours purchased.

The only thing I would caution you about is this. Keep an eye on the cogeneration and self-generation, particularly in areas that have traditionally had low cost for electricity and where suddenly it is becoming high cost. At least, they think it is high cost. These areas include Texas, Louisiana, and Mississippi.

MR. CILIANO: I know you couldn't be precise in terms of quantifying what the level of penetration would be, but there has been much written in many EPRI publications about the R&D that is now going on, and one is led to believe that, by the year 2000, we could see significant penetration on the chemicals industry by some of these technologies. You seem to be a little pessimistic regarding that, yet we seem to be dealing with possibly very large numbers if the penetration could be pulled off.

MR. SHEPPARD: My electricity consumption curve grew at about 1 percent a year, based on the assumption that electricity prices stayed constant in real terms. I understand EPRI's pessimistic scenario had them growing at only 1 percent per year, if I may tease a little bit.

I was saying 20 billion kilowatthours for plasma production of ethylene, and we're talking 132 billion kilowatthours for the industry, so there we are talking a 15 percent increase, spread over 15 years. There is 1 percent from ethylene alone, but it is a big-volume chemical. Maybe it could grow 2 percent or 3 percent a year. But I don't see electrochemistry coming on and replacing the current state of the art in a huge way, not doubling the amount of electricity use.

17

ENERGY CONSUMPTION BY THE PAPER INDUSTRY

Stanley Lancey
American Paper Institute

My remarks today contain both good and bad news for electric utilities and other suppliers of energy to the paper industry. On the one hand, I will demonstrate with statistics and charts that the paper industry is a smokestack industry that has kept pace with the U.S. economy during recent years and for which future growth prospects appear highly favorable. On the other hand, I will also show that the paper industry has done more than almost any other domestic industry to reduce its reliance on purchased energy in the form of fossil fuels.

To begin with, let's examine recent growth patterns for the industry's two principal product lines—namely, paper and paperboard. Production of each totaled slightly more than 30 million tons during 1984, which brought total industry output of paper and paperboard to 68.4 million tons.

PAPER

The overall paper category encompasses four major subgroupings:

1. printing-writing papers, whose end-use markets include magazines, books, stationery, and office reprographics;
2. newsprint, whose principal end-uses are newspapers, catalogs, directories, and advertising inserts;
3. packaging and industrial converting papers; and
4. tissue products.

PRINTING-WRITING PAPERS

Paced by the proliferation of office copy machines, computers and word processors, and by robust growth in magazine and direct-mail advertising expenditures, the printing-writing category has been the paper industry's star performer for more than a decade now. Whereas real GNP rose at the average yearly rate of 2.8 percent between 1974 and 1984, U.S. consumption of printing-writing papers spurted ahead at the average annual rate of 4.7 percent. Nevertheless, not all is rosy for producers of printing-writing papers. Import penetration occasioned by the dollar's lofty exchange rate has taken a toll. Whereas U.S. consumption of printing-writing papers grew 11 percent last year, domestic shipment of these grades advanced at the more gradual pace of 6.2 percent. By reason of that disparity, imports' share of U.S. printing-writing paper consumption rose from 7.6 percent in 1983 to 11.5 percent last year.

In addition to import penetration, another important consideration in the outlook for printing-writing paper demand pertains to the evolution of new information technologies. As I noted a moment ago, the spread of office copy machines and computers has thus far been a boon for printing-writing paper demand. However, because the cost of transmitting and storing information electronically is declining sharply, some analysts suggest that there will eventually be less need for hard-copy documents. Proponents of this view contend that office workers will increasingly create and store text on computer terminals and transmit messages to their colleagues electronically, whether they be down the hall or across the ocean.

However, research on this subject conducted by the American Paper Institute points to a different conclusion. Although it does seem likely that a declining proportion of new information will be recorded on hard copy in the future, the volume of information generated by our society is expected to escalate at an increasing rate. In this case, a smaller slice of a larger pie will probably translate into increased demand for hard-copy documents. Furthermore, even though information will increasingly be transmitted from one location to another electronically, the outstanding image quality offered by printed matter cannot yet be matched by computer screens, suggesting that most readers will still desire printed copies of text transmissions, even if they have the opportunity to initially view the information from a screen. In sum, looking ahead a decade of so, the prospects for printing-writing paper growth appear excellent.

NEWSPRINT

Newsprint is akin to printing-writing papers in one key respect: it is used to communicate information. U.S. circulation of daily newspapers has been on a stagnant trajectory for more than a decade now, while sales of Sunday editions continued to edge ahead at the sluggish pace of 1 percent. Despite these lackluster circulation trends, domestic newsprint consumption has outperformed the U.S. economy by a significant margin during recent years. Between 1981 and 1984, for example, newsprint consumption grew at the average annual rate of 4 percent, versus real GNP growth of 2.7 percent.

The comparative strength of newsprint demand derives

from robust gains in advertising outlays, which have given rise to thicker newspapers, and a proliferation of advertising inserts, many of which are newsprint based. Before leaving this subject, let me stress that newsprint is a North American rather than a U.S. industry. By this I mean that 58 percent of newsprint used in the United States last year originated in Canada. That's down from 64 percent in 1979, but the trend toward greater U.S. self-reliance with respect to newsprint may be entering a period of hiatus, the reason being that no new U.S. newsprint capacity is in prospect for the next three years.

PACKAGING AND INDUSTRIAL CONVERTING PAPER

Not all of paper's end-use markets are keeping up with the economy at large. Demand for packaging papers has trended sideways since 1980, mainly because of stepped up competition from plastic substitutes produced from polyethylene films. To cite some specific examples, film-based products presently account for about half of the merchandise bag market and plastic sack liners have partially displaced kraft papers in selected multiwall sack applications. The largest paper market currently at risk from plastics is the carry-out grocery sack. Some of you have probably witnessed the substitution of plastic grocery sacks for kraft paper sacks at your local supermarket. Plastic's share of the U.S. grocery sack market was negligible just 4 years ago, but has since grown to about 11 percent. Some prognosticators expect plastic grocery sacks to grab an even larger share of the market during the next few years, a view that is partly predicated on the fact that plastic resin prices have been softening lately due to oversupply and weak oil and natural gas prices. Other analysts suggest that paper sacks will continue to dominate the grocery sack market because they can be packed more quickly than their plastic counterparts and are thus more cost-effective to use. Only time will tell which of these views is correct.

TISSUE PRODUCTS

The last of the four major paper categories is tissue, a grouping that encompasses toilet and facial tissues, napkins, towelling, and wiper stock. Tissue production in the U.S. has grown at the average yearly rate of 1.7 percent since 1979, which was six-tenths of a percentage point faster than the U.S. population growth, but slower than real GNP gains, which averaged 2.1 percent for the period. All indications are that per-capita tissue consumption will continue its uptrend because of rising levels of real spending power and the public's ever increasing desire for convenience.

PAPERBOARD

That takes care of paper; now let's focus on paperboard. The two key components of paperboard are **containerboard**, which is used to produce corrugated shipping containers, and **boxboard**, which is used for folding and set-up boxes. Corrugated shipping containers are characterized by a fluted inner material, known as corrugating medium, sandwiched between two plies of linerboard. They are used to package all manner of consumer and industrial products, ranging from foodstuffs to machine parts and appliances. Corrugated containers are most conspicuous in supermarkets, particularly when cans and jars are being unpacked; however, they can also be found in almost any nonfood retail store and in factories and warehouses.

CONTAINERBOARD

Demand for containerboard is responsive to industrial production and consequently tends to be highly cyclical. After abating 8 percent during the 1981-1982 recession, domestic containerboard demand bounced back 9.5 percent in 1983 and then grew an additional 9.5 percent last year. However, 1985 is shaping up as a weak year for this segment of the industry, mainly because U.S. output of manufactured products has remained on a plateau since September 1984.

Like packaging papers, corrugated containers are also facing competition from plastic films. In this case tight-fitting films known as shrink case are being used to contain bottles and cans on corrugated trays. The trays require less containerboard than full-sized cartons, thereby reducing demand. In any case, only a handful of products possess physical properties (i.e., stackability and rigidity) conducive to the use of shrink case, implying that containerboard demand will not suffer the major negative repercussions that have befallen many categories of packaging papers. Indeed, a recent study of the corrugated box market indicates that growth will average approximately 2.5 percent a year between 1983 and 1990, and that competition from shrink case will not pose a major threat.

BOXBOARD

Another leading paperboard grade is boxboard, which is used to produce folding boxes, milk and juice cartons, and rigid boxes such as those used to package shoes and jewelry. Boxboard demand varies cyclically in accordance with the rhythms of the economy, but has exhibited an overall trend that has been essentially flat since 1976.

Paperboard exports, which are chiefly composed of linerboard used for the outer facings of corrugated containers, accounted for 15 percent of domestic paperboard production last year. United States paperboard exports rose at the average yearly rate of 8.4 percent between 1976 and 1983, but subsequently contracted 9.3 percent during 1984, mainly because of the strong dollar.

INDUSTRY TRENDS

Now that we have examined the principal components of paper and paperboard demand, the next step is to look at aggregate industry trends. New supply, which is defined as domestic production plus imports minus exports, measures the tonnage volume of paper and paperboard that is actually consumed or stockpiled in the United States during a given year. United States production is lower than new supply because our imports, which are mainly newsprint, typically exceed our exports. U.S. paper/paperboard consumption grew in tandem with GNP between 1959 and 1973. In fact, real GNP growth and new supply of paper/paperboard both grew at the average annual rate of 4 percent between 1959 and 1973.

The ratio of new supply to GNP dropped sharply during the 1974-1975 recession as users of paper and paperboard worked off excess inventories. Paper and paperboard production that dropped more extensively than GNP (a measure of output of cyclically stable service industries as well as manufacturing) was not expected, although the relative deterioration was sharper than in prior recessions. What was surprising is that the ratio never fully rebounded to its prerecession levels. After sagging from 48,000 tons per billion dollars of real GNP in 1973 to 41,500 tons in 1975, the ratio climbed back to 45,300 tons in 1976 and then began to edge down again, reaching 44,900 tons in 1979. True to form, the 1981-1982 recession brought further deterioration in the ratio of new supply to GNP, the difference being that this time the ratio more than regained the ground it lost after the recession ended. Indeed, by 1984 the ratio had risen to 45,600 tons per billion dollars of GNP, its highest level since the mid-1970s.

While the recent strong performance of paper and paperboard demand is partly cyclical in nature—i.e., demand for the industry's products tends to grow more quickly than the economy during economic expansions and to contract more sharply during recessions—indications are that consumption of paper and paperboard will grow as rapidly or nearly as rapidly as the economy during the next ten years. Underlying this upbeat scenario are the dual assumptions that the computer revolution will remain a positive force with respect to office paper demand and that future penetration of packaging paper markets by plastics will take place gradually rather than precipitously.

In addition to growth of end-use markets, another critical variable for future paper and paperboard production pertains to foreign trade. Imports of paper and paperboard rose from 12 percent of new supply in 1982 to 14 percent in 1984, and remained at that relative percentage share during early 1985. As a result of increased import penetration, domestic production of paper and paperboard gained 5.3 percent during 1984, even though U.S. consumption grew an exceptionally strong 8.2 percent. Much of the recent import penetration stems from the excessively high exchange value of U.S. currency and may not be a major impediment over the long haul. In fact, the dollar has eased somewhat during recent months and a continued softening may benefit paper and paperboard production by promoting exports and reducing imports.

THE PAPER INDUSTRY'S ENERGY CONSUMPTION PATTERNS

We have now examined the prospects for individual segments of the paper industry and for the industry at large. But what of the industry's energy requirements—will they grow apace of paper and paperboard production? If past patterns are any indication of what lies ahead, the answer is a resounding "no."

One reason the answer is "no" is that the paper industry has instituted various processing efficiencies that have kept its energy requirements per unit of output on a nearly steady downtrend since the early 1970s. To cite some specific statistics, the industry's overall energy consumption per ton of output edged down from 32.6 million Btus in 1972 to 27.7 million Btus last year, a cumulative decline of 15 percent in 12 years!

The broad topic of paper industry energy consumption patterns is quite interesting in and of itself, but the focus of today's conference suggests that we devote special attention to electricity. The paper industry purchased 43.7 billion kWh of electricity in 1984, and based on data for 1981 (the latest information available from the U.S. Department of Commerce) ranks third among all U.S. manufacturing industries in terms of purchased electricity consumption. Although the industry's overall consumption of fossil fuel-based energy declined since the early 1970s, its purchased electricity consumption has shown a consistent and significant increase, rising from 27 billion kWh in 1972 to almost 44 billion kWh last year, a gain of 60 percent.

Part of the industry's increased consumption of purchased electricity derives from the use of electrostatic precipitators and scrubbers to meet emission standards set by the Clean Air Act of 1970, as amended in 1977, as well as aerators to treat effluents. Another contributory factor has been rapid increases in the production of thermomechanical pulp, or TMP for short. TMP is a groundwood pulp that first came into use during the early 1970s and presently accounts for about 4.1 percent of all pulp production. Used mainly in newsprint applications, it is stronger than most other groundwood pulps, but highly electricity-intensive to produce. TMP will continue to experience rapid growth according to the American Paper Institute's most recent capacity survey. In particular, the survey indicates that TMP capacity will expand at the average rate of 4.6 percent a year between 1983 and 1987, as compared with 1.4 percent average annual growth for overall pulp capacity.

Not only has the paper industry increased its use of purchased electricity, it is also generating more of its own electricity. As many of you already know, the paper industry is the leading U.S. industry in self-generation and cogeneration of electric power, supplying 37 percent of its own electricity needs and accounting for 29 percent of all industrial cogeneration in the United States today. Cogeneration of electricity can take various forms, but in the pulp and paper industry it is most commonly achieved through the generation of steam at high temperatures for use first in electric power turbines and then in process applications such as pulping and drying. Under the authority of the Public Utility Regulatory Policies Act of 1978 (PURPA), regulations have been issued that encourage cogeneration in the nonutility sector. Partly in response to incentives embodied in that legislation, the U.S. paper industry's self-generation of electricity—i.e., cogeneration and hydroelectric power—increased from 34.2 billion kWh in 1982 to more than 40 billion kWh at the end of 1984, an increase of 270 percent relative to 1982's sales.

I suspect that many of you are interested in learning the regional breakdown of purchased electricity consumption by the paper industry. Here are some key facts: Paper producers located in the Mountain and Pacific regions of the country used the most purchased electricity per ton of paper and paperboard capacity in 1984—an average of 3.8 million Btus per annual per ton. Producers located in the South Atlantic states used the least—1.1 million Btus per ton. Overall, however, pulp, paper, and paperboard producers located in the Mountain and Pacific states accounted for

27.7 percent of the industry's total purchased electricity consumption in 1984, followed closely by the South Central states, which accounted for 27 percent of the total. Next comes the North Central region, which consumed 16.4 percent of the total, and the South Atlantic region, which took 14.6 percent. New England and the Middle Atlantic states accounted for 7.7 percent and 6.6 percent, respectively, of the industry's 1984 purchased electricity consumption.

Reviewing growth trends for the 1979-1984 period, we find that the most rapid gains in the industry's use of purchased electricity took place in New England, where growth averaged 7.1 percent a year, and in the South Central states, where growth came to 4.3 percent a year. Gains in the South Central region of the country can largely be explained by rapid additions to paper and paperboard capacity, but the same explanation does not carry over to New England, where the industry's capacity expanded a relatively sluggish 1 percent a year. Purchased electricity rose from 5.1 percent of the overall energy requirements of New England paper producers in 1979 to 7.1 percent last year. Over that same time span, however, this geographical segment of the industry reduced its use of residual fuel oil from 54.7 percent of its total energy consumption to 33.4 percent, and increased its use of self-generated energy from a 37.3 percent share to 51.4 percent.

Because of declining paper and paperboard capacity, the paper industry's use of purchased electricity in the Middle Atlantic states ebbed at the average rate of 0.9 percent between 1979 and 1984. Although capacity growth and increases in purchased electricity consumption do not necessarily go hand in hand, it is well to note that for the next few years growth of paper and paperboard capacity will be most rapid in the South Central states (2.1 percent a year) and slowest in the Middle Atlantic region of the country 0.7 percent a year.

All that I have said here today can be summarized in just one key sentence: Growth prospects for the paper industry appear quite good, but if past patterns are a reliable guide, the industry will continue to reduce its overall use of purchased energy through conservation and self-generation. However, if recent trends are maintained, despite the growth of cogeneration by the paper industry, it will continue to rely heavily on purchased electricity.

QUESTION-AND-ANSWER SESSION FOLLOWING PRESENTATION

MR. AHMAD FARUQUI, EPRI: I would like to know what you expect to see in terms of the future share of electricity in total energy consumption, given the historical trend that you were just showing.

MR. LANCEY: It will go up, but, exactly how much, I can't forecast. I can't give you an exact number.

MR. FARUQUI: Can you comment on the role of cogeneration in terms of future development?

MR. LANCEY: Well, the only thing I am at liberty to say, after consulting with our Energy Department at The American Paper Institute—I am more in the marketing end of it—is that cogeneration will not increase enough to offset that trend of increasing electricity consumption, and, furthermore, the paper industry growth overall will probably average slightly below that of GNP or maybe even match GNP growth. So, we have an industry that is growing apace with the economy, and purchased electricity will probably continue to increase its share of our overall energy consumption. Exact numbers, I don't know.

MR. SAM SUGIYAMA, Bonneville Power Administration: We have a significant pulp and paper industry in the Pacific Northwest. I heard you mention an increase in thermomechanical pulping. What are the range of scenarios that you are aware of for the increase in percentage of thermomechanical relative to all pulping activities?

MR. LANCEY: Well, right now, thermomechanical accounts for about 4.1 percent of the total, as I mentioned, and it will continue to probably go up to about 6 percent of the total over the next three years. So, after that, it is somewhat questionable, because it depends on how much newsprint is produced in the United States versus Canada. Right now, there are no new newsprint mills, per se, scheduled to be built over the next three years, and thermomechanical pulp is mainly used in newsprint, so we may see a leveling off in its growth. We may, say, three years out. In the short term, what is coming on stream for the next three years is strong growth in thermomechanical pump. When we get out closer to 1989, then that may slow.

MR. SUGIYAMA: In terms of my question, relative to the range, it sounds like there is a narrowing. What do you expect about that?

MR. LANCEY: Well, because what is going to happen during the next three years is already in the pipeline, we know exactly. We survey our members and we know what commitments they have and what they say they are going to produce. If you survey out more than three years, then there are no hard figures.

MS. CONNIE CLONCH, Pacific Gas and Electric Company: You mentioned thermomechanical. What I want to ask is in some of the other processes, such as the drying, are there any new technologies coming on that would reduce or that would increase the efficiency?

MR. LANCEY: Of overall energy consumption?

MS. CLONCH: Yes, especially in the corrugated containers.

MR. LANCEY: Corrugated right now is made mainly from linerboard, which is one of the least energy-intensive processes in the paper industry, and, moreover, it uses a very high percentage of spent pulping liquors as an energy source, so that is one sector that is very, very low, in terms of purchased energy. Whether further efficiencies will be achieved, I don't know.

MR. BOB CILIANO, Dun and Bradstreet: Is it only in the quality printing paper that you see significant import penetration? You quoted a lot of statistics, and I hope they are in your paper.

MR. LANCEY: That is where it has been mainly taking place. We have a minority in the newsprint market; the Canadians have the majority. We account for something like 48 percent. I forgot the exact number. In the containerboard market, we are an exporter. We export around 2 million tons a year. And we are vulnerable to displacement in our export markets but not imports. In packaging papers, we are losing market share in plastics to a weak market, so there is no opportunity for imports. So, all of the imports are taking place in the fine papers, mainly coated paper and uncoated ground wood. We hope that it is a temporary phenomenon deriving from the dollar.

We are the world's low-cost producer, and some companies are optimistic enough to think that they are not only going to take back the market share lost to imports, but also that we are going to increase our exports. This may be hopeful thinking, but that is what I think paper producers truly do think. We have the forests over here, and there is no country in the world that has the natural resources, in terms of timberland, that could match ours. So, long term, we think we are going to recover that market share.

MR. DAVID GOLDFARB, Georgia Power Company: The question I have is in regard to the regional differences. Do you have any explanation for why the kilowatts used per ton is lower in some of the other regions, like the Southeast versus the Pacific Northwest? And will that have any bearing on future self-generation or cogeneration in those two areas?

MR. LANCEY: Well, it depends on cost. Electricity is cheaper in the Northwest, so companies tend to be more profligate with their use of electricity. In the South and the Northwest—it is a big lumber-producing region also, so companies rely more on residues. In the Northeast, you don't have that. The companies are heavily dependent on imported fuel oil.

MR. GOLDFARB: Did you say the Northeast was the highest?

MR. LANCEY: In terms of electricity, the Pacific Northwest, but that is because of cost. It is cheaper.

18

THE U.S. ELECTRONICS INDUSTRY IN THE WORLD ECONOMY

John A. Alic and Martha Caldwell Harris
Office of Technology Assessment

Robert R. Miller
University of Houston

INTRODUCTION

Today, the world's electronics industry looks vastly different from that of the late 1960s (1). In the aftermath of World War II, U.S. companies had been in a better position to turn emerging technologies into commercial products than manufacturers in other countries. American engineers developed semiconductors and integrated circuits; in most parts of the world, American firms became the chief suppliers of computers.

Despite this head start, by now about half the consumer electronics products sold in the United States are imported. European firms are on the defensive virtually across the board. South Korea makes more black-and-white TVs than any other nation. We hear that Japanese semiconductor firms have taken more than half the world market for some types of random access memory circuits, that their fifth-generation computer project foreshadows an assault on the United States' dominance in information processing. In 1985, after repeated threats, the U.S. semiconductor industry began to file formal trade complaints against Japanese companies.

What has changed since the 1960s? Consumer electronics technologies have become relatively standardized, thus available to manufacturers in newly industrializing countries (NICs) as well as advanced economies. Japanese corporations, which took over product leadership from their American and European counterparts during the 1970s, already find themselves under severe pressure from manufacturers elsewhere in Asia (Table 18-1). Even in more demanding products, such as video cassette recorders, the NICs are moving rapidly ahead. With Japanese firms refusing to license VCR technology to Korean manufacturers, the latter have developed their own product designs—a task that American companies declined to undertake. For no other reason, Japanese manufacturers would have been forced to move up the technology ladder into microelectronics and computers. The fears expressed by many Europeans in the 1960s—that the United States would benefit from long-lived if not permanent technology gaps (2)—have hardly been realized.

Now the questions for Europe as well as the United States are: How to compete in standardized goods with the NICs? How to compete in high technology electronics with the Japanese? Japan has its own problems: Having largely caught up with the West in technology, can it find strategies that will help it take the lead? For the NICs, many of whom face shortages of engineers and technicians, the critical issue becomes: How to develop a store of home-grown technology sufficient to move into higher value-added products and services.

TABLE 18-1
Production and Exports in Asian Electronics Industries (1979)

	Total Electronics Production (millions of dollars)	Exports as a Percentage of Total Electronics Production (percent)
South Korea	$3,300	70
Taiwan	3,200	80
Hong Kong	2,000	90
Singapore	1,850	90
Malaysia	990	75

Source: International Competitiveness in Electronics. Washington, DC: Office of Technology Assessment, November 1983, p. 128.

INTERNATIONAL PATTERNS OF PRODUCTION AND SALES

CONSUMER ELECTRONICS

Televisions typify the manufacture and sale of relatively standardized products. As Table 18-2 shows, TV sales account in value terms for more than a quarter of the total U.S. consumer electronics market. Foreign-owned firms now take many of these sales through imports or local assembly; today 11 of 15 companies with TV plants in the United States are under foreign ownership. Many of the parts and subassemblies for TVs produced here, by U.S. as well as foreign firms, come from overseas.

Import competition came first from the Japanese, who earlier had turned the transistor radio into a mass-market product. As they moved from black-and-white TVs to color production, Japanese firms like Matsushita and Sony increased their output at very high rates. From less than

TABLE 18-2
U.S. Sales and Imports of Selected Consumer Electronics Products (1982)

	U.S. Sales (millions of dollars)	Imports (millions of dollars)	Import Penetration (percent)[a]
Color television	$4,253	$ 546	12.8
Black-and-white TV	507	344	67.9
Video cassette recorders	1,303	1,032	100.0[a]
Home and auto radios[b]	1,579	1,207	76.4
Stereo systems[c]	1,754	1,342	76.5
	$9,396	$4,471	47.6

Source: Electronic Market Data Book, 1983. Washington, DC: Electronic Industries Association, 1983, pp. 6, 19, 23, 31.

[a]Because many items imported in a given year are not sold until the following year, dividing imports during a given calendar year by sales in that same year may give only a rough indication of import penetration; for instance, all video cassette recorders sold in the United States are imported, even though 1982 sales figures exceed 1982 import figures.

[b]Including auto tape players.

[c]Including audio tape units and other component equipment.

100,000 color TVs in 1965, by 1970 Japan was turning out more color sets than any other nation; half the small-screen color TVs sold in the United States came from Japan. Imported TVs, built by large integrated firms which produced many of their own components, were inexpensive. Japan's manufacturers gained scale economies by moving aggressively into markets, not only in North America, but in many other parts of the world; Toshiba, for instance, entered Costa Rica as early as 1971. American consumers found the imports attractive and reliable. Fierce price competition—including instances of dumping proven under U.S. law—drove many American companies from the business. Since 1970, a dozen U.S. manufacturers have disappeared or been bought out by foreign enterprises.

Today, European consumer electronics firms are succumbing fast to similar competitive pressures. While South Korea's output of consumer electronics grew by almost 50 percent per year during the 1970s, aided by low wage rates and supportive government policies, many European manufacturers were losing money and falling behind in research and development and production technology. Telefunken, one of the largest, entered bankruptcy in 1982. Companies like Philips, the Dutch multinational, and Thomson, which benefits from the active support of the French government, seem determined to continue in consumer markets. Smaller European manufacturers, facing fragmented markets and stiff competition from companies based in the Far East, will no doubt continue to lose ground. Already, five Japanese firms assemble TVs in Great Britain through joint ventures or wholly-owned subsidiaries.

HIGH-TECHNOLOGY PRODUCTS

Could the picture painted above, which seems so grim for many of the advanced industrial nations, be replicated in high-technology electronics—e.g., semiconductors or computers? In some ways it already has; in other respects, the patterns are quite different. European companies, with a few exceptions such as the computer manufacturer Nixdorf, are slipping with respect to Japanese as well as American rivals. Japan's major electronic firms, several of which make computers and microelectronics as well as consumer products, have mounted a strong challenge to U.S. leadership in integrated circuits. In computer systems, Japanese companies already possess excellent hardware technology, but have a long way to go in software, marketing, and customer support and service. Meanwhile, several of the Asian NICs are seeking to emulate Japan by moving rapidly from components and consumer products into microelectronics and computer equipment; a pair of locally owned companies in Hong Kong began producing integrated circuits during 1982.

In contrast to U.S. consumer electronics firms, American suppliers of semiconductors and computers began to produce and sell overseas many years ago. American semiconductor manufacturers come in two types: so-called captive producers, typified by IBM, which consume most or all of their output internally, and merchant firms which sell primarily to other companies. Many of the semiconductor producers in Europe and Japan resemble U.S. captives more closely than merchant firms. As Table 18-3 indicates, large and diversified firms like Philips and Siemens, as well as Hitachi and Toshiba, get relatively small fractions of their revenues from semiconductor sales.

For either type of American firm, multinational production is nothing new. Intense competition among entrepreneurial merchant suppliers—many with headquarters in the Silicon Valley region of California—led to rapid transfers of labor-intensive production offshore, mostly to low-wage countries in the Far East. In Europe, American companies built factories to avoid trade barriers and be close to major markets. By 1970, when foreign competition was still nonexistent, U.S. semiconductor manufacturers operated more than 70 overseas plants.

Taking advantage of technical leadership and overseas investments, merchant and captive U.S. producers for many years accounted for about two-thirds of world output. While Japanese entrants have recently taken well over half of merchant sales for several varieties of integrated circuits—notably random access memory chips—this success has not as yet been repeated in products like microprocessors, where market acceptance depends heavily on creative engineering design. Japan's strength has resided in mass production of relatively conventional devices.

In computers, desktop to mainframe, the United States' position has hardly been challenged (**3**). American-owned multinationals dominate markets in most of the world through exports and overseas production. Few companies outside the United States make money on computers. Europe's largest suppliers, only recently surpassed by Japanese firms (Table 18-4), have computer sales not even

TABLE 18-3
Total Sales and Merchant Semiconductor Sales for Selected Firms (1982)

	Headquarters	Worldwide Sales (1982) (millions of dollars) Total	Semiconductor[a]	Semiconductor Sales as Percentage of Total	Percentage of Semiconductor Production Used Internally
Texas Instruments	U.S.	$ 4,370	$ 1,300	30	9[b]
Motorola	U.S.	3,790	1,219	32	5[b]
National Semiconductor	U.S.	1,150	673	59	10[b]
NEC (Nippon Electric)	Japan	5,500	990	18	24
Hitachi	Japan	16,260	800	5	19
Toshiba	Japan	10,150	740	7	8
Philips	Netherlands	16,090	494	3	NA
Siemens	West Germany	16,960	328	2	NA

Source: International Competitiveness in Electronics. Washington, D.C.: Office of Technology Assessment, November 1983, pp. 131, 133, 138, 142.

NA = Not Available.

[a]Outside (merchant market) sales only. IBM, with total 1982 sales of $34,400 million, produced semiconductors valued at $2,080 million, consuming all internally. The company is the largest purchaser in the merchant market, as well as the world's largest producer of semiconductors.

[b]1978.

half those of Digital Equipment Corporation. (Note that the noninclusive list in Table 18-4 omits many American firms that are larger than the foreign firms included.)

The same forces that led companies like General Motors to invest first in captive semiconductor facilities and then to purchase Electronic Data Systems (EDS) are blurring the bounds of the computer industry. With smart electronics—microprocessors, microcomputers, minicomputers—central to their products and their manufacturing processes, companies producing consumer, industrial, and military goods of many types are joining computer and communications suppliers in designing and fabricating custom chips, hiring software specialists, and otherwise reorienting their product

TABLE 18-4
Selected Computer Manufacturers Ranked by 1984 Sales

	Headquarters	Worldwide Computer-related Sales (1984) (millions of dollars)
IBM	United States	$ 46,940
Digital Equipment Corp. (DEC)	United States	6,230
Fujitsu	Japan	3,500
NEC (Nippon Electric)	Japan	2,800
Siemens	West Germany	2,790
Groupe Bull	France	1,560
ICL	United Kingdom	1,220
Nixdorf	West Germany	1,150

Source: P. Archbold and J. Verity, "The Datamation 100," Datamation, June 1, 1985, p. 50.

and process technologies around digital electronics. Captive semiconductor manufacturers ranging from Western Electric and IBM to Hughes Aircraft (also recently purchased by General Motors) have made major contributions to the technology base; continued growth of captives and diversification of firms from other industries should help the United States maintain its standing.

NEW APPROACHES TO RESEARCH AND DEVELOPMENT

Even so, the United States needs to look to its technology base. In the early years, military spending—mostly direct purchases, although research and development funds also contributed—spurred developments in both microelectronics and computers. Military sales totaled half the U.S. semiconductor market in 1960, perhaps 10 percent today. As late as the middle 1960s, the federal government accounted for nearly half of IBM's domestic computer sales. But the ties between military and civilian electronics have long since been broken; seldom does defense spending now lead to commercial innovations. The Defense Department's Very High-Speed Integrated Circuit (VHSIC) program, initiated in 1979 as a response to the technology lag in military equipment, illustrates the shift. VHSIC goals are to catch up—to speed insertion of advanced microelectronics technologies into defense systems.

No longer in a position to benefit greatly from military spending, and for reasons that include growing international competition, escalating research and development costs, and shortages of engineers, the commercial side of the industry has sought new approaches to research and development. American Telephone & Telegraph's reorganization has also changed the research picture. For many years, Bell Laboratories provided a substantial share of the technology base for U.S. electronics firms. As the mission of Bell Labs has shifted, reflecting new business realities, basic research has been de-emphasized; the flow of technology from Bell Labs to the rest of the U.S. (and world) industry is diminishing.

New approaches to research and development include extensions of past practices in joint ventures and cooperative research. For instance, several companies may divide the design and development tasks for a family of integrated circuits. Agreements with Japanese firms provide a striking instance of shifting competitive balance. Once plainly inferior in technology, a company like Oki Electric now cooperates with National Semiconductor in the development of memory chips. Further signs of change include the Semiconductor Research Corporation, many new industry-supported research centers at universities, and Microelectronics & Computer Technology Corporation, the widely publicized joint venture that seems to be in competition with IBM as well as the Japanese.

Could the need for stronger technical foundations, together with a swelling U.S. defense budget, tip the balance back to military leadership in electronics technology? This is not likely. The commercial side of the industry is too big and too dynamic—electronics is spreading too swiftly into products and manufacturing in other sectors of the economy. But if military needs remain specialized and well removed from civilian applications—real-time signal processing being the common example—the balance may nonetheless shift in a more limited way.

THE FIFTH GENERATION

Military spending on "intelligent" systems is the most likely source of renewed spinoffs. With a $600 million, six-year program under way—and prospects for more money in the future—the United States may now seem to have a challenger to Japan's fifth generation and supercomputer projects. Directed at expert systems and other forms of artificial intelligence, and at architectures and the theory of computing (**4**), the program could have real if modest impacts on the civilian side of the industry.

How is this so, given that funding averaging $100 million per year comes to less than half the annual research and development spending of even the smaller U.S. computer firms? The reasons are the same as in earlier instances of spinoffs from military projects. The Defense Advanced Research Projects Agency (DARPA) will be paying for basic research of the sort that few corporations conduct. Not all of this research will be at a far remove from civilian needs. This sets DARPA's strategic computing program apart from most current military research and development. Among the goals in expert systems, a field that seems ripe for rapid advance, are "personal assistants" for military commanders. At least in broad outline, and at the conceptual level, some of the problems faced by designers of software for battle management may not be foreign to strategic and organizational problems in business and government. Research and development directed at autonomous robot vehicles, for another example, might eventually find applications in collision-avoidance systems for passenger cars.

It would be foolish to assume that DARPA's programs—or the very ambitious Japanese efforts (**5**)—will reach all their objectives. Enthusiasts working in artificial intelligence have been perenially over-optimistic. On the other hand, anyone using personal computers, word processors, or the current generation of engineering work stations has first-hand experience of the rate of advance in software aids for relatively routine tasks. Software for use by people with little or no specialized training remains an infant technology. This is one reason military and civilian needs have not yet diverged. Military spending can thus exert a good deal of leverage, as in the early days of semiconductors or computer hardware.

Needless to say, DARPA is not alone in seeking computers that will interact in new and helpful ways with people. This is precisely the goal of government-supported research and development programs in Japan and several of the Western European nations—all of which, however, are pointed at commercial rather than military applications. Will they, as a result, be more successful in stimulating civilian industries?

Given equal success in meeting technical goals, the answer has to be yes. Furthermore, Japan plans to spend at least as much money on its joint government-industry programs as DARPA. Most important, Japan's objectives extend beyond technology _per se_. The Ministry of International Trade and Industry, along with other government agencies, hopes to aid Japanese firms in leapfrogging the United States in computing. This accomplished—easier said than done—Japanese corporations would seek to apply their technical advantages broadly, taking the lead in smart machines and their myriad applications throughout post-industrial societies.

The task may seem daunting. After all, while the Japanese computer industry has been growing rapidly, firms like Fujitsu and Nippon Electric remain puny compared to IBM, especially in international markets. Still, Japan can now bring to bear a good deal of experience in organizing and managing long-term joint research and development, while rough hardware parity already exists. But even if the Japanese make rapid progress technically, they face severe handicaps in marketing computers worldwide. Japan's firms are striving to gain the kind of international business experience that American suppliers like DEC, IBM, and Hewlett-Packard have long since accumulated. In selling information systems, Japan faces problems quite different from building an internationally competitive consumer electronics or automobile industry. Multinational production, rather than exporting, will be called for. Software geared to people's needs, farflung support and service networks, customer hand-holding and make the computer market much different from that for semiconductors, where purchasing agents will switch suppliers to save a few cents per chip. Still, who in the U.S. electronics industry—or in the automobile or steel industry—would continue to underestimate Japan's major corporations?

NATIONAL INDUSTRIAL POLICIES

Japan, with its fifth generation and supercomputer projects, is far from alone in seeing—or wishing to see—electronics at the heart of its future economic structure. Industrial policy programs in a number of other countries have been designed to subsidize and stimulate both direct production and applications of high-technology electronics.

Best viewed as generic, the term industrial policy encompasses the array of instruments—investment grants and subsidies, regulations (including antitrust), research and development funding, government procurement, education and training, and trade policies—that affect a nation's industries, directly and indirectly. While all governments choose from a more or less standard list, national industrial policies differ in goals, timing, and comprehensiveness.

In the past, the industrial policies of several European nations aimed at building an independent capability in technologies such as computing for reasons of national security and prestige. France, fearing that domestic manufacturers would be overwhelmed by American competition, tried in the 1960s with Le Plan Calcul to engineer a "national champion" capable of matching IBM. The French have continued to support local companies in microelectronics as well as computers and communications; through several shifts in government, the bureacracy has encouraged mergers, paid for research and development, aided the training of engineers and technicians, and helped French companies acquire American technology. West Germany, the United Kingdom, and others among the advanced industrial nations have typically set more modest goals. In recent years, governments in countries like Taiwan and South Korea have encouraged shifts from heavy industries (steel, shipbuilding) into knowledge-intensive sectors like electronics, emulating Japan more than the Europeans.

In contrast, neither business nor government in the United States has seen much need for an industrial policy more comprehensive than we have traditionally had; after all, American companies for years remained the leaders in electronics and high technology generally. Lack of consensus on industrial policy matters shows in disputes over appropriate levels of government regulation, export controls intended to restrict technology transfer to the Soviet Union and its allies, and trade protection for troubled firms and industries (where consumer electronics, with a complex history of trade complaints dating back to 1968, has been a primary example). These policy debates have usually been conducted in either/or terms, accompanied by much posturing. Meanwhile, state governments have devised their own strategies for attracting high-technology business development—the competition over Microelectronics & Computer Technologies Corporation which was won by Texas is one recent example; that for General Motors' Saturn site is another.

U.S. skepticism concerning industrial policy comes in part from observations of the European experience. France wanted a strong computer industry but did not get it—nor did West Germany, which pursued a more market-oriented approach. Still, if industrial policies in the United States remain fragmented, attitudes in some cases have begun to shift—due in part to recognition of slipping international competitiveness. Antitrust enforcement under the Reagan administration, for instance, represents a marked break with past practices.

Further changes may be in store for the United States, but European governments, facing the prospect of falling decisively behind in electronics, are searching out new approaches more actively. In many respects, Europe has learned to live with U.S. investments and exports. With recent successes by Japanese companies coming at the expense of European more than American manufacturers, the Europeans have shown renewed interest in joint ventures and other forms of cooperation—e.g., the ESPRIT program (European Strategic Program of Research in Information Technology), organized under the auspices of the Commission of the European Communities. Past attempts to cooperate across European boundaries, such as the Unidata consortium, have been conspicuous failures; indeed, Europe's electronics companies have cooperated more readily with American or Japanese firms than with each other. ESPRIT has had a rocky start; still, Europe has by now lost so much ground that governments and corporate executives may finally be forced to put aside their differences and learn to work together. Even so, given that many European electronics firms have traditionally had good technology, while falling down in commercialization, ESPRIT—geared as it is to relatively fundamental research—could be a success without having much impact on competitive trends.

As pointed out above, government policies in electronics have often sprung from needs viewed as broader than the industry and its technology alone. In Japan, where concerted effort to build an "information economy" goes back to the 1960s, government has looked to electronics as the centerpiece of the nation's future economic structure—a structure emphasizing knowledge-intensive and research-intensive products and services. The comprehensive nature of Japanese industrial policies, exemplified by the "visions" of the Ministry of International Trade and Industry, distinguishes Japan's approach; neither the choice of policy tools nor the money spent on subsidies differs greatly in comparison with other countries.

Elsewhere in the Far East, governments have learned from Japan how valuable a weapon technology can be. Industrial policies in several of the Asian NICs, particularly those such as Taiwan with considerable pools of technical manpower, are aimed at moving away from dependence on imported technologies. Nations farther down the ladder of development, including many in the Middle East, have different objectives. Rather than technological self-sufficiency, their first priority has been to acquire the skills needed to operate and maintain imported equipment. Countries such as Saudi Arabia, which may never produce electronics for export, stress manpower training and technical services in their dealings with more advanced nations.

INTERNATIONAL TRADE

In trade policy, the "rules of the game" have become the issue of the 1980s. The body of laws and regulations governing international trade evolved with objectives such as protecting local industries from imports, particularly those "unfairly" traded, while encouraging domestic companies to export. In the past, governments could draw sharp lines between "their" companies and foreign competitors. Export subsidies flowed to the former; tariffs and anti-dumping laws were directed at the latter. Today, lines blur. Three-quarters of U.S. semiconductor imports consist of intracorporate shipments by American-owned firms. In Japan, IBM's local subsidiary remains the leading exporter of computers.

In such a world, questions of national interest can easily become ambiguous. Policies directed at foreign producers may hurt domestic firms, as the United States found when quotas limiting imports of color TVs from Taiwan cut into shipments by Zenith and RCA, both of which had extensive operations there. Should the U.S. government decide to restrict imports of semiconductors from Japan, it would quickly find Japanese companies shifting production, not only to the United States, but to export platforms in the same countries favored by American firms for offshore assembly.

A second quandry for trade policy arises as more countries adopt relatively explicit industrial policies. Often pursued with goals at least ostensibly domestic, which tends to remove them from the reach of traditional trade remedies, industrial policies nonetheless have distinct impacts on trade patterns. Research and development subsidies are typical; they can help local companies design and develop products that will meet with success overseas as well as in the home market. (For years, other nations have complained that U.S. defense spending subsidizes commercial production.) As joint ventures and multinational production spread in industries ranging from electronics to petrochemicals to automobiles, with companies competing aggressively in some parts of the world while cooperating elsewhere, the policy environment grows more tangled. Adapting national and international trade laws and regulations to this environment poses a major challenge. Given the consequences of a breakdown in the world trading system, the stakes are high.

EMPLOYMENT: IMPACTS OF TECHNOLOGICAL CHANGE AND MULTINATIONAL PRODUCTION

Do foreign investments by American electronics firms cost large numbers of U.S. jobs? How about technological change as exemplified by cheap computing power? Will manufacturing continue to migrate to developing nations, leaving advanced economies with increasingly stratified labor forces—engineers and managers near the top of a wage pyramid dominated by poorly paid service-sector workers? Will underemployment rise, as societies struggle to find new ways to help people occupy their time? Such questions have no simple or direct answers, but they are among the most critical facing governments both here and abroad (**6**).

In the past, technological change created more jobs in the aggregate than it destroyed: The motor car displaced carriage makers and stable hands, but automobile production grew to employ more than a million Americans, with many more finding jobs making steel or tires, or building roads and suburbs. The entertainment industry expanded with the advent of radio and television; people found work in broadcasting and advertising as well as manufacturing, sales, and service. Will such trends hold in the future? Unfortunately, there can be no guarantees.

During the 1950s, people worried that computers would take over the workplace, creating massive unemployment. So far that has not happened—but such fears have resurfaced, fueled in part by the publicity given robots.

Certainly, advances in electronics have led to vast changes in the way we work. Quartz-crystal technology plus competition from the Far East decimated Switzerland's watch industry. Keypunch operators have virtually disappeared. Moreover, it is plain that high-technology employment—e.g., in computer manufacturing—will not rise as fast as output itself; this is true even in the absence of overseas production. Part of the reason is simply that labor productivity tends to grow rapidly in the technologically dynamic industries. When productivity increases, jobs will be lost unless output goes up even faster. In the U.S. color TV industry, the number of blue-collar workers dropped by half between 1971 and 1981, even though output doubled. As productivity rises in rapidly expanding high-technology industries, total employment may go up, but not as rapidly as output. When slowly growing industries such as textiles and apparel automate, employment typically declines.

International dispersion of production adds another dimension. Today, American semiconductor firms employ many more production workers overseas than at home. Multinational production is becoming the rule in sectors like automobiles as well. For such reasons, the United States finds itself losing unskilled and semiskilled jobs at a time when overall unemployment is already high. Unemployment always concentrates in the ranks of those with low skills. Where will they find work in the future? One point seems clear: The United States must reinvigorate its approach to education and training, including retraining, if people are to find satisfying jobs and the labor force is to keep pace with the demands of new technologies.

The United States is not alone in these dilemmas. Most of the European nations have experienced steadily rising unemployment for a decade and more. As developing countries become cost-competitive producers of goods that were once mainstays of the advanced economies, the latter—with their high wage rates and living standards—have to adjust at a pace that strains both people and institutions. Meanwhile, developing nations worry about their own ability to move upscale, to learn from Japan and the West while moving away from dependence on imported technology and capital. Some feel exploited by foreign investors and foreign governments; others will take jobs any way they can get them. Most have learned to drive hard bargains with overseas companies and governments. Is it any wonder that doomsayers forecast trade wars, renewed economic isolationism, even a collapse of the world economy?

CONCLUSION

What then of the future? Are the doomsayers likely to be right? Certainly if we have learned one thing from the postwar history of the electronics industry, it is to be wary of predictions. Neither the people who claimed the total market for computers would be a dozen or two, nor those only a few years later who said that computers would take away the jobs of many Americans, have been close to the mark. Electronics has brought new industries, new forms of entertainment, new ways of waging war—to all of which we have so far managed to accommodate ourselves.

Other countries are striving to catch up with the United States, or to avoid being left behind. Competitive pressures from both Japan and the NICs indicate that, if the United States is to maintain its position, we must continue to push for technological advantages (7). This has been our edge. The burden rests primarily on the private sector, but within government a more coherent approach to policies affecting industry—embracing human resources development (education, training, and retraining), support for technology development (including cooperative efforts and diffusion to U.S. industry), and economic adjustment—would help in coping with change. To implement such a policy effectively, the United States must take a strategic view of technology in its role as a driving force for the nation's economy.

REFERENCES

1. International Competitiveness in Electronics. Washington, D.C.: Office of Technology Assessment, November 1983. (This report is the primary basis for the paper.)

2. Gaps in Technology: Electronic Computers. Paris: Organization for Economic Cooperation and Development, 1969.

3. J. A. Alic and R. R. Miller. "Export Strategies in the Computer Industry: Japan and the United States." Strategic Computing: Defense Research and Computer Technology. Edited by P. Edwards and R. Gordon, forthcoming.

4. "Strategic Computing." Washington, D.C.: Defense Advanced Research Projects Agency, October 28, 1983. (This section of the paper also draws heavily on Reference 3.)

5. "Outline of Research and Development Plans for Fifth Generation Computer Systems." Tokyo, Japan: Japan Information Processing Development Center, Institute for New Generation Computer Technology, May 1982.

6. J. A. Alic and M. C. Harris. "Employment Lessons from the U.S. Electronics Industry." Stuttgart, Federal Republic of Germany: Proceedings of the 2nd International Conference on Human Factors in Manufacturing, June 1985, p. 9.

7. "Development and Diffusion of Commercial Technology: Should the Federal Government Redefine Its Role?" Washington, DC: Office of Technology Assessment, Staff Memorandum, March 1984.

QUESTION-AND-ANSWER SESSION FOLLOWING PRESENTATION

MR. JOE WHARTON, EPRI: You mentioned the fifth generation. Do you think the Feigenbaum idea that the Japanese, by this coordinated government-supported effort, will beat us to the fifth generation is not the case for this particular technology? Could you comment on whether that is going to be the case and whether that kind of effort, in general, has proven to be important to the centralized planning source and whether it is necessary in the future.

MR. ALIC: The Japanese have, for roughly 20 years, been pursuing joint government/industry research and development projects, of which the fifth generation is one of the more recent and certainly one of the most heavily publicized. Some of these have been pretty successful; some of them have been failures. The Japanese have been, in essence, going down a learning curve, and they are doing a better job today than they were of making these things work. So, I don't want to say the fifth-generation project is going to be unsuccessful.

I think there is a substantial probability that it will have an important stimulating effect on the Japanese industry, even if it does not reach its technological goals. Because one of the things the Japanese do with their joint government/industry research associations is this: They try to overcome some of the bottlenecks in their own economic structure, and those bottlenecks have to do with things like limited personnel mobility. They don't have the people moving from company to company or simply communicating with one another informally as is characteristic of the technical communities in the United States. They have, I think, more difficulty in managing some of the transfers from research to product development. They have used these joint government/industry associations to overcome a lot of specific problems.

For example, they had to make a transition from integrated circuits intended for consumer products to digital ICs for computers back in the '70s, and they did that, in part, with the help of a project called the VLSI project—the Very Large-Scale Integrated Circuit project. That was successful, not so much for the technology it developed, but for its more indirect effects in pushing the knowledge out to the companies and encouraging them, giving them indirect support. There is a lot of cheerleading involved.

So, I think the Japanese have a system that works pretty well for them. The message I would leave you with, in terms of the fifth-generation project, is that its budget is dwarfed by the budgets of individual Japanese computer companies, much less the individual American computer companies; that the objectives, in a technical sense, are much the same as those of other countries in other parts of the world; and that there are few secrets in technology development and commercialization, including the results of projects like that in Japan.

MR. DAVID GOLDFARB, Georgia Power Company: We seem to have a virtual monopoly of the software industry, at least up to this point. Do you see any future competition from either Europe or Japan in this area?

MR. ALIC: That is, of course, one of the areas that the fifth-generation program in Japan is pointed at. They are well behind in software; they need to get better, because that is important for selling machines, and software will be more important in the future than it is now and has been in the past. I think this is a field where it is wrong to expect that we can stay ahead forever, just as it was wrong in microelectronics to think that the United States could stay ahead, at least across the board, forever. So, I would expect that our margins of superiority in software will shrink in the future. But that does not mean that there aren't areas, particularly the more advanced kinds of software, for things like artificial intelligence, expert systems, and so on, or maybe a certain kind of networking, where we cannot maintain advantages. Both the Europeans and the Japanese, just in terms of the structure of their software industries, will have to start doing things differently, because they are still producing almost all of their software on a custom programming basis. That's grossly inefficient. So, it is almost inveitable that they will move away from custom software toward standardized packages, and, hence, that our advantages will decrease. But, again, I think this is an area where we should be able to keep useful margins over the longer term.

MR. RALPH FERRARO, EPRI: I wonder if you could comment on two recent announcements in The Wall Street Journal, where AT&T announced a megabit chip being available sometime next year for prototype samples. At the end of the same article, they announced that Toshiba would have commercially available one-megabit chips within the next six months.

Your presentation focused mostly on microelectronic semiconductors. The second comment in The Wall Street Journal was on power semiconductors and the joint venture being established by Westinghouse, General Electric, and Mitsubishi Electric Corporation to take on the power semiconductor device industry.

MR. ALIC: There are a couple of things that strike me in your first question about the megabit chip. The first point I would make is: I think we are going to see continued very intense competition in that part of the market, and that is a part of the market, by the way, that AT&T has been trying to crack for several years and really hasn't had too much success so far. They may make it and they may not.

But, even more than that, the point that I would emphasize is this: The real competition isn't only in chips, it is in utilization. It is in things like software systems and it is in what you do with chips. But system architectures are far, far more important—you can have the biggest and best chip—and systems applications is an area in which I think the United States is, generally speaking, ahead and will continue to be ahead, although there are some exceptions—perhaps factory automation.

Now, on your second question, I think the most interesting aspect is that it is part of a much broader trend in which American companies are cooperating technologically and on a commercial basis with foreign companies. There is evidence, first of all, that we are not way ahead in our technology, we are more or less on a par with our technology, and so other countries have something to bring to the table, so to speak. We aren't selling technology, in all cases, as we once were. We may sell it, but, in many cases, now we are trading it, or we are bringing it to a joint venture, to a coproduction deal, technology exchange, technology sharing.

You can see that in many, many industries. Again, it is part of the broader picture of structural change in the world economy, and I can't resist commenting that we are also seeing some businesses in some industries filing trade complaints on the one hand and cooperating with those same rivals on the other hand.

MR. AHMAD FARUQUI, EPRI: To what extent is the "foreign trade problem" facing the electronics industry due to exchange rate imbalance versus real cost competitiveness types of issues? I ask that in the context of some of the other presentations where the impression some of the speakers left with us was that once the exchange rates—macroeconomic types of problems—were corrected, then

American industry would be as competitive as any other in the world.

MR. ALIC: That is a very important question. And I'm sorry I haven't been here for the other presentations or I probably would have tried to address it earlier.

The kinds of structural changes, competitive shifts, and competitive problems that I have been outlining can, in most cases, be traced back 15 to 20 years. The patterns are present when the dollar has been strong and when the dollar has been weak. Certainly, the strong dollar has aggravated the competitive difficulties of many American industries, virtually all American industries, but the situation we are in is a new one. We aren't going to return to happy days when the dollar drops. That's the message. This is a much, much deeper set of changes.

That is just inarguable. All you have to do is look at what's happening in the industries themselves instead of focusing on economic statistics and it comes through loud and clear.

19

SECTORAL SHIFT AND INDUSTRIAL ENERGY DEMAND: WHAT HAVE WE LEARNED?

Hillard G. Huntington[1]
Stanford University

John G. Myers[1]
Southern Illinois University

INTRODUCTION

Energy use per unit of output in manufacturing has fallen substantially since the 1973 oil embargo. This trend holds for total energy, for fossil fuels, and for electricity. In the case of fossil fuels, this decrease represented an

[1]We would like to acknowledge the insightful discussions of the EMF 8 working group that provided the basis for this paper. We have benefitted from the detailed comments and suggestions of Gale Boyd, Ahmad Faruqui, Peter Lilienthal, Bob Marlay, Marc Ross, Blair Swezey, Paul Werbos, John Weyant, David Wood, and Frances Wood.

acceleration of the pre-embargo trend of declining fuel intensity. For electricity, on the other hand, the post-embargo trend represented a reversal of the increasing electrification observed before 1973.

A significant portion of this decline was due to shifts among sectors of the economy, among industries (measured at various levels of disaggregation), and among products. At a minimum, these changes in the composition of economic output, or product mix, account for at least one-third of the decline in energy use per unit of output in manufacturing since the embargo. We will refer to such shifts among sectors within manufacturing as the sectoral-shift effect.

In addition, energy use per unit of output within each sector has decreased because firms have adopted new processes (e.g., the electric-arc furnace in steel production) or have improved their operation and maintenance of existing facilities. We will refer to trends in the intensity of energy use within sectors as the change in sectoral energy intensity.[2] Together these two effects determine the trend in the energy-output ratio in manufacturing as a whole, or aggregate energy intensity.

This paper focuses on what is currently known about the

---

[2]Since we choose to divide all changes in aggregate energy intensity into two categories, "sectoral shifts" and "changes in sectoral intensity," it is not surprising that only one of the terms can be used concisely. The focus here is on sectoral shifts. Some changes in aggregate energy intensity may be labeled "changes in sectoral intensity" that one would not necessarily categorize as such. In particular, sectoral shift depends on the definition of "sector." Any shifts within sectors at aggregation levels that are finer than those used by the analyst would be mislabeled as a "change in sectoral intensity." Examples would be the shift away from aluminum within the broadly defined primary metals group or the shift toward higher-valued products within the chemicals group. Thus, the use of the term "changes in sectoral intensity" should be viewed as a convenient catch-all for factors not explicitly quantified as "sectoral shifts."

sectoral shift effect and its influence on fuel and electricity use patterns in the United States. Our goals are twofold. First, we want to document the importance of this effect in the post-embargo period by reviewing a number of studies conducted on this topic. And second, we want to highlight the role of sectoral shift in forecasting and understanding future energy demand trends in manufacturing. Although much of the Energy Modeling Forum (EMF) working group's efforts on industrial energy demand has been directed toward comparing different projections of future energy use, the study has also provided a forum for probing what happened during the previous decade as well.

While a consensus has not yet developed on the causes of the decline in major energy-using sectors, it appears clear that future energy demand will be strongly influenced by trends relating to the transformation of the economy. It is important to recognize sectoral shift's distinct role in shaping aggregate energy intensity. If current economic trends remain intact, shifts within industry will tend to keep growth in energy demand low. However, low energy prices, reduced capital costs, and more rapid growth, if they occurred together, could improve the prospects for important energy-using industries, partially countering the energy trends observed during the previous decades. Moreover, changes in the mix of goods produced could alter energy use relatively rapidly, compared to the slower process of replacing the capital stock within individual industries.

We begin in the next section by explaining the concept of sectoral shift and why it is important. Section 3 describes a methodology for separating this effect from changes in sectoral intensity. Section 4 summarizes the findings of previous studies on the importance of this effect during the post-embargo period, including a recently completed study by Argonne National Laboratory that was prepared concurrently with the working group's efforts. Section 5 highlights the role of sectoral shift in the EMF projections of industrial energy use. We conclude this paper with a section on further disaggregation of the reductions in energy use per dollar of output that are associated with changes in sectoral energy intensity, as projected by the models participating in the EMF study.

DEFINING THE SECTORAL-SHIFT EFFECT

Sectoral shift is a change in the relative economic importance of the individual sectors that constitute industry in the aggregate. Importance here is measured by the value of output after adjusting for increases in prices over time (i.e., output measured in constant prices). For example, within the aggregate "manufacturing" sector, the rising importance of electrical machinery has been accompanied by the waning importance of primary metals in terms of output produced in each industry.

Shifts in the mix of industries are important because they account for a substantial fraction of the decline in the energy-output ratio since 1973, regardless of the level of aggregation one uses to represent this effect. As a lower-bound estimate, this effect appears to account for at least one-third of the decline in fuel use per dollar of output and well over one-half of the decline in electricity use per dollar of output.

In general, the finer the decomposition of an aggregate, the larger the fraction of the decline in the energy-output ratio that can be attributed to sectoral shift. Thus, a study that disaggregates all industry into many sectors or even individual products can generally be expected to show a larger industrial shift effect because more of the shift away from energy-intensive goods is being incorporated by the sectoral shift measure. However, even if one decomposes all industry into only two sectors, a group of industries producing materials for other industries and all remaining sectors, this effect remains important.

Shifts between sectors can raise as well as lower the aggregate energy-output ratio in manufacturing. While this effect has reduced the energy-output ratio in the post-embargo period, new conditions could emerge that would reverse these effects over time. Thus, it is inappropriate to consider these trends as permanent structural shifts in our economy that cannot be reversed. We have avoided the term "structural change" for precisely this reason.

Reductions in the energy-output ratio are sometimes considered to be improvements in energy productivity because more output is being produced with a given amount of energy. Energy productivity in this context is an elusive concept that can occasionally mislead one about important trends. As the previous discussion suggests, improved energy productivity can occur simply as the composition of output shifts from more to less energy-intensive sectors. Careful measurement of the sectoral shift effect can clarify what fraction of the gains in energy intensity are due to changes in processes or technologies within individual sectors as opposed to shifts among industries.

Energy intensity is also an elusive concept for other reasons, which are not directly addressed with better measures of industrial shift. Some reductions in energy use may result only because industry is using more labor or capital. For example, a firm could hire an extra worker to turn off lights to conserve energy. If output is unchanged, energy intensity would be reduced but labor intensity would increase. In this case, an improvement in energy productivity leads to a deterioration in the productivity of labor. A similar situation arises when firms install more expensive capital equipment that also uses less energy.

For this reason, decreases in energy intensity, either for aggregate manufacturing or for individual sectors, should not be assumed to be improvements in energy efficiency. This latter issue should be examined within the context of identifying the gains in efficiency related to all factors. (Total factor productivity change can be defined as the increase in output after adjusting for the contributions of all factor inputs.)

MEASURING THE SECTORAL-SHIFT EFFECT

Conceptually, the sectoral shift effect can be measured by allowing the output mix in manufacturing to change but holding constant the energy intensity within each industry (i.e., energy use per dollar of output). Any observed decline in the aggregate energy-output ratio would then reflect shifts among sectors rather than changes in production processes within the individual sectors. For analogous reasons, the change in sectoral intensity can be measured by allowing energy use within sectors to vary but holding the output mix constant.

Index number computations permit the decomposition of changes in a time series into its components. To decompose changes in energy consumption for aggregate manufacturing, an analyst could weigh output in each sector by its energy use, before summing across sectors. Variations in this energy-weighted output series would reflect changes in the relative importance of the more energy-intensive industries; change in energy use patterns within sectors would not contribute to movements in this index.

MEASUREMENT OF OUTPUT

Several problems deserve explicit considerations. Energy-output trends will be sensitive to how one measures output in the individual sectors. One source of the differences in the estimated sectoral shift effect during the post-embargo period has been that the studies have used different measures of economic activity. Three widely used measures of activity—value added, gross output, and Federal Reserve Board (FRB) production indices—have different implications for energy-output ratios. Moreover, all three concepts may differ substantially from pure physical measures of output, e.g., tons of steel.

MEASUREMENT OF ENERGY

Inputs must also be measured carefully in representing these trends. Electricity is a higher-valued energy form with many uses that cannot be met by oil and gas. When electricity replaces oil and gas in the aggregate energy picture, an additional BTU of electricity will generally be more productive than the BTU of fossil fuel that is "replaced," i.e., it can sustain a higher value of economic activity for each BTU. If energy is aggregated on the basis of BTUs, all energy sources are being viewed as equivalent replacements for each other in terms of BTUs. This procedure ignores the greater productivity of a BTU of electricity. A similar problem exists when oil or gas replaces coal.

Aggregating energy on the basis of relative prices is usually preferred when one wants to adjust for differences among fuels in terms of their economic value. When this is done, the decline in energy intensity during the post-embargo period is noticeably less than that revealed when energy is aggregated on the basis of BTUs.[3] This result

---

[3] Analyses conducted by Hogan (1) and the Energy

reflects the fact that the price-weighted measure gives more importance to electricity, the use of which rose relative to other energy sources during the post-embargo period, and less importance to coal, which decreased during the post-embargo period.

Foreign trade can also affect the measurement of energy in this ratio. When semi-processed materials are imported, the direct energy use does not appear in the accounting for U.S. industry. The energy-output ratio would decline even though processes within each industry have not changed.

CHOICE OF INDEX

And finally, there exists a technical problem in constructing indices that will accurately decompose the two effects of sectoral shift and changing sectoral intensity. For many indices, it is possible that the results will depend upon whether the sectoral shift effect is measured directly or whether it is calculated as a residual after first measuring the contribution from the change in sectoral intensity. This situation arises when there are important differences in sectoral intensity trends among the various industry groups.

Economists have found the Divisia Index to be a useful tool for eliminating the above decomposition problem as well as for aggregating across diverse components of an input like energy. This index is a weighted sum of growth rates. In the case of aggregate energy intensity, the growth rates are those of energy intensity in each sector and the relative economic importance of each sector. The weight in each sector will depend upon whether one is disaggregating the energy in terms of BTUs or expenditures. In the former, the weight is the sector's share of energy use in BTUs, while in the latter it is the cost share of energy for that sector. This index was chosen by the Argonne researchers in a study

Information Administration (2) clearly demonstrate this result. Even if energy is measured in BTUs, the computation of energy-output ratios can be greatly affected by whether energy is being measured in primary or end-use BTUs. The electrification trend can result in either a decrease or an increase in electricity intensity, depending upon which measurement is used.

discussed in the next section as well as for reporting the EMF projections on this issue. This approach is discussed in greater depth in the appendix.

THE SECTORAL-SHIFT EFFECT DURING THE POST-EMBARGO PERIOD

All past studies investigating this issue have concluded that the sectoral shift effect has been important in determining the observed energy-output ratio during the post-embargo period. Table 19-1 summarizes the studies' findings regarding the importance of shifts in the composition of output as a determinant of the decline in the ratio. It should be emphasized that no two studies are strictly comparable to each other because they frequently are based upon different methodologies, data bases, etc. Studies can differ from each other by employing different levels of industrial aggregation, different indices for aggregating the sectors, or different time periods. In addition, some studies analyzed fuels and electricity separately while others combined them as simply energy.

TRENDS IN ENERGY INTENSITY CHANGES

Despite these fundamental differences in approach, there appears to be a consensus on the approximate magnitude of the effect. Given the results reported in Table 19-1, many of the studies suggest that perhaps one-third of the decline in the energy-output ratio can be attributed to this effect. And under certain conditions, the relative importance of this effect is considerably larger.

The sectoral shift effect was relatively minor during the few years just prior to the embargo, as revealed by the first entry for Myers and Nakamura (**3**) in Table 19-1. In recent years, however, the role of sectoral shift in energy-output trends has increased dramatically, especially since 1980. This trend is observable for the Marlay (**4, 5, 6**) results, which are based upon a very detailed disaggregation of industry, as well as for several studies conducted by Werbos (**7, 8**).

The estimates in Table 19-1 refer to the relative importance of shifts between sectors within manufacturing. In

TABLE 19-1
Relative Importance of Composition Effect (Sectoral Shift)[a] in Previous Studies of the Post-Embargo Period

Study	No. of Sectors	Decline in Energy Intensity (% change) (1)	Composition Effect (% change) (2)	Relative Importance of Composition Effect (%) (3)=(1)/(2)
Myers & Nakamura	2			
Energy 1967-76		10.8	0.6	6
Energy 1974-76		4.2	1.4	33
Samuels, Vogt, & Evans	448			
Energy, 1975-80		18.8	4.4	23
Fuels, 1975-80		20.9	5.1	24
Electricity, 1975-80		8.6	0.7	8
Marlay	475			
Energy, 1972-80		16.5	5.9	36
Energy, 1972-84		34.4	18.5	53
Fuels, 1972-84		42.4	21.2	50
Electricity, 1972-84		15.0	10.0	67
Werbos	18			
Energy, 1974-81[b]		17.2	5.7	33
Jenne & Cattell				
U.K. Energy, 1973-78	9	12.5	4.2	34
	104	12.5	6.8	54
Boyd et al.	20			
Fuels, 1972-81		20.5	7.0	34
Roop & Belzer	NA			
Energy, 1972-77		21.0	9.7	45
Fuels, 1972-77		21.4	9.6	45
Fujime	NA			
Japanese Energy, 1974-83		NA	NA	33

Source: See Boyd et al. (1985) for all estimates except Marlay (1985), Roop & Belzer (1986), and Fujime (1985).
NA = Not available.

[a]Based upon sectoral shifts and energy-intensity changes within manufacturing. Additional shifts may occur between manufacturing and other sectors.

[b]Fuel-switching effects have been netted out and feedstocks are excluded. Composition effect increases to 59 percent of total effect when shifts between manufacturing and other sectors (including services) are included for the 1974-81 period. When feedstocks and shifts within industry (i.e., including agriculture, construction, and mining) are included, the sectoral-shift effect increases to 75 percent of the total effect for the 1974-82 period.

addition, there will be compositional shifts between manufacturing and the rest of the economy. Werbos (**7, 8**) has found this latter shift to be important. When it is included, his estimates of the sectoral shift effect increases to 59 pecent of the total decline in energy intensity between 1974 and 1981. Moreover, its relative importance increases to 75 percent when 1982 is added and feedstocks are included.

CHANGES IN THE TREND FOR FUEL INTENSITY

Sectoral shift accounts for an even larger proportion of the change in the trend of the energy-output ratio from just before to after the embargo. Since all studies did not report estimates of both pre- and post-embargo trends, we will discuss this point with reference to only one of the studies, that by Boyd et al. (**9**). This effort was conducted at Argonne National Laboratory concurrently with the EMF effort on comparing industrial energy models. It represents one of the more recent attempts to decompose the sectoral shift and technological change effects. It employed the Divisia Index described previously to analyze separately fuels and electricity in BTUs for the 1967-81 period. Its conclusions are similar in character to many of the other studies.

The fossil fuel-output ratio was already falling by about 1 percent per year during the years immediately preceding the embargo (1967-74).[4] Table 19-2 shows that this decline incorporates a 1.9 percent per year decline due to the change in sectoral intensity, but that this effect was offset partially by a 0.9 percent per year increase in aggregate energy intensity due to the output shift among two-digit SIC industries. During the post-embargo period (1974-81), the decline due to the change in sectoral intensity became more

---

[4]Some researchers have found the 1967-74 trends in energy intensity to differ from trends during previous periods. (Personal correspondence from Blair Swezey on a forthcoming EPRI study not yet completed.) For convenience, we will continue to refer to the 1967-74 results of Boyd et al. as pre-embargo, although the reader should keep this caveat in mind.

TABLE 19-2
Trends in Specific Energy Intensity and Product Mix (Percent per Year) (Industry Disaggregated at Two-Digit SIC Level)

	Fuel	Electricity
Specific Energy Intensity		
Pre-embargo[a]	-1.9	+2.0
Post-embargo[b]	-2.9	0
Difference	-1.0	0[c]
Sectoral Shift		
Pre-embargo[a]	+0.9	+0.8
Post-embargo[b]	-1.0	-1.3
Difference	-1.9	-2.1

Source: Argonne National Laboratory (1985).

[a] 1967-1974.
[b] 1974-1981.
[c] No precise estimate is available but the effect is negative.

pronounced (2.9 percent per year), which was reinforced by a 1.0 percent decline due to sectoral shift. Aggregate fuel intensity, the sum of these two effects, declined by 3.9 percent per year during this period.

Comparing the two periods, aggregate fuel intensity declined by 2.9 percent per year more during the post-embargo period than in the pre-embargo period. Sectoral shift accounted for about 66 percent (1.9 percentage points) of this difference in energy-output ratios between periods, as its contribution to aggregate intensity moved from being positive to negative. This result contrasts with trends for the post-embargo period (not the change between periods), during which sectoral shift represented about 26 percent (1.0 percentage points) of the annual decline of 3.9 percent in the aggregate energy-output ratio.[5]

[5] The results in Table 19-2 on the change in intensities between periods were based upon estimates where output was measured in physical terms rather than value added, so the estimates differ somewhat from those in Table 19-1.

CHANGES IN THE TREND FOR ELECTRICITY INTENSITY

The trends in electricity per dollar of output in Table 19-2 are somewhat different. Both effects together were causing aggregate electricity intensity within manufacturing to increase by a total of 2.8 percent per year during the pre-embargo period. Sectoral shift accounted for a yearly increase of 0.8 percent per year, while the remaining 2 percent was attributable to changes in sectoral intensity. After the embargo, aggregate electricity intensity reversed directions. The sectoral-shift effect became negative, contributing to a decline of 1.3 percent per year. The effect of the change in sectoral intensity was about neutral, causing little additional change in the aggregate electricity-output ratio for this period. Thus, virtually all of the decline in aggregate electricity intensity during the post-embargo period was attributable to sectoral shift.

In comparing the two periods, sectoral shift alone was causing the electricity intensity within manufacturing after the embargo to fall 2.1 percentage points per year below its pre-embargo trend. Changes in sectoral intensity within individual industries were also contributing to a slower growth in electricity intensity for all manufacturing, although the Argonne researchers were reluctant to estimate the magnitude of this effect because the estimates fluctuated widely from year to year. However, if one assigns the 2 percent implied in Table 19-2 to the sectoral intensity effect, the sectoral-shift effect would account for at least 50 percent of the 4.1 percentage-point decline in the aggregate ratio between periods.

A TWO-SECTOR DECOMPOSITION

The Argonne study also concluded that for fossil fuels, about 80 percent of the composition effect was attributed to shifts bewteen two major manufacturing sectors. Five industries (paper, chemicals, petroleum refining, primary metals, and stone, clay, and glass) share the common features of being highly energy-intensive as well as producers of raw materials for other industries. These five materials industries were separated from the rest of manufacturing in a two-sector analysis of the decline in aggregate fuel and electricity intensities. This finding suggests that detailed disaggregation of manufacturing may not be necessary to incorporate much of the sectoral-shift effect for fossil fuels. Therefore, one may not require a detailed set of economic projections by industry for forecasting fossil fuel consumption in manufacturing. A less expensive, more aggregate economic projection differentiating between the two major sectors may be sufficient for many purposes.

With regard to aggregate electricity intensity, the disaggregation into materials-producing and all other industries accounted for substantially less than the full sectoral-shift effect at the two-digit level. This result once again emphasizes the difference between the energy-intensity trends for electricity and fossil fuels.

CAUSES OF SECTORAL SHIFT

There has been substantially more research on documenting the extent of sectoral shift and its importance for energy consumption than on the underlying factors causing this shift. This issue should be given much more emphasis than has been afforded it in the past. Until this is done, it will be impossible to reconcile the wide range of economic projections that are currently available.

Some of the studies cited previously have offered possible explanations for the shift among sectors, although few have investigated the relative importance of these effects. These factors include: (1) higher energy prices, (2) a higher cost of capital, (3) slower economic growth, (4) shifting consumption due to changes in income, (5) technological advancements that reduce the use of materials, and (6) recent exchange rate shifts and increased foreign competition.

After testing several competing hypotheses, Werbos (**8**) concludes that the direct effects of energy price changes on output mix are small compared to the effects of a more slowly growing economy, in which investment and the demand for energy-intensive capital goods are lower. The Wharton projections discussed in a later section are also consistent with this explanation. In contrast, however, Jorgenson (**10**) finds a relatively large effect of energy price changes on shifts between major sectors of the economy for 1972-76, although these results were not disaggregated by industry within manufacturing. Given these disparate results, more effort should be directed toward assessing the relative importance of energy price changes and other factors in shaping the product mix within manufacturing.

SECTORAL SHIFT IN THE EMF PROJECTIONS

Four of the five models evaluated in the EMF study require external information about the economic growth of different industries in projecting industrial energy demand. The EMF working group sought to standardize assumptions about economic growth, including output mix effects, so that differences among models reflected differences in model behavior rather than in input assumptions. For this reason, economic scenarios from the Wharton annual model were used to provide the necessary inputs on gross output by industry in the reference case, as well as high and low oil prices and high and low GNP cases.

ECONOMIC PROJECTIONS FOR THE REFERENCE CASE

The Wharton projections of sectoral output reveal a continuing trend toward less energy-intensive sectors, although at a noticeably slower rate than during the historical period since the embargo. For example, four large energy-using industries (paper, chemicals, petroleum refining, and primary metals) experienced a decline of 3.3 percentage points in their collective share of manufacturing gross output between 1974 and 1981. This share declined only 0.5 percentage points during the first ten years in the Wharton projections covering the 1985-95 period.[6] The trends for the six key sectors plus aggregate industry that are projected by Wharton for the reference case are shown in Figure 19-1, where the 1985 level for each sector has been

[6] Wharton's economic series show the output share of the energy-intensive sector to decrease from 20.3 to 18.5 percent over the 1985-85 period. Thus, most of the decline from its 1981 share has already occurred by the beginning year of the projection horizon.

FIGURE 19-1
Gross Output by Industry, Wharton Reference Case (Gross Economic Output) Indexed to 1985

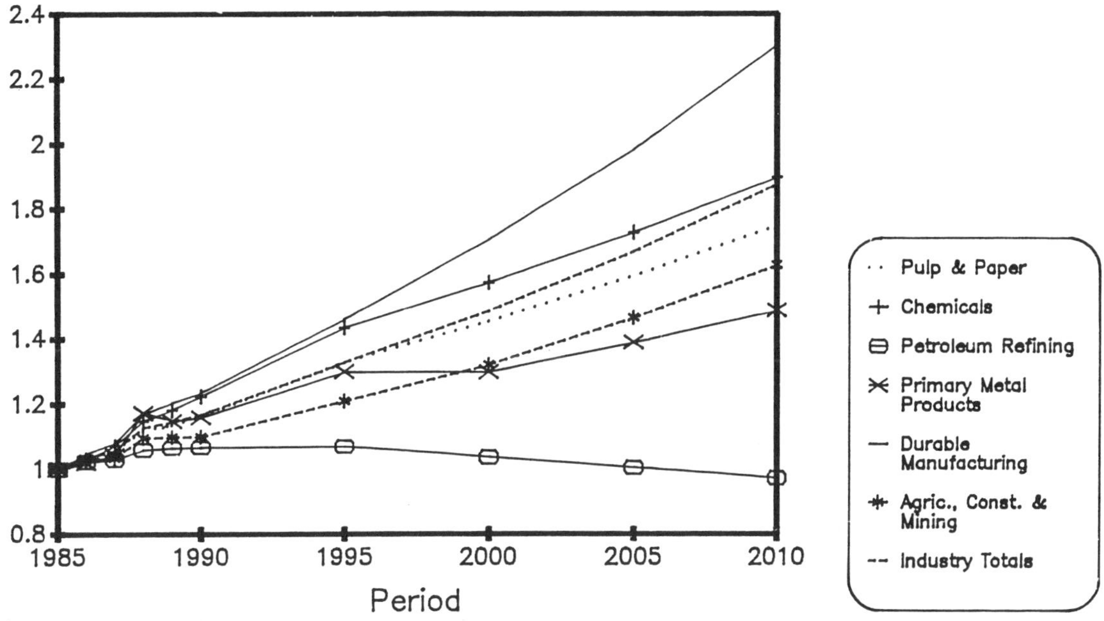

indexed to equal one. Only durable manufacturing, chemicals, and miscellaneous manufacturing (not shown), grow more rapidly than aggregate industry in these projections.

The Wharton economic projections assume relatively stable or smoothly growing oil prices, in stark contrast to the experiences of 1970. They also assume that the high value of the dollar does not continue through the period.

REFERENCE ENERGY DEMAND PROJECTIONS

Figures 19-2a through 2e show that aggregate fossil fuel use per dollar of output is projected by the five models to fall by 40 to 50 percent in the reference scenario over the next 25 years. These trends refer to purchased heat and power in manufacturing, which excludes the agriculture, construction, and mining sectors that are included in measures of

FIGURE 19-2
Change in Aggregate Fossil Fuel Intensity (Reference) and Change Due to Sectoral Shift (2 and 6 sectors)

FIGURE 19-2-a
PURHAPS Manufacturing Fossil Fuel (Div. by Gross Econ. Output)
Indexed to 1985

FIGURE 19-2-b
ORIM Manufacturing Fossil Fuel (Div. by Gross Econ. Output)
Indexed to 1985

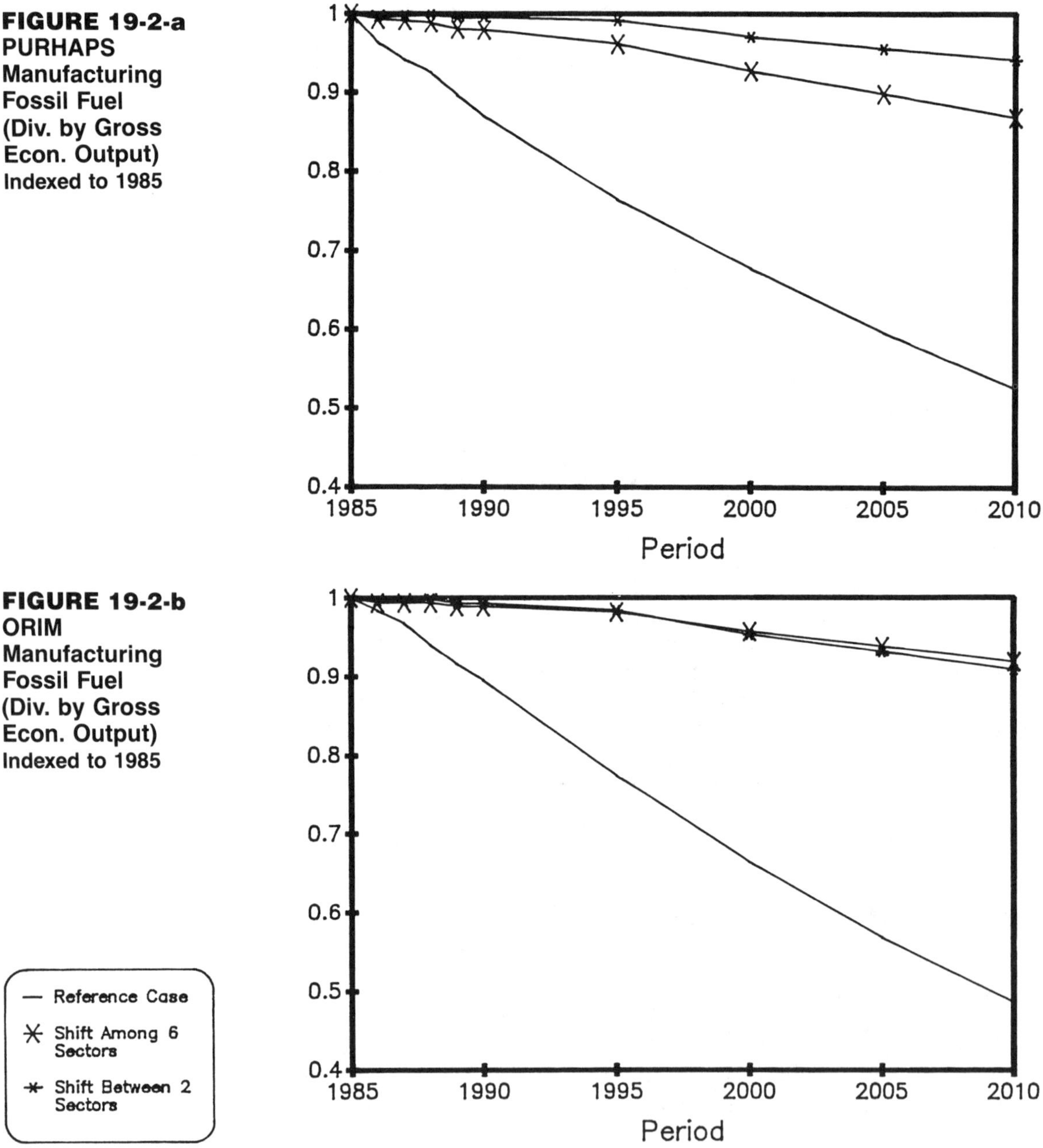

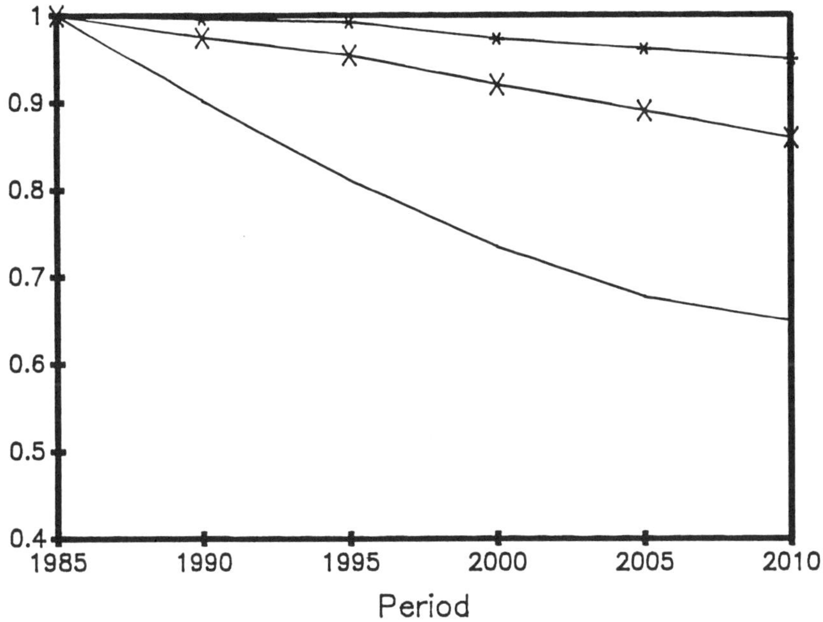

FIGURE 19-2-c
AES/ISTUM-I
Manufacturing
Fossil Fuel
(Div. by Gross
Econ. Output)
Indexed to 1985

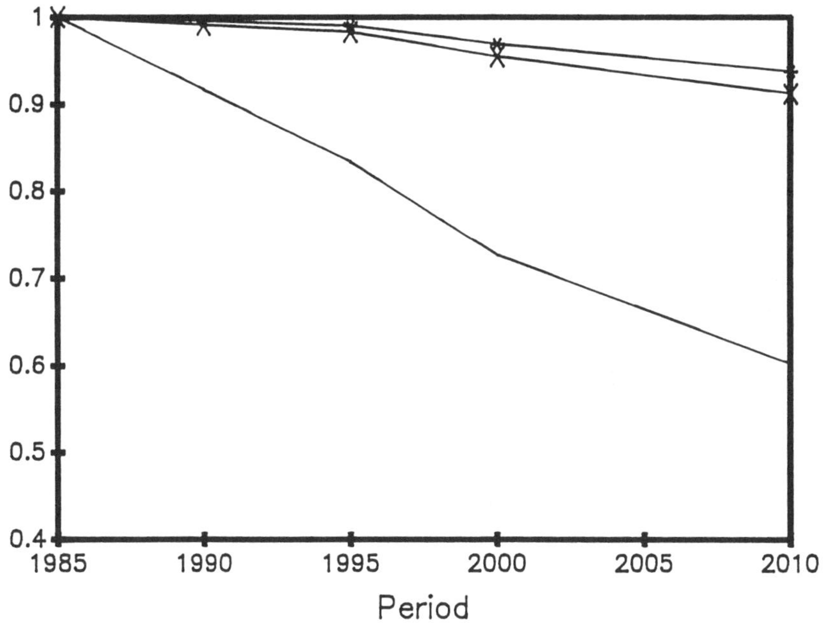

FIGURE 19-2-d
ISTUM-II
Manufacturing
Fossil Fuel
(Div. by Gross
Econ. Output)
Indexed to 1985

FIGURE 19-2-e
INFORUM Manufacturing Fossil Fuel (Div. by Gross Econ. Output, Indexed to 1985)

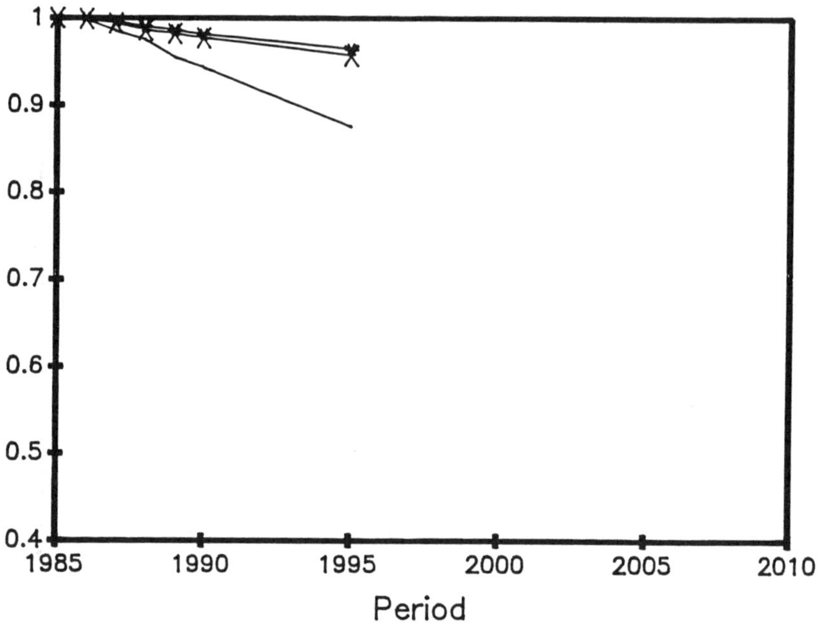

industrial energy use. The energy-output ratio for each model has been indexed to its 1985 value.

These figures also reveal that the decline in aggregate energy intensity attributed to the compositional shift in output between the seven major sectors[7] is much more modest—approximately 10 percent. Thus, sectoral shift accounts for about 20 to 25 percent of the decline in the projected energy-output ration in the EMF study. Moreover, the figures also represent the sectoral shift effect when manufacturing is disaggregated into only two sectors: the four major energy-using industries identified previously and all other industries. For most models, this more aggregate measure of sectoral shift incorporates most of the effect represented in the measure based upon the seven industry groups.

By contrast, Figures 19-3a through 3e show electricity use per dollar of output either rising or falling slightly during the next 25 years. Even when declining, however,

---

[7]Sectoral shift has been measured with the Divisia Index in these figures.

FIGURE 19-3
Change in Aggregate Electricity Intensity (Reference) and Change Due to Sectoral Shift (2 and 6 sectors)

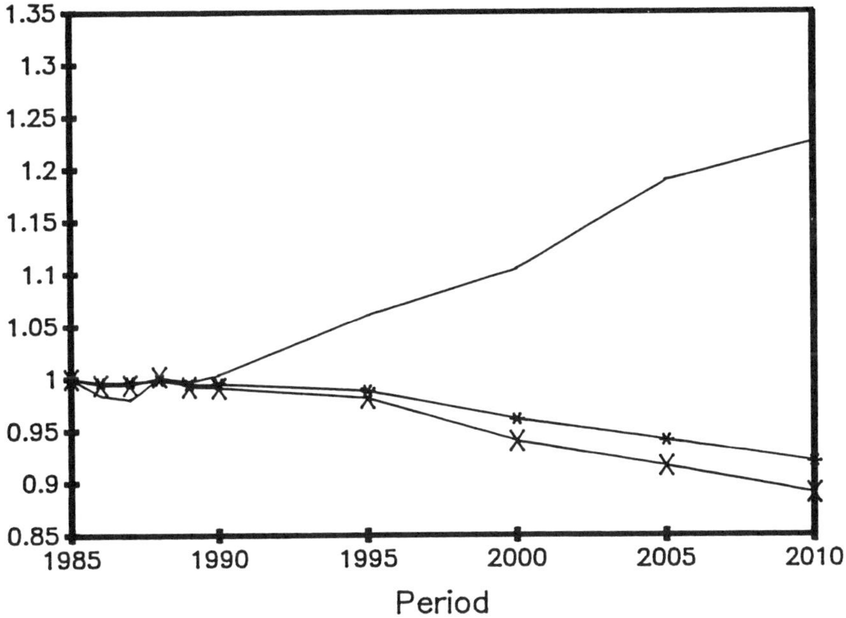

FIGURE 19-3-a
PURHAPS
Manufacturing Electricity Cons. (Div. by Gross Econ. Output)
Indexed to 1985

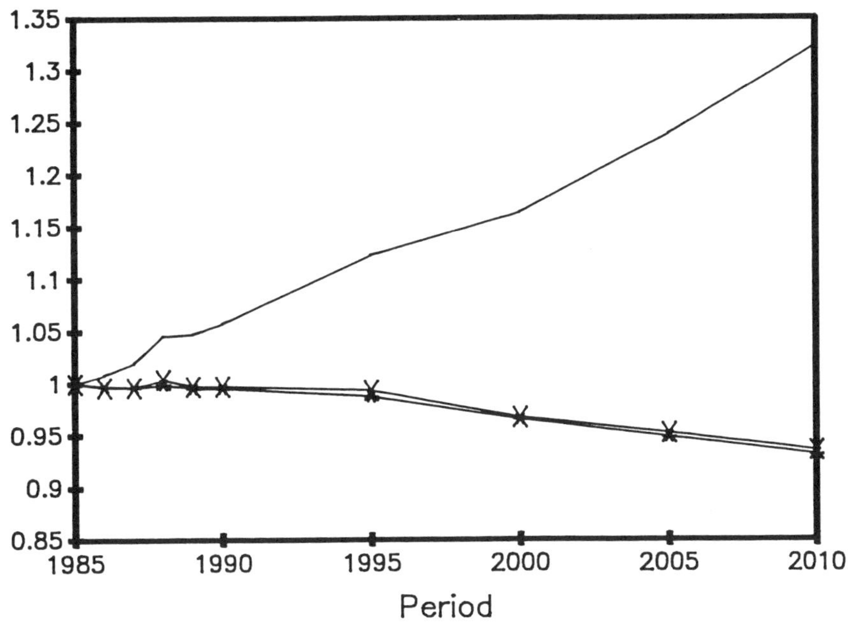

FIGURE 19-3-b
ORIM
Manufacturing Electricity Cons. (Div. by Gross Econ. Output)
Indexed to 1985

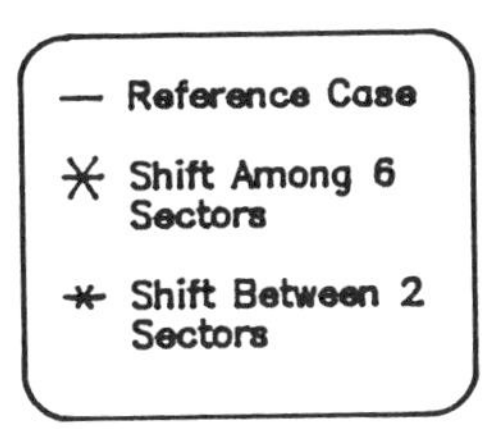

FIGURE 19-3-c
AES/ISTUM-I
Manufacturing
Electricity Cons.
(Div. by Gross
Econ. Output)
Indexed to 1985

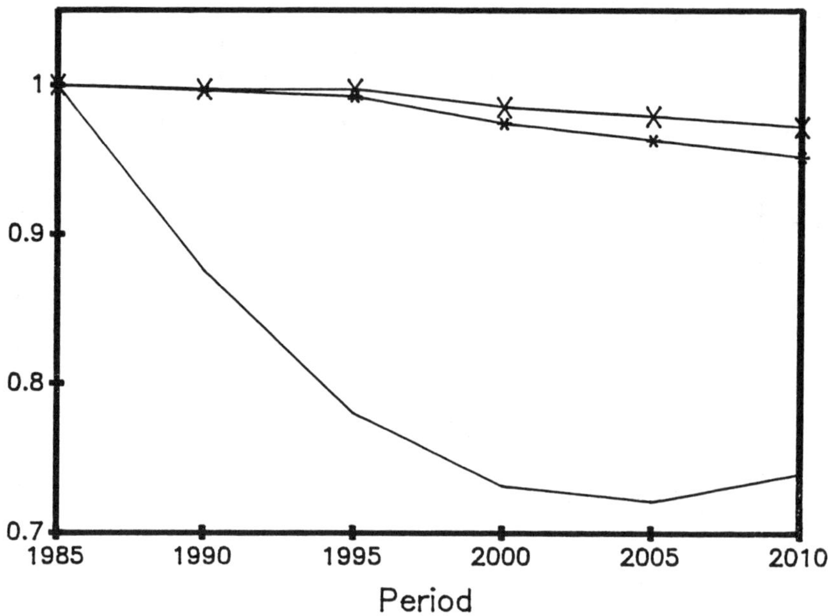

FIGURE 19-3-d
ISTUM-II
Manufacturing
Electricity Cons.
(Div. by Gross
Econ. Output)
Indexed to 1985

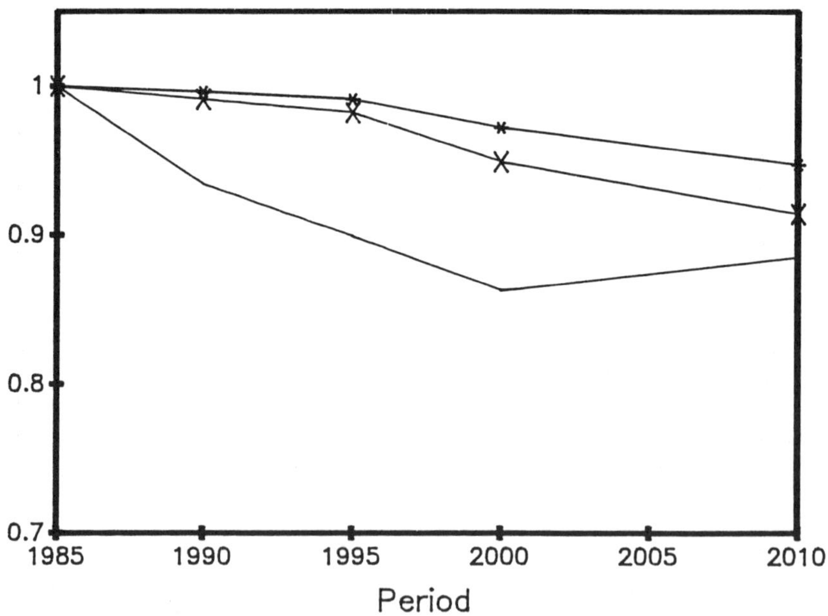

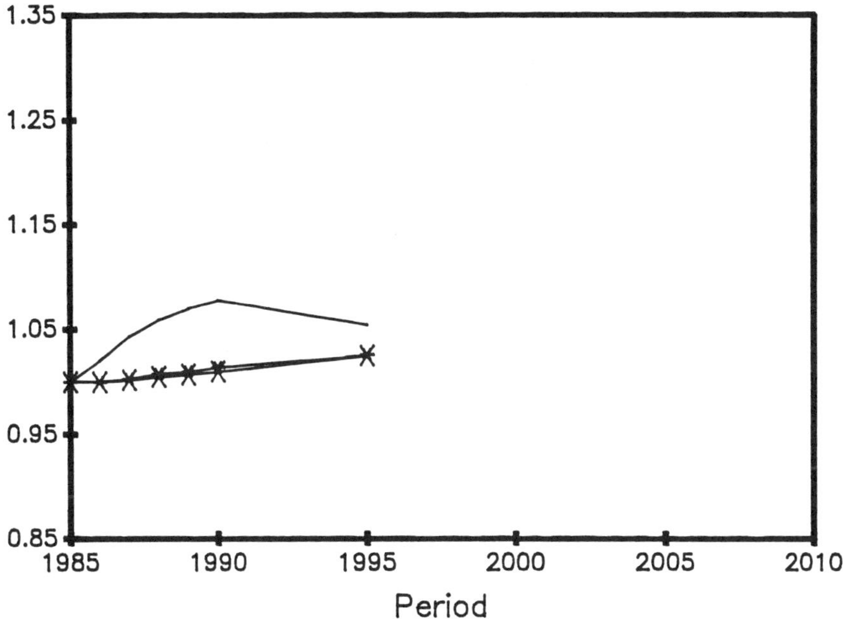

FIGURE 19-3-e INFORUM Manufacturing Electricity Cons. (Div. by Gross Econ. Output) Indexed to 1985

electricity intensities fall much more modestly than their fossil-fuel counterparts. The electricity-output ratios for ORIM and PURHAPS increase by about 25 to 30 percent over the full period, averaging about 1 percent per year. Some of the decline for the two process models, ISTUM-1 and ISTUM-2, reflects the penetration of cogeneration, which reduces purchased electricity but not necessarily consumed electricity. Even so, these models show electricity consumption per unit of output declining, as discussed in the main EMF report (forthcoming, 1986). The declining electrification trend in the process models occurs because increased efficiency in electric equipment more than offsets the penetration of new applications for electricity within manufacturing.[8]

All models indicate that the sectoral-shift effect, either at the six- or two-sector level of aggregation, operates to

---

[8] Analyzing several major electric-intensive industries during the 1973-81 period, Ross (**11**) found that increasing efficiency has offset the trend toward electrification, resulting in a slightly negative electricity-output ratio when output is measured in physical tons.

retard slightly the electrification trend. This compositional effect reduces the electricity output ratio by less than 10 percent over the period in all models.

ALTERNATIVE ECONOMIC PROJECTIONS

Despite its importance, there is little consensus on the causes of the significant compositional shift between industries. Similarly, it is uncertain to what extent this process will continue in the future. The resolution of this issue will clearly have important implications for future industrial energy demand.

The Wharton economic projections appear to be relatively insensitive to alternative energy price paths, provided that differences in energy prices occur gradually. Figure 19-4a reveals the effects of high oil prices on gross output in the six major sectors plus aggregate industry. Figure 19-4b shows the comparable effects of low oil prices. Except for petroleum refining, the trends in gross output in the high or low oil price cases appear almost indistinguishable from the reference case.

The small effect of gradual changes in energy price on the structure of the U.S. economy is not necessarily inconsistent with the experiences of the 1970s, during which both high oil prices and significant sectoral shift were observed. The energy price changes of the previous decade were both sudden and large; these price shocks also had significant indirect effects on the economy's structure through decreased economic growth and a rise in the interest rate that retarded capital accumulation and the premature obsolescence of the capital stock in different industries. In contrast, these indirect effects are not present in the Wharton projections of the economy with gradual changes in energy prices. Smooth energy price increases will still have direct effects on the output mix by raising the relative prices of more energy-intensive products and thereby reducing the consumption of these goods.

By contrast, the Wharton projections for industrial mix are much more sensitive to the state of the economy. A

FIGURE 19-4
Change in Gross Output by Industry, High and Low Oil Price Cases

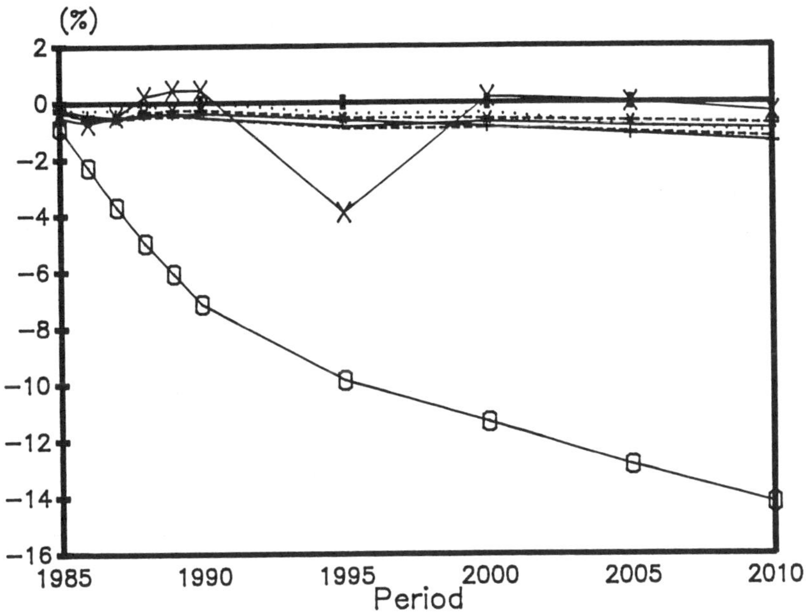

FIGURE 19-4-a
WHARTON
High Oil Price
(Pct. Diff. w.r.t. Reference Case)
Gross Econ. Output

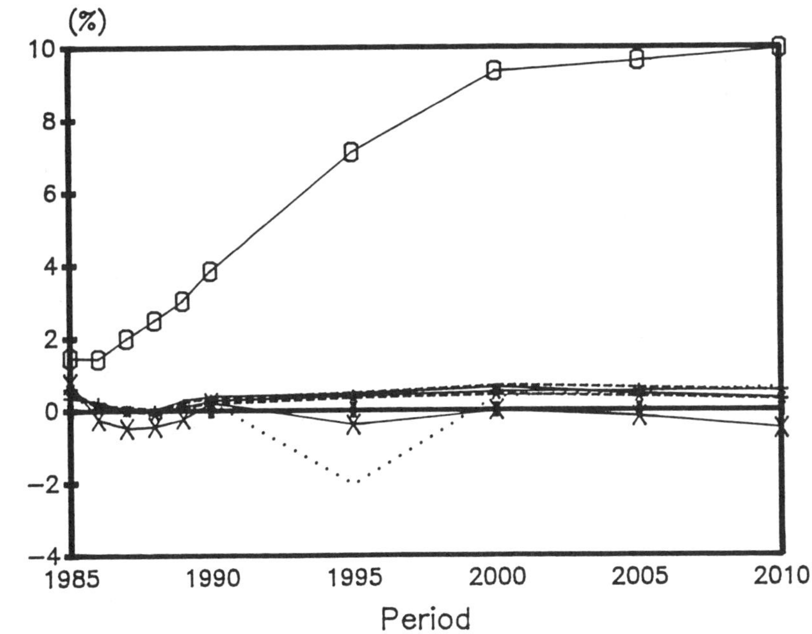

FIGURE 19-4-b
WHARTON
Low Oil Price
(Pct. Diff. w.r.t. Reference Case)
Gross Econ. Output

high GNP scenario produces a noticeable shift toward energy-intensive sectors (Figure 19-5a); a low GNP path is associated with a shift away from these sectors (Figure 19-5b). The primary explanation for this result is that faster growth requires a more than proportionate increase in investment, which raises the demand for energy-intensive capital goods. Moreover, the macroeconomic conditions conducive to a high-growth case result in a lower real interest rate, and hence lower capital costs, in the Wharton projections. Lower capital costs stimulate growth in capital-intensive sectors, which are also important industrial energy users.

ALTERNATIVE ENERGY DEMAND RESULTS

These shifts between sectors, however, would not obviate the conclusions about declining energy-output ratios in manufacturing derived from the reference case. Figures 19-6a through 6d show that fossil-fuel use per dollar of output falls for the three models shown, even in the high GNP case.[9] Differences in the ratio across scenarios (high GNP, reference, and low GNP) are not very pronounced. These figures do not show the energy-output ratios in the high and low oil price case because they are virtually indistinguishable from the reference case path.

---

[9]In simulating the alternative GNP scenarios, the EMF 8 modelers held capital costs and fuel prices constant. However, these assumptions do not appear to alter appreciably the conclusion about energy-output ratios. The effect on energy demand of lower capital costs in the high GNP case would vary depending upon the model. Based upon results from the low capital-costs scenario, energy demand would be higher in PURHAPS, lower in ORIM, and remain essentially unchanged in ISTUM 2. In addition, if higher GNP were accompanied by higher energy prices, energy demand and the energy-output ratio would be lower than reported in the EMF study.

FIGURE 19-5
Change in Gross Output by Industry, High and Low GNP Cases

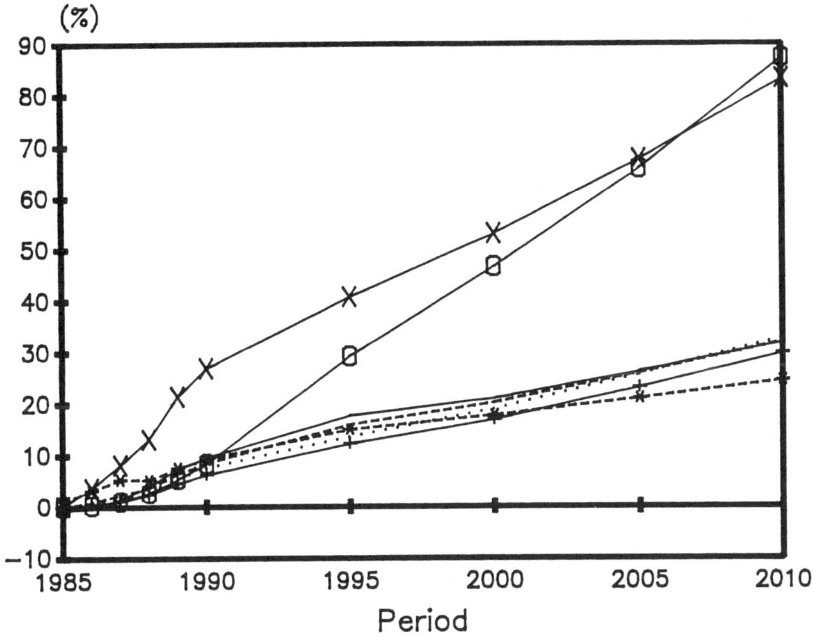

FIGURE 19-5-a
WHARTON
High GNP
(Pct. Diff. w.r.t. Reference Case)
Gross Econ. Output

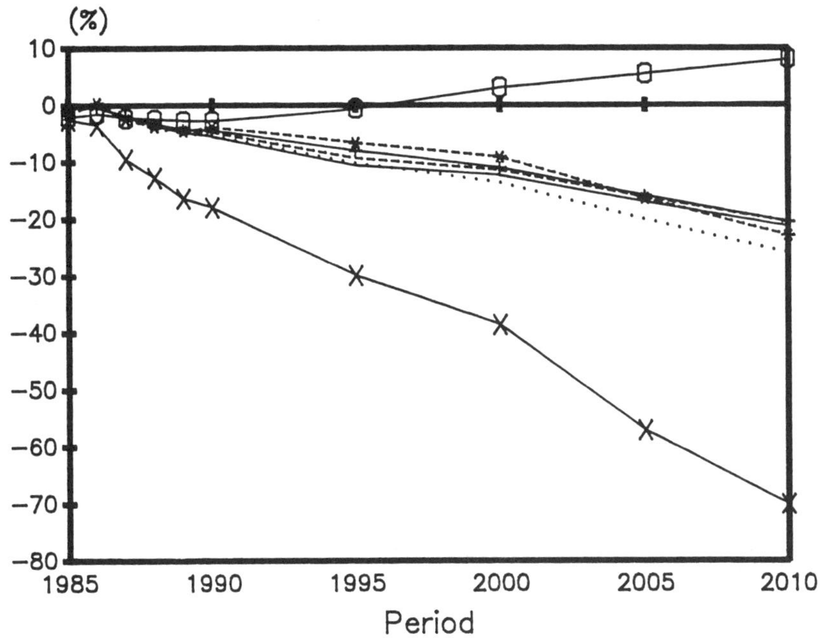

FIGURE 19-5-b
WHARTON
Low GNP
(Pct. Diff. w.r.t. Reference Case)
Gross Econ. Output

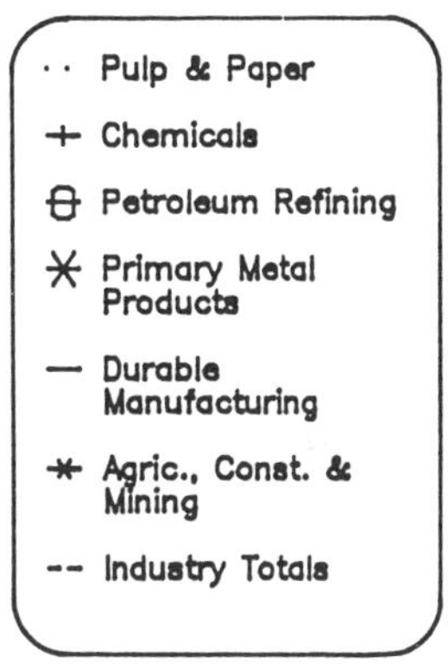

FIGURE 19-6
Aggregate Fossil Fuel Intensities in Selected Scenarios

FIGURE 19-6-a
PURHAPS Manufacturing Fossil Fuels (Div. by Gross Econ. Output)
Indexed to 1985

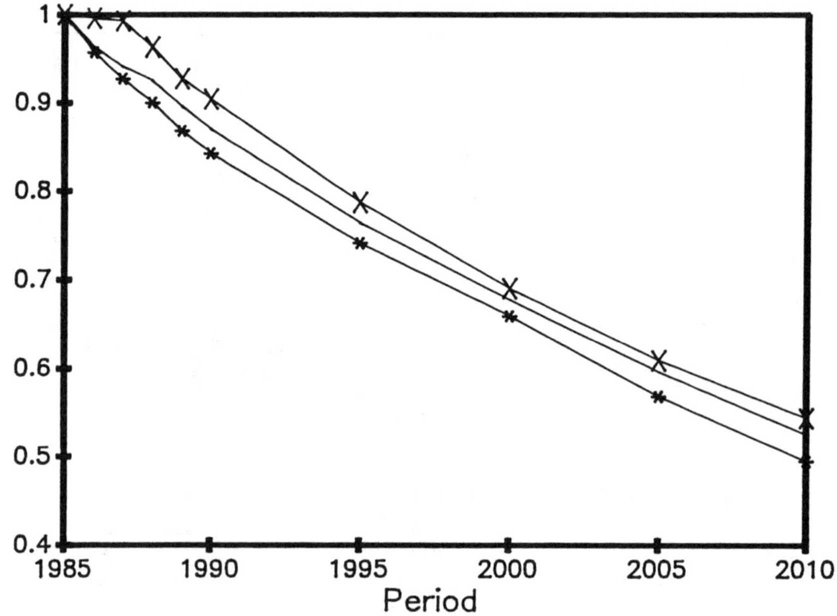

FIGURE 19-6-b
ORIM Manufacturing Fossil Fuels (Div. by Gross Econ. Output)
Indexed to 1985

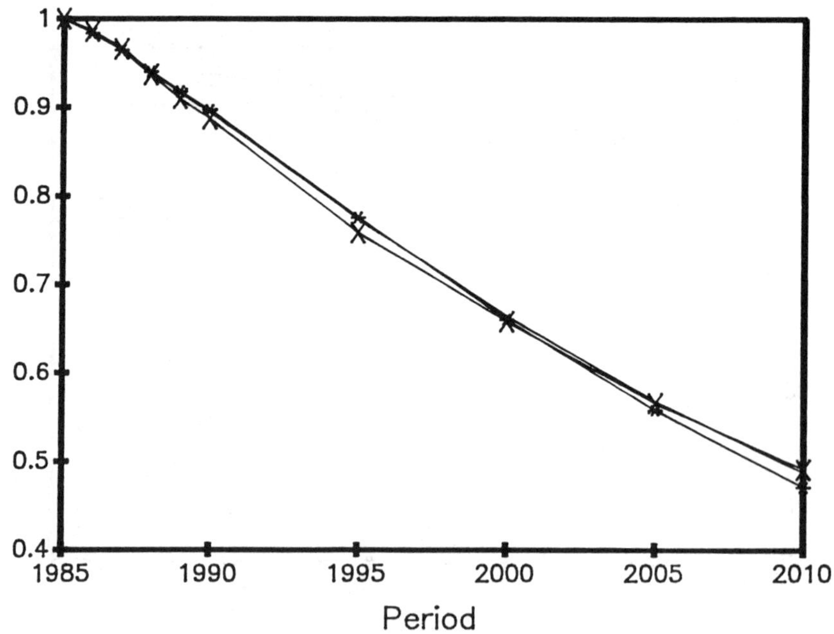

— Reference Case
✕ High Gnp
✱ Low Gnp

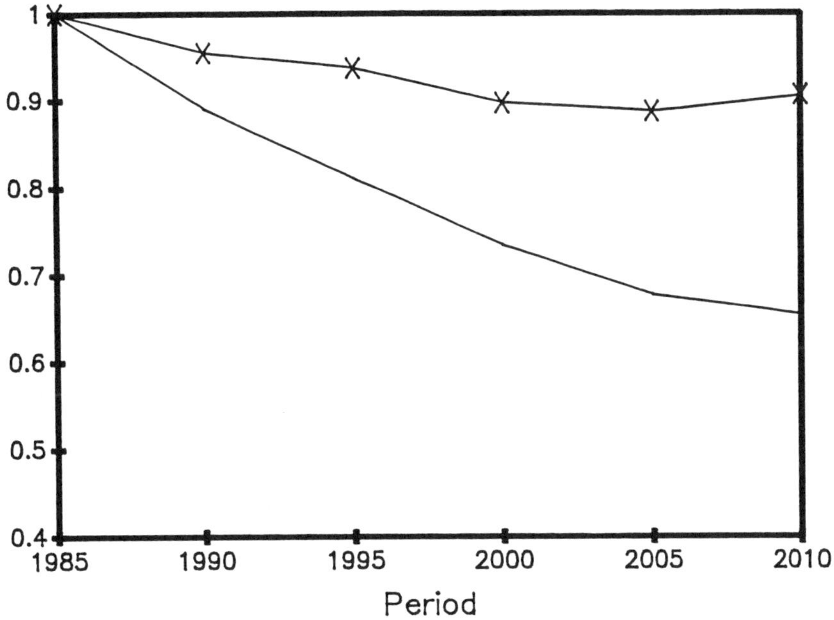

FIGURE 19-6-c
AES/ISTUM-I Manufacturing Fossil Fuels (Div. by Gross Econ. Output)
Indexed to 1985

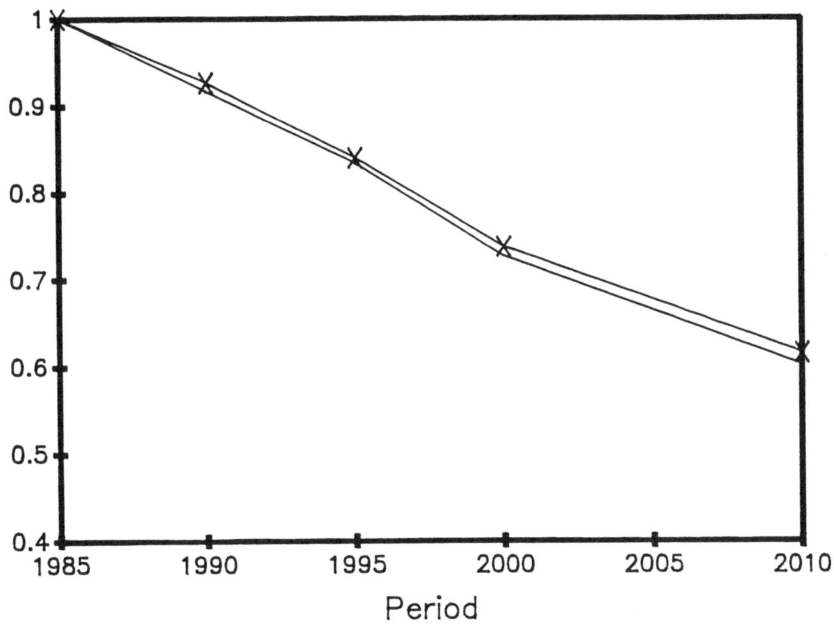

FIGURE 19-6-d
ISTUM-II Manufacturing Fossil Fuels (Div. by Gross Econ. Output)
Indexed to 1985

FIGURE 19-7
Decomposition of Aggregate Intensity Changes into Sectoral Shift (6 sectors), Product and Process Shifts, and Efficiency and Maintenance Improvements

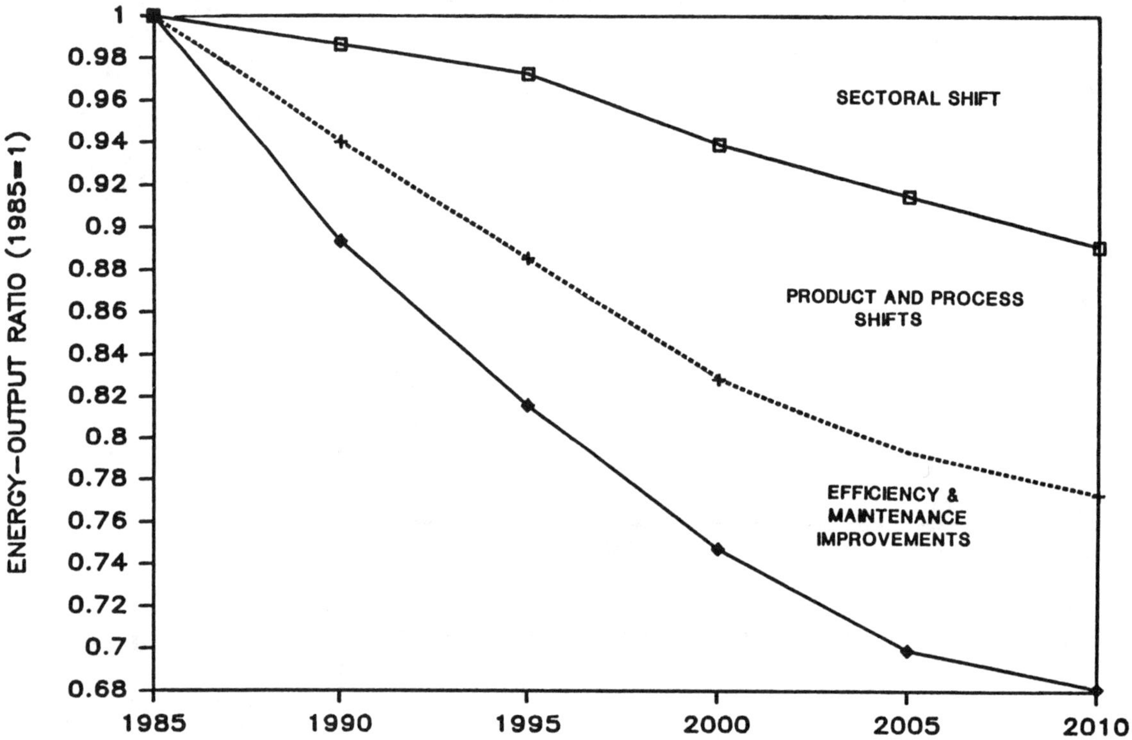

FIGURE 19-7-a
Sectoral and Technological Shifts, AES/ISTUM-I Reference Case

DECOMPOSITION OF TECHNOLOGICAL CHANGE

The effects of changing sectoral intensity on energy use per dollar of output can also be disaggregated further in some EMF models, especially the process models. Changing sectoral intensity can be decomposed into changes in the product mix within an industry (e.g., from low to high value-added chemicals), changes in the processes employed to produce a specific product (e.g., replacing open-hearth with electric-arc steel production), and changes in the equipment

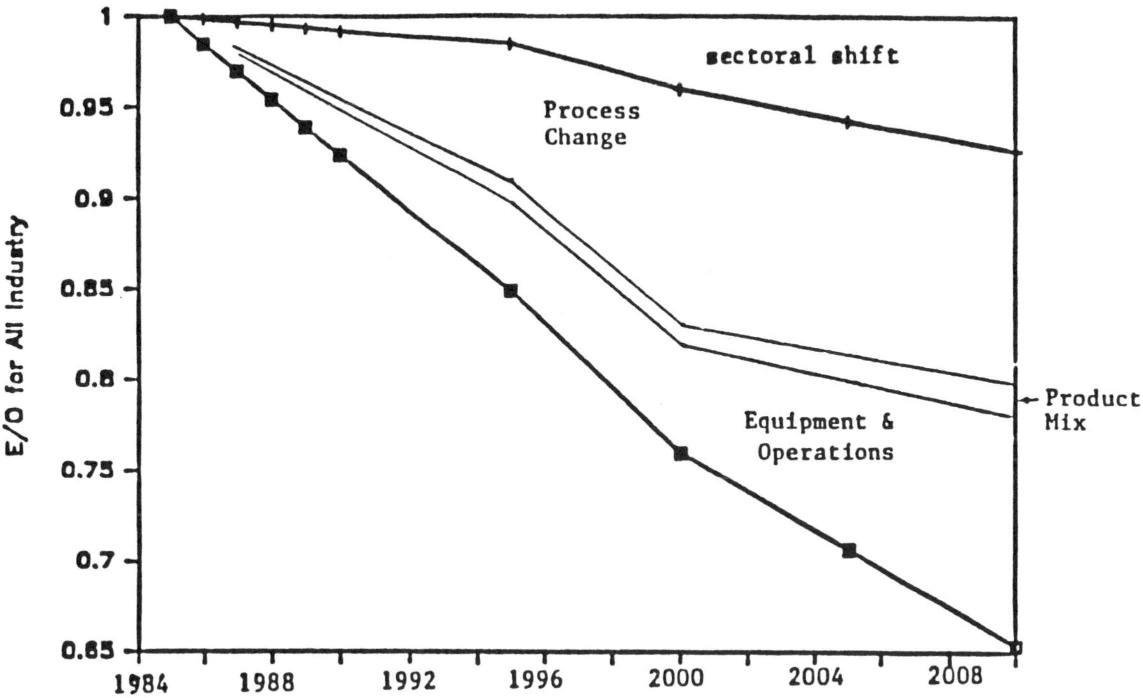

FIGURE 19-7-b Sectoral and Energy-Output Ratio, ISTUM-II Reference Case

and maintenance procedures associated with a particular process (e.g., adding or deleting work crews).

This decomposition is summarized for the reference case for the two ISTUM models in Figure 19-7a and 7b. Both figures separate the contribution of newer, more energy-efficient processes and products from that of adopting more efficient operation and maintenance of existing processes. It should be emphasized that the first effect represents the decline in energy intensity due to the increasing penetration of existing, more energy-efficient processes or products; no new technologies awaiting commercialization are imbedded in these results. This effect has been calculated from the mix of processes and products selected by the model, as well as their respective energy intensities, based upon assumptions in the reference case.

The two components appear to account for approximately equal shares of the decline in sectoral energy intensity in the EMF reference case. And for ISTUM 1, the total decline in aggregate energy intensity appears to be comprised of three almost equal factors: sectoral shift, product and process shifts, and efficiency and maintenance improvements. (The reported ISTUM 2 results are preliminary and may be revised.)

While changes in sectoral intensity are not explicitly decomposed in econometric models, certain key assumptions lie behind the forecasts generated by such systems. In PURHAPS, the energy-output ratio will be reduced most in those industries—chemicals and durables—where there has been a strong trend that way in the past, based on a shift to high value-added products. In metals, on the other hand, there will be more energy-intensive alloys. A change in the trend in chemicals toward less feedstock growth and more energy growth is expected because of the future growth in more specialized plastics. The proprietors of PURHAPS believe these changes in sectoral intensity are very important, even if they are too complex to be captured explicitly by the model.

CONCLUSIONS

Shifts among economic sectors within manufacturing have had significant effects on industrial energy demand since 1973. At least one-third of the reduction in energy intensity for fossil fuels over this period can be attributed to sectoral shifts. Moreover, since compositional shifts among sectors prior to the embargo increased rather than decreased the energy-output ratio, much more of the change in energy intensities between the 1967-74 and 1974-81 periods was accounted for by this reversal in the sectoral shift effect—perhaps as much as two-thirds of the total decline in aggregate fossil-fuel intensity.

Virtually all of the decline in aggregate electricity intensity since the embargo was produced by sectoral shifts. Changes in electricity intensity within sectors showed no noticeable general trend during this period. About half of the change in aggregate electricity intensity between the 1967-74 and 1974-81 periods can be traced to changes in the sectoral-shift effect between periods.

Shifts between economic sectors will continue to be an important source of uncertainty in forecasting industrial energy demand. Unfortunately, analysts know substantially less about the causes of these shifts than what their trends have been. Thus, one must continue to be cautious about interpreting any set of industrial energy demand projections.

The EMF projections are based upon Wharton economic scenarios that anticipate substantially less shift away from energy-intensive industries than has been observed during the 1970s. As a result, sectoral shift does not exert as great an influence in these projections as has been the case recently. While the assumptions behind these economic projections are reasonable, they have not been tested in this study to the same degree as other factors affecting industrial energy demand.

APPENDIX: THE DIVISIA INDEX

The derivation of the Divisia Index for studying energy intensity is given in the appendix of the Argonne study by Boyd et al. (9). The index is considered attractive on theoretical grounds because it is not sensitive to where in the time period you begin computing the index or in which direction. This property is called path independence. Additionally, when decomposing a trend, the individual components will have the same magnitudes regardless of which one is calculated first.

In this appendix, we provide an example of this index, emphasizing its principal advantages, although empirically we show that the measured sectoral-shift effect between 1984 and 1981 is relatively insensitive to whether or not a Divisia Index is used to decompose aggregate energy intensity in physical (BTUs) terms. However, it should be emphasized that the Divisia Index may still be preferred when measuring energy in terms of expenditures rather than BTUs. Past studies have shown this distinction to be important for studying aggregate energy intensities.

ENERGY INTENSITIES

Table 19-3 shows the trend in energy intensity for all manufacturing, for the four major energy-consuming sectors within manufacturing as a group, and for the remaining sectors collectively between 1974 and 1981. Gross output

TABLE 19-3
Energy/Output in Manufacturing, 1974 and 1981

	1974	1981
All Manufacturing		
Energy	12819	11369
Output	815.9	987.9
E/O Ratio	15.7	11.5
Energy Intensive		
Energy	8212	7186
Output	192.8	200.4
Share (%)	23.6	20.3
E/O Ratio	42.6	35.9
Other		
Energy	4607	4182
Output	623.1	787.5
Share (%)	76.4	79.7
E/O Ratio	7.39	5.31

Notes:
Energy is measured in trillions of BTUs.
Gross output is measured in billions of 1972 dollars.
Energy-intensive sector includes paper, chemicals, petroleum refining, and primary metals.
Source: Department of Energy and Wharton Econometric Forecasting Associates.

shares for the two broadly defined sectors are also shown. The table emphasizes the dramatic difference in energy intensities between the two sectors; the energy-output ratio in one sector is six to seven times its counterpart in the other sector. Thus the decline in the energy-intensive sector's share of gross output from 23.6 to 20.3 percent over this period had an important effect on aggregate energy intensity within manufacturing.

DECOMPOSITION WITH FIXED-WEIGHT INDICES

From the information on output shares and sectoral intensities in Table 19-3, the aggregate energy intensities for manufacturing in the two years are defined as:

$$E_0 = \Sigma e_0 s_0 = (.236)(42.6) + (.764)(7.39) = 15.7$$

$$E_1 = \Sigma e_1 s_1 = (.203)(35.8) + (.797)(5.31) = 11.5,$$

where e represents the energy intensity within a sector, s is the sector's share of gross output, and the subscripts 0 and 1 denote the years 1974 and 1981, respectively. Thus, aggregate energy intensity falls from 15.7 to 11.5, or by 26.8 percent from its 1974 level.

The contribution of sectoral shift can be estimated by holding the sectoral intensities at their levels for either 1974 (a Laspeyres Index) or 1981 (a Paasche Index). The two estimates, however, will not necessarily be the same. Denoting the Laspeyres and Paasche indices as L and P, respectively, they can be written as:

$$L = \frac{\Sigma e_0 s_1}{\Sigma e_0 s_0} = \frac{(.203)(42.6) + (.797)(7.39)}{15.7} = \frac{14.54}{15.7} = 0.926$$

$$P = \frac{\Sigma e_1 s_1}{\Sigma e_1 s_0} = \frac{11.5}{(.236)(35.8) + (.764)(5.31)} = \frac{11.51}{12.5} = 0.920$$

The two indices are pretty similar to each other because the trends in energy intensities within the sectors are not very different. The Laspeyres Index shows that 27.6 percent of the decline in aggregate energy intensity can be attributed to sectoral shift; the sectoral-shift index falls by 7.4 percent compared to a total reduction of 26.8 percent for aggregate energy intensity. The Paasche Index reveals a slightly greater sectoral-shift effect; the 8 percent decline in that index represents 29.9 percent of the total change in aggregate energy intensity.

DECOMPOSITION WITH DIVISIA INDICES

The Divisia Index is a weighted sum of growth rates. For energy intensity (in BTUs), the formula for the index (d) becomes:

$$\Delta \ln d = w_i \ln e_i + w_i \ln s_i$$

where w_i is the share of total energy use (in BTUs) in sector i, e_i is the sector-specific energy intensity, s_i is the share of gross output in sector i, and the percent change for each variable has been represented as the change in logarithms (Δln). The first set of terms on the right-hand side of this equation represents the effect of changing sectoral intensities, while the second set represents the effect of the shift between sectors.

When applied to the previous example, this formula provides a superior decomposition of the aggregate energy intensity into sectoral shift and changing sectoral intensities. The energy share weights could be either arithmetic or geometric averages for the two years. Using simple means, we note that the energy use in the major energy-consuming manufacturing sectors averaged 63.6 percent for the two years, 1974 and 1981. Thus, the energy weights are .636 and .364 in the formula, which now becomes:

$$\Delta \ln d = [(.636)(-.171) + (.364)(-.331) +$$

$$+ (.636)(-.151) + (.364)(.042)]$$

$$= -.229 - .080 = -.309, \text{ or an index of } 0.734.$$

Indices (1974=1) for changing sectoral intensity and sectoral shift are .795[exp(-.229)], and .923[exp(-.080)], respectively. These results are quite similar to the sectoral-shift effect estimated using the Laspeyres and Paasche indices in this case. Sectoral shift accounts for 28.9 percent (.007/.266) of the decline in aggregate energy intensity, compared to a range of 27.3 to 29.9 percent using fixed-weight indices. Thus, while theoretically superior, the choice of index appears to be relatively unimportant in decomposing aggregate BTUs per dollar of output over this period.

Note that the product of the sectoral shift and changing sectoral-intensity effects (.734) implies a total decline of 26.6 percent, which is very close to the proportional decline in aggregate energy intensity. This result would be maintained even if the trends in energy intensities within individual sectors varied more dramatically from each other. This property is an important advantage of Divisia indices applied in a more general setting.

REFERENCES

1. W. Hogan. "Patterns of Energy Use." Discussion Paper Series. #E-84-04. Kennedy School of Government, Harvard University, May 1984.

2. Energy Information Administration. "Energy Conservation Indicators, 1984 Annual Report." DOE/EIA-0441(84). Washington, DC: U.S. Department of Energy, 1984.

3. J. Myers and L. Nakamura. Saving Energy in Manufacturing. Ballinger, 1978.

4. R. Marlay. "Industrial Energy Productivity." Unpublished Ph.D. dissertation, MIT, May 1983.

5. R. Marlay. "Trends in Industrial Use of Energy." Science, December 14, 1984, pp. 1277-83.

6. R. Marlay. "Industrial Electricity Consumption and Changing Economic Conditions." In A. Faruqui and J. Broehl (eds.), Forecasting Industrial Structural Change and Its Impact on Electricity Consumption. Battelle Press, 1986 forthcoming.

7. P. Werbos. "Industrial Electricity Demand: Prospects and Uncertainties." In Proceedings: Forecasting the Impact of Industrial Structural Change on U.S. Electricity Demand. EA-3816. Palo Alto, CA: Electric Power Research Institute, December 1984.

8. P. Werbos. "Industrial Structural Shift: Causes and Consequences for Electricity Demand." In this book, Section 6.

9. G. Boyd, J. MacDonald, M. Ross, and D. Hanson. "Effects on Energy Demand of the Changing Composition of U.S. Manufacturing Production." Argonne, IL: Argonne National Laboratory, 1985.

10. D. Jorgenson. "Econometric and Process Analysis Models for Energy Policy Assessments." In R. Amit and M. Avriel (eds.), Perspective of Resource Policy Modeling: Energy and Minerals. Ballinger Publishing Co., 1982.

11. M. Ross. "Trends in the Use of Electricity in Manufacturing." In Technology and Society. Forthcoming March 1986.

12. A. Faruqui, S. Braitwait, C. Gellings, and L. Andrews. "Sources of Variation in Industrial Energy Demand: United States Manufacturing, 1970-81." In Industrial Structural Change. Palo Alto, CA: Electric Power Research Institute, forthcoming 1986.

13. K. Fugime. "Japan's Energy Supply-Demand Achieved in 1984 and Background." In Energy in Japan. Tokyo: Institute of Energy Economics, 1981.

14. C. Jenne and R. Cattell. "Structural Change and Energy Efficiency in Industry." Energy Economics, April 1983, pp. 114-123.

15. J. Roop and D. Belzer. "Changes in Energy Use in the U.S. Economy, 1972-77: An Input-Output Analysis." Richland, WA: Pacific Northwest Laboratory, February 1986.

16. G. Samuels et al. "Shifts in Product Mix Versus Energy Intensity as Determinants of Energy Consumption in the Manufacturing Sector." In Proceedings: Forecasting the Impact of Industrial Structural Change on U.S. Electricity Demand. EA-3816. Palo Alto, CA: Electric Power Research Institute, December 1984.

QUESTION-AND-ANSWER SESSION FOLLOWING PRESENTATION

EDITORS' NOTE

At the seminar, Hillard Huntington gave a presentation entitled "Findings from a Comparison of Industrial Energy Demand Models." This presentation was based upon some preliminary results from an ongoing Energy Modeling Forum (EMF) study. At that date, the working group for the study had not had an opportunity to review the final results. Thus, we have not included his presentation in this proceedings. Instead, a related EMF paper coauthored by Huntington and John Myers has been included. The findings of the EMF study will be published in a report issued in 1986. Preliminary drafts of this report are available from EMF (406 Terman Engineering Center, Stanford University, Stanford, CA 94305).

EMF studies provide a forum for high-level experts from government, industry, and academia to study important issues of common interest. In each study, teams of specialists in the major competing methodologies prepare projections of key variables that are compared and evaluated by experts on the relevant industries, markets, and regulatory regimes.

The study on industrial energy demand (EMF8) is examining the implications of various alternative scenarios and sensitivities on fuel demand in that sector through the year 2010. It is focusing on a number of key energy-using industries that tend to dominate energy consumption in manufacturing. The working group brings together energy producers and users and is chaired by Daniel M. Greeno, a former group vice president at Stauffer Chemical Company.

The question-and-answer session that followed Huntington's presentation raised some interesting discussions and is reproduced below.

MR. WERBOS: My comments [on the difference between the econometric and process models used in the EMF study] are certainly going to be prejudiced, but I will try to compensate a little. I can't help being prejudiced because I am responsible for one of the econometric models.

The first is a very minor comment. When we get all of this calibration stuff done, I think that ORIM, perhaps, would be a lot closer. But I think the pattern will still be there; the econometric models will be fairly close going one direction and the process models will be going in another.

Before EIA developed an econometric model of industrial demand, we, in fact, used ISTUM and ORIM in the process models. Those were our bread and butter. Some of you may remember an infamous thing called the PIES Study—Project Independence Evaluation System. I think two years ago we were going to be free of all other imports, or something like that. And somebody, at some point, said,

"You are being a little bit too optimistic." And that experience has conditioned our choice of modeling philosophy quite a bit here.

We asked ourselves, "If you had to do a process model five years ago—let's say, to predict energy use in the motor vehicle industry—what would you have projected?"

We certainly would have projected an improvement in the energy efficiency of all of the relevant technologies—motors and so on. You would not project technology going backward. One of the things that we projected in Project Independence was an increase in energy per unit of output. But, at the same time, you saw an increase in electricity per unit of output, and in some industries you actually get an energy increase, depending on how you weight electricity.

The question is this: How can you see improvements in technology, improvements in efficiency, in the past decade or the future decade? How is it that, in the past, you could still get more energy intensity? I think the answer has to be the kinds of things that Al Sobey was alluding to earlier about more advanced process technologies. I think things like electrotechnologies have something to do with it; things like printing and writing papers increasing relative to newsprint.

In the historical period, we find the energy per unit of output is different than the improvement in efficiency in specific things like motors. Therefore, when we project the future, if we do it based on history, we continue those trends, because the point is this. There won't be just improvements in machine drive, there will also be these subtler changes in technology in the future, as there have been in the past.

It is very hard to represent these subtler changes in technology in a process model. ISTUM II is, without doubt, the best process model in existence for industry. The level of detail is incredible. And yet, at the same time, the amount of detail you would need to represent minor improvements in robotics in the manufacturing sector is beyond the scope of any model in existence in the world today.

So, my very prejudiced point of view is that the fundamental reason for the divergence between these two models is that one is reflecting a historical trend that includes two kinds of factors, whereas the other is projecting forward, based on one kind of factor.

The process modelers would say that there are more opportunities for conservation now than there were ten years ago. Maybe the improvement is going to be more rapid. Maybe, as capital becomes more available, things that were being projected, that never came on line, will finally start to come on line. I would say that we have a much better case in the example of co-generation, because we have only been forecasting the electricity shares through 1995. We did this out to 2010 just for the Forum.

A lot of the trend in the last ten years has been away from self-generated electricity in some industries. My impression is that that is particularly true for paper and chemicals, just from a glance at the ASM [Annual Survey of Manufactures] data. And if the trend away from self-generated electricity slows down, then you would expect a certain amount of slowdown in the trend toward electrification. How much, I don't know.

But I would say, in terms of the energy graph, I feel a lot of confidence in the historical approach, in terms of the electricity per unit of output beyond 1995. You know, things may not be quite as rosy as they look in these graphs. But who knows what is coming on line after 1995? None of us have advanced ceramics in our models, and none of us really fully capture robotics in our models.

MR. HUNTINGTON: I think Paul gave a very elegant defense of econometric models. I only wish I had the ISTUM folks. They would give you an equally defensible and perhaps impassioned view of the benefits of the technology process models.

Something I didn't point out here is that a lot of the models look pretty similar in the period 1985 to 1990. However, ORIM is actually showing less energy conservation in that early period. Maybe it is not significantly less, but clearly less improvement in energy efficiency. And then over time it starts falling below the econometric models on this graph, indicating greater energy efficiency.

One of the ways I interpret the econometric results is that the improvement in energy efficiency—attributable mostly to turnover in capital stock—begins to slow down over a period of time. In contrast, in some of the end-use technology models, some significant improvements are still taking place out in the later period. So, they differ more in terms of the long-run view of what is happening rather than in the short run.

MR. KAHANE: I think this is going to be a fascinating study, and I can already think of 10 or 20 questions, but I think we should give you time to write your report.

But I think that the point that Mr. Werbos was making is something that you really have to keep in mind when you are writing this report. What we are really interested in is what story you can tell about the different lines; not just the numbers you can come up with, but what picture of the future is different in the different models that accounts for the different results.

MR. HUNTINGTON: I agree a hundred percent.

MR. KAHANE: I know this is the EMF approach. The reason I think that is important is this: If you do that for the outputs of the different models, we will be able to see where the real disagreements are and where we need to work, where we need to do more basic research in modeling. To a certain extent, the industrial analysis that all of the modelers are doing draws on a rather small amount of actual information about the industrial sector. A lot more needs to be done. I hope that, by comparing these model results, we can see where the real unanswered questions are, and not just about which number is higher or lower.

MR. HUNTINGTON: I agree. As I prefaced my comments—and maybe I will just reemphasize that one more time—these are preliminary comments. The working group really hasn't had a chance to probe these very much. At our meeting, in a month and a half, these are the kinds of issues that we are going to be thrashing about, and Mr. Werbos is going to be there throwing in his ideas. I believe this applies not just to industrial energy demand but is appropriate for a lot of different issues as well. Lots of times you take a model and say, "I have just spent the last year looking at this model and now I am going to run some sensitivity analyses on it." And you run the sensitivities. Then you say,

"Okay. Things are going to be plus or minus 10 percent." You say, "That is the range of uncertainty."

I say, "Perhaps." That may or may not be. In some cases it may be appropriate. In other cases, just by the methodology you have selected, you may have constrained yourself to a narrow range of uncertainties that aren't warranted. That is the point I was really trying to make with my first graph. And I agree with you, in terms of the report and in terms of what the working group can contribute to the discussion about industrial energy demand, explaining the differences between the results is really what it is all about.

But I can tell you frankly, at this point, I myself do not know enough about the different results to begin explaining those results. I think that this will only come out over the next few months.

MR. KAHANE: The point I was trying to make is that I think you will find a lot of the differences between the models at our present state of knowledge are due to different assumptions. And when we know which of the assumptions are really critical, we can know which assumptions we have to research more.

MR. HUNTINGTON: Yes. I don't have any problem with that.

20

INDEPTH LEVEL I RESULTS: ECONOMETRIC FORECAST MODELS FOR 20 INDUSTRIES

Laurel M. Andrews
Synergic Resources Corporation

As electric utilities continue to expand and refine their strategic planning to meet the challenges in today's marketplace, the need for electricity load forecasting models that provide reliable forecasts and support policy analysis is increasing. To provide tools to meet this need, the Electric Power Research Institute (EPRI) has sponsored the development of forecasting models for the residential, commercial, and industrial sectors. The residential and commercial models are complete. This paper discusses a portion of the forecasting models that are being developed for the industrial sector in an EPRI project titled "Industrial End-Use Planning Methodology (INDEPTH)."

As in the residential and commercial sectors, both econometric and end-use (broadly defined) models are being used for forecasting the demand for electricity in the industrial sector. The INDEPTH project draws on the strengths of each of these approaches in the three levels comprising the project. These levels and their applications are:

- Level I, Econometric Models—useful primarily for forecasting.
- Level II, Process Models—useful both for forecasting and policy analysis.
- Level III, Equipment Models—useful for policy analysis.

This paper focuses on the development of the econometric models, Level I. These models may be used independently by a utility or may be used in combination with the process and equipment models, Levels II and III. A description of the INDEPTH system is provided in (1).

The goals of Level I of the INDEPTH project are to develop econometric forecast models for 20 two-digit SIC Code industries covering the manufacturing sector, SICs 20 through 39; develop state-specific models; and test dynamic model specifications. To meet these goals we reviewed a number of econometric models reported in the literature and used by electric utilities, and examined the availability of data to support model estimation. We then chose specifications to test and are currently estimating these models. This paper discusses the econometric models examined and our final choice. A discussion of the data base development follows. Preliminary results and a discussion of forecasting using these models are then presented. The paper concludes with a discussion of lessons learned from the project and of the project's future.

APPROACH

MODEL OVERVIEW

Electricity is one of a number of factors of production (or inputs) used to create products in the manufacturing sector. Thus, the starting point for modeling the demand for electricity is the production function used by the firm. This holds whether we are discussing an econometric model or an engineering model. The broad conceptual framework is the same. Specifically, to produce a product a firm chooses between various production processes and inputs into those processes. Generally, it is assumed that the choices of production process and inputs depend primarily on the available technology, costs of the production process, and the costs of the inputs. More specifically, a firm chooses the production process and inputs that minimize the cost of production of a particular level of output. Figure 20-1 illustrates this model of a firm's behavior.

The firm has a production function comprised of specific processes that can be used to produce the output. This production function, along with the prices of the factors of production or inputs (i.e., capital, labor, energy, and materials), and the desired quantity of output Q_O are inputs into the firm's decision-making process. Using this information, the firm chooses the production processes and inputs that

FIGURE 20-1
Conceptual Model of the Industrial Producton Process

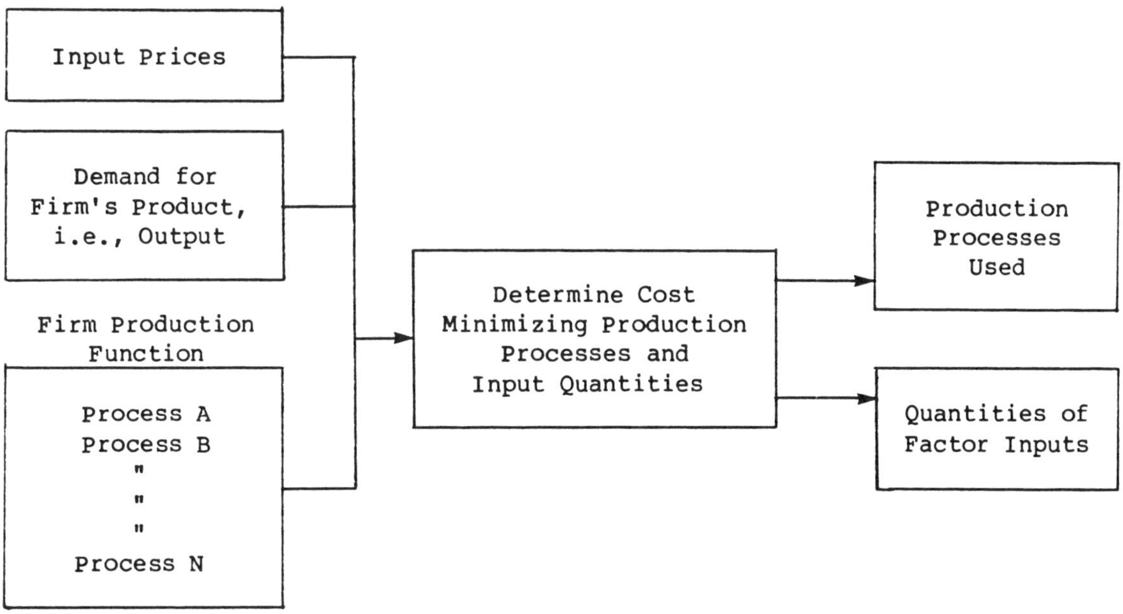

produce Q_0 at the lowest cost (and, thus, providing the highest level of profits). Once the decision is made it is possible to identify the quantities of each input used in production.

The implementation of the conceptual model to model industrial electricity demand can take a variety of forms. Engineering models tend to retain a large portion of the detail shown in the conceptual model. A process model, for example, explicitly considers a number of processes that are available and derives the cost-minimizing set of production processes and inputs using life-cycle-costing techniques. These models are generally based on engineering calculations along with data collected from actual firms. INDEPTH project Level II, the process model, is of this type. The equipment model, Level III, is also a kind of engineering model.

Econometric models are based on the same conceptual framework, but take a different approach to modeling. The detailed information contained in the firm's production function and explicit cost-minimization calculations is summarized into two functions: cost function and factor demand equations. These functions reflect choices made regarding equipment and production processes as input prices and output quantities change, but do not explicitly model these choices. Thus, econometric models can estimate the responsiveness of a factor of production, such as electricity, to changes in its price. Some models also provide information on the type of change resulting from a rise in electricity price, such as movement toward more capital-intensive (or labor-intensive) processes. However, these models do not provide information on which capital equipment is now being used, as engineering models do. Figure 20-2 illustrates the econometric model.

The summarization, or aggregation, of information inherent in econometric models provides one advantage over engineering models—they can be estimated using historical data. Thus, instead of using data on how firms say they behave, or how engineers think they behave, actual data on actual behavior is utilized.

Within the classification of econometric models a number of alternatives exist that vary in level of detail and historical data needed for estimation. They are discussed in the following sections.

FIGURE 20-2
Industrial Economic Models

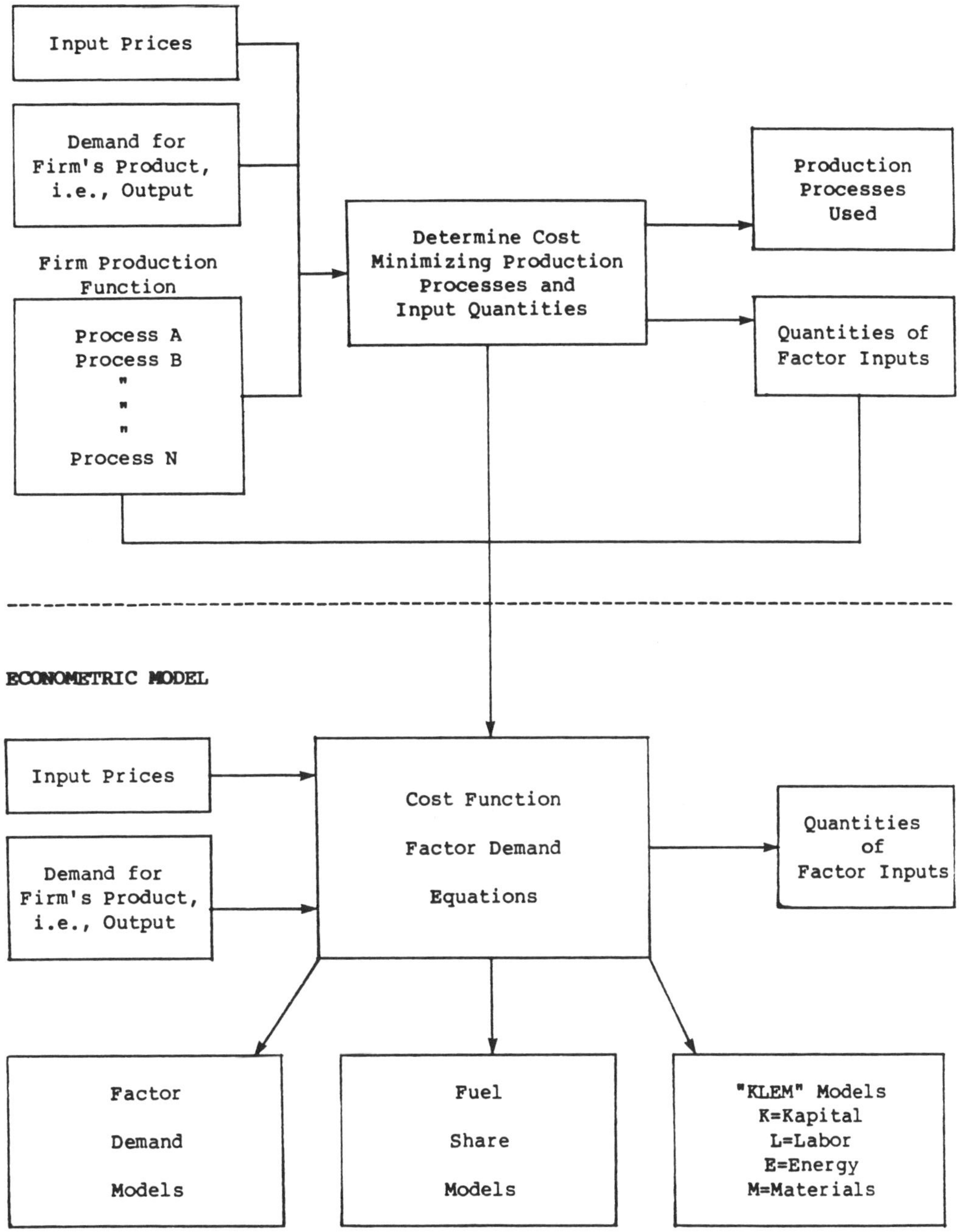

ECONOMETRIC MODELS

Three classes of econometric models have been developed to model the demand for electricity. These models vary in the level of detail and whether nonenergy production inputs are included. The types of models are:

- Factor Demand Models.
- Fuel Share Models.
- Flexible Functional Form or "KLEM" Models.

We examined each class of models before choosing the KLEM models for this project. A brief description of each model type, along with our reasons for choosing the KLEM models, follows.

Factor Demand Models. Factor demand models are the simplest models to estimate, and, generally, the first type utilized by utilities. These models describe the demand for a factor, such as electricity, as the function of the price of electricity, price of other inputs, and the quantity of output. The major advantage of this specification is that it has a small set of data requirements that are generally easy to satisfy.

The major limitation of this approach stems from the fact that the demand for one input is estimated in isolation from other inputs. As a result, the estimated demand for the input under study will not be consistent with the choices of other inputs used in production. The model parameters provide statistical relationships between the demand for electricity and other variables derived from a well-behaved cost function.

Fuel Share Models. Fuel share models have been developed to incorporate interrelationships between energy input demands. Generally these models consist of an aggregated energy demand equation and logit share equations for the individual fuels. The advantage of this approach is that the demands for individual fuels are made to be consistent with total energy demand. The major problem with these models is that they are not derived from economic theory and thus, like the factor demand models, they provide only statistical relationships between quantities demanded and other variables. They do not uncover the true functional relationships between factor demands and variables affecting production costs.

KLEM Models. KLEM models can be viewed as a set of simultaneous factor demand equations with an imposed structure derived from the economic theory of cost-minimizing behavior. The equations model the entire cost function, which is dual to an underlying production function that would be explicitly modeled in an engineering model. Thus the KLEM model provides estimates of functional relationships between factor demands and variables affecting costs and is consistent with economic theory. The models derive their name for the four general categories of production inputs: capital (K), labor (L), energy (E), and materials (M). Selected studies conducted using KLEM models are contained in **(2)**, **(3)**, and **(4)**. Table 20-1 shows the dependent and independent variables for a typical KLEM model.

The KLEM models are meant to approximate the true cost function and a variety of model specifications are possible. These are developed by combining choices for the following:

- Functional form—Recent research has focused on flexible functional forms which are second-order approximations to the cost function and thus are less restrictive than the traditional Cobb-Douglas and CES specifications. Flexible functional forms include the Translog, Generalized Leontief, and Generalized Cobb-Douglas specifications.
- Static or dynamic model—A static model assumes all inputs are variable and optional adjustments to price changes are made instantaneously. Dynamic models assume some factors are fixed in the short-run and adjustment of these quasi-fixed factors to optional levels is a dynamic process that occurs over time.
- Separability of inputs—A nonseparable model assumes all input decisions are made simultaneously and thus all input prices must be included in all equations. Strong separability of energy inputs from other inputs implies input choices can be made in two stages. In the first stage, total expenditure on aggregate energy and nonenergy inputs are chosen. In the second stage

TABLE 20-1
Illustrative KLEM Model
(static, translog model, nonseparable, biased technological change, constant returns to zero)

Equation	Dependent Variable	Independent Variables
Cost	Total costs	Price of capital Price of labor Price of electricity Price of natural gas Price of coal Price of oil Price of materials Output Time
Share of each input input	Share of total cost of each input	Price of capital Price of labor Price of electricity Price of natural gas Price of coal Price of oil Price of materials Time

the shares of the individual fuels of aggregate energy are chosen. The second stage choices are not dependent on the price of nonenergy inputs. The possibilities for aggregating various inputs is limited only by the data set and numerous forms of separability can be tested in model estimation.

- Returns to scale—Various assumptions about returns to scale can be imposed on the model.
- Technological change—Neutral technological change lowers total cost over time and is not biased toward any inputs. It is captured by adding a time trend to the total cost equation. Biased technological change

results in increased or decreased use of the relative quantities of inputs. It is captured by adding time trends to the factor share or demand equations as well as to the cost equation.

One of the major advantages of KLEM models is that they model the entire production process and, thus, capture functional relationships between electricity demand and the demand for other factors of production. These models are consistent with economic theory, allow estimation of the impact of technological change in a straightforward manner, and provide a dynamic specification derived from economic theory. The major disadvantages of this model class are that it is computationally demanding to estimate, more difficult to interpret the results, and requires a larger data set than the other model types.

We chose the KLEM models for the INDEPTH project because we believe that the advantages of this model class outweigh the cost of constructing the larger data set required for model estimation. We are testing a number of specifications for the INDEPTH project using the Translog function. The major specifications being considered include:

- Alternative assumptions regarding separability
  —Energy separable
  —Nonseparable
- Imposing constant returns to scale
- Alternative assumptions regarding technological change

These models are being estimated using a combined state and national data base, discussed next. This allows tailoring of the model to each state for use by local utilities.

Forecasting with KLEM Models. We also examined the feasibility of using KLEM models to develop utility-specific forecasts of electricity load before proceeding with model estimation. We determined that the computations required were reasonable and that the forecast series needed could be obtained, possibly at some cost. Our experience with developing forecasts for the test-case utilities, discussed next, confirms this.

DATA BASE FOR KLEM MODELS

The data base needed to estimate KLEM models is extensive. Data elements required are the prices of all inputs (capital, labor, electricity, natural gas, oil, coal, and materials), the expenditures on all of these inputs, and a measure of output. Since state-specific models are needed, state-level data for each of these variables are needed to support model estimation. In addition, a long-time series with sufficient observations to allow model estimation is required. The data must also be reliable and available at a reasonable cost.

We examined a number of alternative data sources for use in this project. Our first choice was the Longitudinal Establishment Data (LED) file under development by the Census Bureau. This file contains individual establishment data used to create the Annual Survey of Manufactures and the Census of Manufactures. The large number of observations, and regional detail, made this data base desirable. However, to use the data we needed a public-use tape. This tape was not available within the time frame necessary for this project. We then examined state and national data bases that contained the necessary data. We found complete national data for years 1958 to 1981. However, state data were available for 1967, 1971, and 1974 to 1978 only. This is a relatively short time-series and does not incorporate the second OPEC oil price shock. Thus, we decided that state data alone could not be used for model estimation. The national data, on the other hand, provided the time-series needed but did not support estimation of state-specific models. We solved this dilemma by developing a KLEM model specification that allowed using both state and national data to estimate the model. This specification assumes constant-returns-to-scale holds and provides a model with common slope coefficients and state-specific intercept terms.

The state portion of the data base was the most time-consuming to develop. State data are available from one source, the Annual Survey and Census of Manufactures. Some of these data are available on computer tape. Some are not. We keypunched and verified most of the data not available on tape. Some of the data needed were provided on tape by Dr. Barry Field. And the data were cleaned. We discovered a number of questionable values for variables, negative numbers, and even a keypunching error on a census tape. While time-consuming, we believe that our model results were improved because of the care we took in cleaning the data.

The major activity in developing the national data base was to determine which data set to use. Three national data bases are available:

- Office of Business Analysis, Department of Commerce
- Department of Energy
- Annual Survey and Census of Manufactures, Census Bureau

We reviewed the three alternatives and chose the Office of Business Analysis (OBA) data because they were more reliable, better documented, and contained estimates of energy data based on the work by Jack Faucett and Associates. In addition, the methodology for estimating capital stock used by OBA is quite comprehensive in this data base and this variable is carefully constructed. (See (5) for details.) These data were available for the years 1958 to 1981.

Our combined state/national data base consisted of national data for the following years: 1958-1966, 1968-1970, 1972-1973, and 1979-1981. State data for states with data for all variables and a "residual state" for the balance were used for the years 1967, 1971, and 1974-1978.

Compiling and cleaning the data base was a major effort that took approximately three months. In addition to the tasks undertaken to process the state data base discussed above, we processed the national data tapes and aggregated them to the two-digit SIC code level. The national data were checked for quality and consistency. Very few problems were discovered. The major problem was unreliable LPG data. However, data on this fuel were not available at the state level, so it was treated as a material in the analysis. Usage of this fuel is generally very small.

After the state and national data were processed and cleaned, the variables needed for model estimation were constructed and expressed in constant dollars. Estimates of capital stock by state and industry were made using a methodology we developed to approximate the approach used to construct the national capital stock estimates.

Once this was complete, model estimation began. The preliminary results are discussed next.

ILLUSTRATIVE RESULTS

MODEL PERFORMANCE

Preliminary results from model estimation demonstrate that KLEM models can be successfully estimated using a combined national/state data base. While some modifications are still needed, the models estimated to date are generally well behaved and produce reasonable estimates of the price elasticity of electricity.

The data used to estimate these models show that electricity costs are a small share of total costs. We found that electricity as a percentage of total costs varies from 0.3 percent to 1.4 percent across industries. Other inputs have a much larger share of total costs, as shown below:

Input	Share of Total Costs
Capital	15% to 55%
Labor	10% to 30%
Materials	30% to 70%
Electricity	0.3% to 1.4%

On the other hand, electricity is a large share of energy cost, more than 50 percent in most industries.

During the first round of model estimation we tested both nonseparable and energy-separable specifications for both static and dynamic models. Static models for 9 industries and dynamic models for 12 industries worked sufficiently well during this round; they required only minor modifications to be used for forecasting. At this time, efforts are under way to develop satisfactory models for the remaining industries and test modifications in the specifications suggested at the meeting held with the Project Review Committee for the INDEPTH project held in September 1985.

In addition, models are being checked to ensure that they satisfy "curvature constraints" (ensuring negative demand elasticities) and are modified until they satisfy this condition. These models, when completed, will be available for utilities to test for their service area on an "experimental" basis, and will be part of the INDEPTH forecasting system, with user-friendly software for its use to be developed in Phase III of this project.

FORECASTING METHODOLOGY

While slightly more complicated than generating a forecast from factor demand models, the algorithm used by a utility to implement these models is not difficult. The steps to generate a forecast are as follows:

1. Calculate total costs of production using the estimated cost equation and forecast series for the utility.
2. Calculate share of cost for electricity using estimated share equation and forecast series for the utility.
3. Calculate cost of electricity as:
   —Total cost x Share of Electricity
4. Calculate quantity of electricity as:
   —Cost of Electricity/Price of Electricity.

We are developing the needed software to make these calculations as part of this project.

The one crucial element needed for a utility to use these models is the data set of forecast series specific to the utility. The forecast series needed are:

- Price of electricity, natural gas, oil, coal
- Price of capital (static model) and capital stock (dynamic model) by two-digit SIC
- Wage rate
- Price of materials
- Output by two-digit SIC

It is desirable to have industry-specific prices (two-digit SIC) as well.

We are in the process of testing the econometric forecast models at five test utilities. Collection of forecast inputs from the five utilities is complete. Our experience demonstrates that compiling the needed forecasts is feasible, although it does require devoting some resources to the project. We found that energy prices and wage rates were readily available to utilities. Capital and material price forecast were not as available, but can be obtained from national sources.

The two most difficult variables to obtain were output and capital stock by industry. A forecast, sometimes a crude one, of output by industry was developed for all five utilities. Only one utility had a forecast of capital stock. We believe that devoting resources to forecasting output is warranted because it is a far superior measure of activity in the industrial sector compared to employment, a variable commonly used. For example, it is possible to observe declining employment and rising electricity usage and output. Thus, allocating resources to forecasting industrial output is a worthwhile investment for a utility to make. We also believe that forecasts of capital stock can be developed at the utility level. Further research into a methodology for doing this is needed.

CONCLUSIONS

This paper describes the econometric models under development for Level I of the INDEPTH project. Estimation of the KLEM specification using a combined state/national data base is under way, with preliminary results lending support to this choice of model specification. The final product of Phase II of the INDEPTH project will be econometric models for each of the 20 industries comprising the manufacturing sector. Phase III of the project, if funded, will provide user-friendly software for using all of the forecasting models developed for this project as well as additional testing of the forecast models.

We have learned a number of lessons from this project. First, KLEM models are a viable specification for electricity load forecasting models. A significant number of industries have poorly performing energy-separable models, but had energy-nonseparable models that performed well. This implies that including nonenergy inputs in the econometric models can be important and that, for these industries, simple factor demand and fuel share models would be inferior model choices. Second, development of a state/national data base to support estimation of KLEM models is time-consuming and expensive. Special care had to be taken with capital stock and capital service price variables. Third, we found that the state data base had to be cleaned prior to usage. This took time but, we believe, improved the results from our model estimation.

Finally, we found that it is possible to construct the needed forecast series for a utility to use the KLEM models estimated to generate electricity load forecasts for its service area. We are working with five test-case utilities to test the models on data for their service areas. We found that the necessary model inputs could be developed for the static model, with a forecast of output (in dollars) by industry for the service area being among the hardest variables to forecast. A forecast of capital stock by industry, needed for the dynamic model, was often not available for the test-case utilities. This lack of data restricts use of the dynamic model for forecasting purposes until this problem is solved.

The econometric models under development for Level I of the INDEPTH project are proceeding well and will provide a viable and useful tool to utilities for use in forecasting the electricity load for their industrial customers.

REFERENCES

1. Electric Power Research Institute. Industrial End-Use Planning Methodology. Palo Alto, CA: Electric Power Research Institute, EA 4019. February 1985.

2. E. R. Berndt, C. Morrison and G. C. Watkins. "Dynamic Models of Energy Demand: An Assessment and Comparison." In E. Berndt and B. Field (eds.), Measuring and Modelling Natural Resource Substitution. Cambridge: MA: MIT Press, 1981.

3. R. Halvorsen. "Energy Substitution in U.S. Manufacturing." Review of Economics and Statistics, Vol. 59, November, 1977, pp. 381-388.

4. C. Harper and B. Field. "Energy Substitution in U.S. Manufacturing: A Regional Approach." Southern Economic Journal (Vol. 50, No. 2), October 1983, pp. 385-395.

5. Bureau of Labor Statistics. U.S. Department of Labor. Capital Stock Estimates for Input-Output Industries: Methods and Data. Bulletin 2034. 1979.

QUESTION-AND-ANSWER SESSION FOLLOWING PRESENTATION

Mr. SAM SUGIYAMA, Bonneville Power Administration: What exactly is the structure of the dynamic model relative to the static translog?

MS. ANDREWS: Instead of including the price of capital as an independent variable, we included the quantity of capital; and for the cost equation, we used the variable cost. The cost of materials, energy, and labor, and not the cost of capital, is included in variable costs.

MR. SUGIYAMA: Were you able to do a short-run and a long-run, in the elasticity situation?

MS. ANDREWS: Yes. We haven't made the calculation for the long run yet, but we have the parameters to do it.

MR. SUGIYAMA: We tried a number of different specifications using our regional data and ran into some problems. In terms of the forecast, it was just too high. For example, pulp and paper, which won't grow at 3 percent, was projected to grow at 3 percent given the input forecasts that we received.

If you find some problems with your forecast, do you have an alternative to using the KLEM specification? The capital stock was a problem for us.

MS. ANDREWS: We understand that, and that is why our finished product will be both static and dynamic models, because a model can fit well and not provide a reasonable forecast.

MR. PAUL WERBOS, U.S. Department of Energy: It seems as if we have been taking very similar approaches, and it is possible some of our experience might be useful.

While we haven't tried estimating KLEM models at the state level, because of the lack of good capital and materials data at the state level, we have tried doing the same thing at the national level using Fawcett data extended by some material from Dale Jorgenson in the annual survey. And our results with the translog were extremely poor at the national level. We also did some investigations, actually at Massachusetts Institute of Technology, on the state level, before we did this, and that helped convince us we shouldn't.

MS. ANDREWS: Poor in terms of fit or forecast?

MR. WERBOS: Well, poor in a number of ways. Our documentation is available for free, and I think there are about ten pages on it that nobody would want me to reproduce here. But, if I had to make a quick summary, I would say that if we used Dale Jorgenson's original estimates of the most recent updated data that met the correct concavity constraints, we found 50 percent error, systematically, across industries in the postembargo period. When we relaxed the curvature constraints, we found we could get to about 6 percent error, but the problem was the elasticities; if you calculated them for different time periods, they were just ridiculous. They were just not believable.

We found that we could get 1 percent error with the KLEM structure model, if we used a certain kind of dynamic model, which we are now using. And, as I say, our mean absolute percentage error is 1 percent with the KLEM structure, but not a translog structure. And what we are doing is sort of seat of the pants, to be honest, because we are interested in energy, we are not forecasting labor.

We are basically assuming constant elasticity kinds of things. And the way we get to 1 percent error is by putting in a different kind of dynamics from the quasi-fixed kind that you are talking about. The quasi-fixed formulation basically assumes there are problems in adjusting the quantity of capital to the optimal quantity of capital; whereas, what we hear on the engineering side is that conservation is lagging because of quality of capital, because of things embedded in the capital. And it is when we take a capital-embedded price response that we can get our errors down to 1 percent with the kind of KLEM specification that we are looking at.

Now, I understand that there are some generalized Leontief versions that have a capital-embodied price response, and I know of a couple of people at Stanford who have been playing with that. I am not aware of it ever having been applied to this kind of a model, and, as I say, we haven't because we didn't need to spend the extra money to predict labor shares. It would have been very good, from a theoretical point of view, to do that. If I were at Wharton, I would be looking at those sorts of things, but 1 percent error was fine for us.

One very minor thing is that we do have all of that state data you were looking for, with the fill-ins, available on one tape, if anyone wants to avoid these kinds of hassles in the future.

MS. ANDREWS: We talked to you. We got your tape and it didn't have all the inputs. Maybe there was a misunderstanding.

MR. WERBOS: There might have been. The inputs were in the form of SAS data sets, so you can't read those particular files if you can't handle a SAS data set, and I seem to remember some problems with being unable to read IBM files. But, at any rate, all of this is available from the National Energy Information Center in documents DOE AI0420, which they will send on request.

MR. JOHN BROEHL, Battelle Columbus Division: My question is really twofold. If a utility was interested in forecasting, how would I use the models that you are developing; and, second, what kind of inputs do I have to provide?

MS. ANDREWS: The models, although they are a lot more complicated than maybe simple OLS models a utility might be using now, are operated the same way. We have some code written and the utility makes an input file, loads the input file into the program, the code trunks through, and out comes the forecast of quantity of electricity. In the intervening time, it first calculates the total cost; then it calculates the share of electricity of the total cost. Then these two quantities are multiplied together to obtain the dollars spent on electricity. This is divided by the price of electricity, to yield the quantity of electricity. I have more detailed equations written out if anybody is interested.

The inputs that are needed are the prices of capital, labor, energy, and material, or the prices of all of the production inputs. Capital price, as I mentioned, can be difficult to forecast, although I believe, with one coordinated effort, we could solve that problem. Labor price is a forecast of wage rates, which are generally available. Energy price is a forecast of the energy prices, which are also generally available. The price of materials is more difficult and, again, could be done at the national level. The hardest variable to forecast is output for the service area. Many utilities don't have it. They have a forecast of employment. We know that employment can go down while output goes up, and you need a forecast of output to run these models. That is basically it.

MR. DEREK H. HOWELL, Northeast Utilities Service Company: What is the interpretation of these elasticities?

MS. ANDREWS: The static elasticities are long-run demand elasticities for electricity; the dynamic are short-run.

MR. HOWELL: With respect to price?

MS. ANDREWS: Yes. Change in quantity of electricity over change in price of electricity.

MR. HOWELL: Does anyone else in this room think that they are kind of high, that 0.43 is a very high price elasticity for industrial? I do. I just thought you might want to know that before the Review Committee sees them.

MR. BOB CILIANO, Dun and Bradstreet: Did you indicate that initially you tried to go to Commerce and get establishment-level data?

MS. ANDREWS: Right.

MR. CILIANO: And, failing that, you had to settle for state- and, in some cases, national-level data.

In your professional opinion, to what extent do you think that that has somewhat compromised the validity and value of what you have done here, if at all?

MS. ANDREWS: It depends on why the industries differ at the state level from the national level. They can differ because there is a different underlying production technology, or because there is a different mix of industries at the two-digit SIC. For example, SIC 20 might be all potatoes in one place and all corn somewhere else, and it might be a very different production process. Third, it could differ because there are different prices. If what we have is one underlying technology in the whole country, but we see different technologies because the prices are different, that is not a problem for the econometric models.

If the difference is because of the different prices, and so different industries in different parts of the country have different production technologies because of the different prices, this model is okay because is assumes one production function and you can be anywhere along it. If the difference is due to different production technologies, perhaps because of different times the firms were started and what was available at the time, or a different mix of industries in the two-digit SIC, the econometric models aren't going to work as well.

That is why we think you have to combine the econometric models with the process models. For the two or three big industries in the utility service area, like aluminum to BPA, you have to model it separately; you have to learn about that industry; you have to do engineering analysis. For the other ones that are smaller, there may be enough similarities between a service area and the nation that the econometric approach will work. Obviously, the LED data would have been better, but I think this is workable.

MR. CILIANO: Lastly, to what extent do you think you can compensate for the lack of an extensive time-series if you had literally thousands of data points on a cross-sectional basis?

MS. ANDREWS: If we hadn't had the second OPEC oil price shock, I would be more comfortable. The state data were available in 1967, 1971, and 1974 through 1978, which we did not feel was enough of a data series. The LED data are available for ten years up through 1981, so it is not really a problem when you get to that level. I don't know if that answers your question.

MR. CILIANO: No, I was really trying to get your opinion as to how important it is to capture some of the effects that would be embedded in a time-series versus the value, in econometric terms, of having literally thousands of points on a cross-sectional basis to measure the types of things you are seeking to measure.

MS. ANDREWS: I can't quite answer the question at this point. We haven't had time to compare the national model results with the cross-section time series model results. I think it is an empirical question.

21

INDEPTH LEVEL II RESULTS

John H. Broehl
Battelle Columbus Division

INTRODUCTION

Future uncertainties over electricity consumption in the industrial sector highlight the need for improved forecasting and have prompted the Electric Power Research Institute (EPRI) to initiate a major project titled "Industrial End-Use Planning Methodology (INDEPTH)," which will enable utilities to analyze the future of electricity sales at the service-area level. The purpose of this paper is twofold: (1) to give an overview of the entire INDEPTH project and (2) to present the structure and some results of the process model component of the INDEPTH system.

OVERVIEW OF INDEPTH

The primary objective of the INDEPTH project is to design, develop, and demonstrate a methodology for forecasting and shaping industrial electricity use at the service-area level. The emphasis of the project is on:

- Long-term forecasting (i.e., a 5- to 20-year forecasting horizon).
- The manufacturing industries (i.e., customers in SICs 20 through 39).
- Development of a forecasting system consisting of a conceptual framework, mathematical equations, computer code, and underlying data bases.
- Development of a flexible system that recognizes the diversity within the manufacturing sector at the service-area level and the different needs of forecasters within the utility industry.

In general, forecasts of industrial electricity consumption must capture the impact of three major influences including changes in: the level of industrial output, mix of industrial output, and electricity use per unit of output (electricity intensity). These phenomena have dramatically affected historical electricity consumption and are likely to continue to do so in the foreseeable future. The level of industrial output is influenced by macroeconomic factors, consumer confidence, and international competition. The mix of output is influenced by the competitive position of various industries, the required factors of production, consumer preference, and competition. Finally, electricity intensity is affected by technology developments, production costs, and investment decisions. Figure 21-1 depicts the numerous factors affecting industrial electricity use and how the INDEPTH system addresses them.

Based upon a review of the industry's forecasting needs, an end-use forecasting and planning system incorporating a hierarchical menu of approaches was designed. The system allows the user to develop energy forecasts for the entire industrial sector, to examine in detail those industries and industrial processes most important to the service area, and to investigate those uses of electricity, such as motors or lighting, that are of interest in demand-side programs. As shown in Figure 21-2, the system operates on three analytically distinct levels:

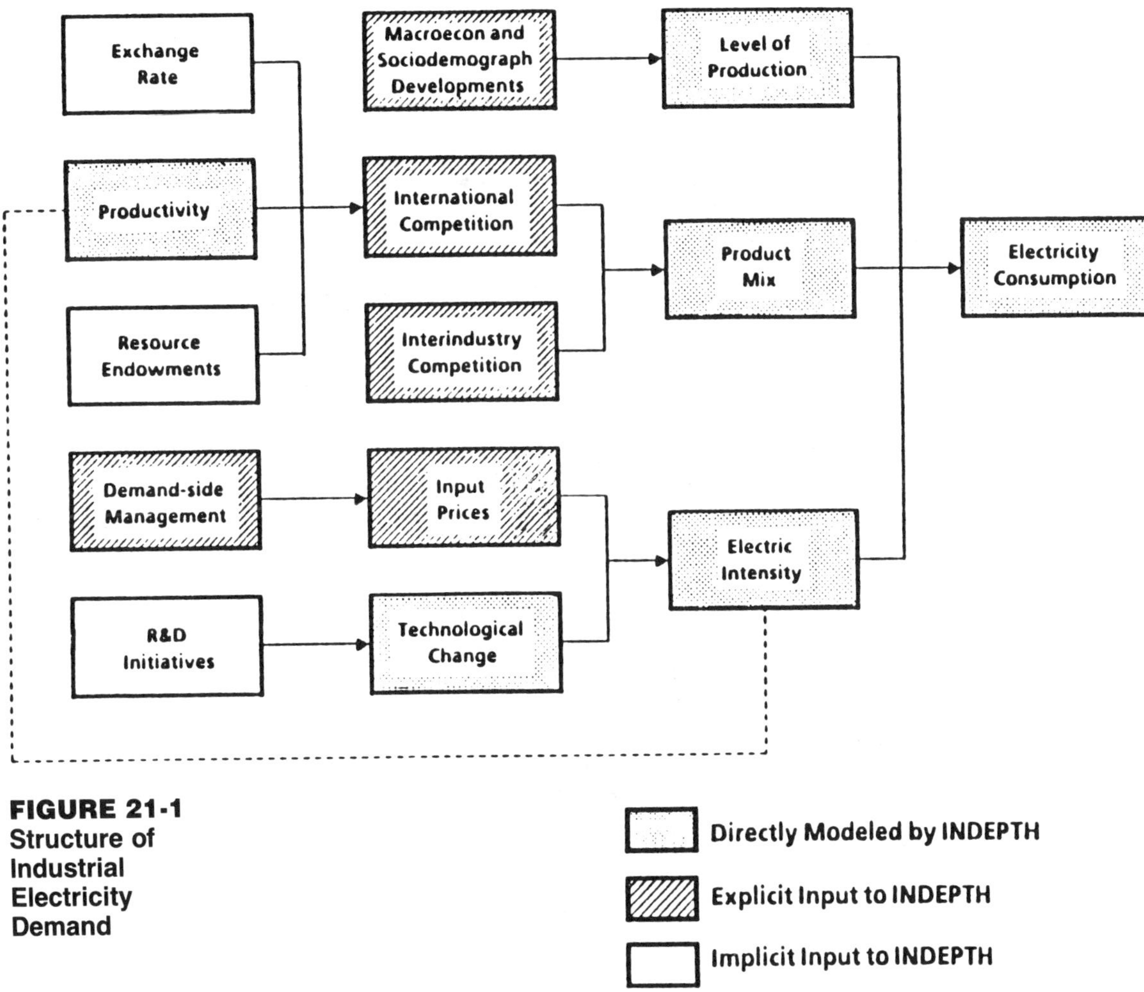

FIGURE 21-1
Structure of Industrial Electricity Demand

1. The econometric model level (Level 1)
2. The process model level (Level 2)
3. The equipment model level (Level 3).

Having described the overall framework of INDEPTH, the emphasis of this discussion is on the description of the structure of the process model (Level 2) and the presentation of preliminary results.

FIGURE 21-2
INDEPTH System Design

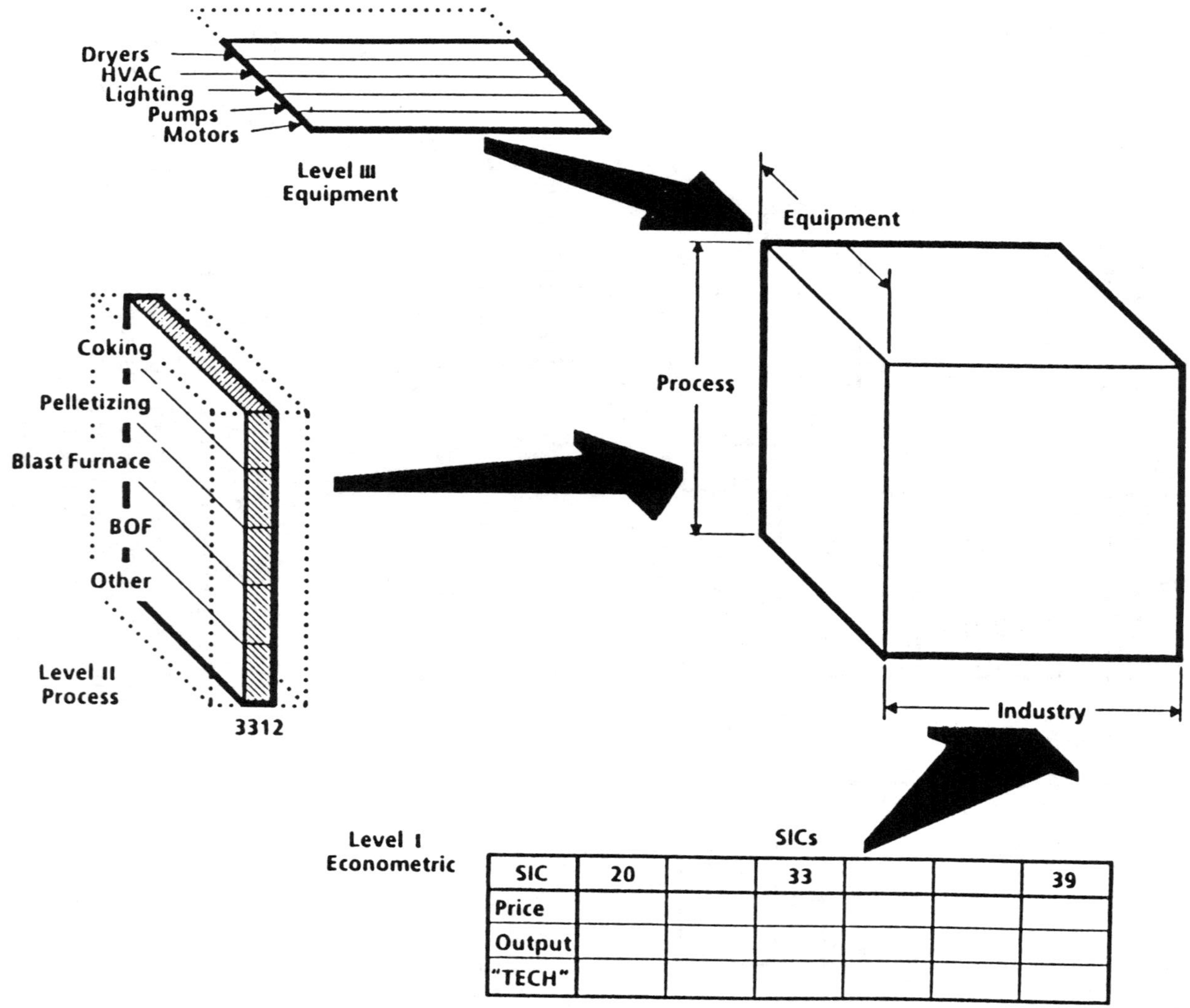

DESCRIPTION OF INDEPTH PROCESS MODEL

Process models estimate consumption by focusing on the major electricity-consuming technologies and processes in a given industry. The key assumption underlying a process model is that an industry has discrete production technologies or processes at its disposal to produce a given slate of outputs. Moreover, the processes are selected for implementation and utilization by the industry on the basis of least cost. The level of production or the amount of output that is being produced by the industry, and to a large extent the mix of that output, are exogenous inputs.

Process models are most appropriate for those industries experiencing or anticipating major technological change. This change takes place when existing production technologies are challenged by new technologies. The challenger technologies are typically newly developed, have a low market share, and offer the industry high potential savings. These savings may be due to reduced operating costs, higher efficiencies, or improved product quality. Process models simulate the competition between challenger and defender technologies to the extent that it is driven by production costs.

MODEL STRUCTURE

The structure of a process model may best be described in terms of a network depicting the major material flows within an industry. The nodes of the network represent the major material transformations within the industry, while the arcs represent the material flows. The level of the material flow and thus the utilization of the nodes is calculated such that the exogenously supplied product demand is satisfied in the least-cost manner. As an example, a network representation of the iron and steel industry is given in Figure 21-3. The network shows that there are basically two processes in iron making: direct reduction and the blast furnace. Moreover the blast furnace may be operated with different levels of major inputs (i.e., pellets, concentrated ore, and coke).

Central to the concept of a process model are the processes used in a given industry. As pointed out above, a process is a distinct step in the production of an industry's output and it is described in terms of its input materials, its

FIGURE 21-3
Network Representation of the Iron and Steel Industry

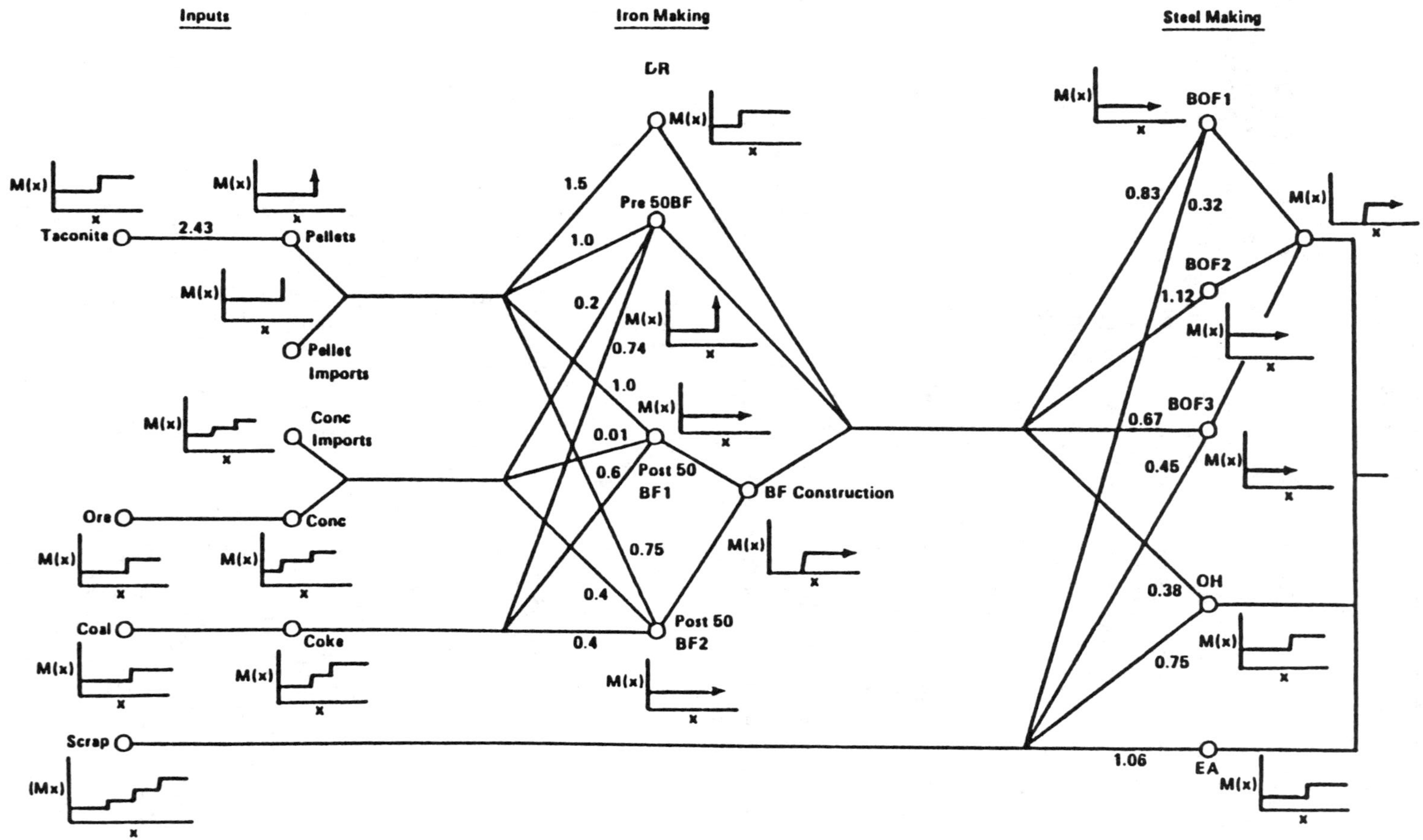

output, as well as production capacity, efficiency, and cost. Figure 21-4 depicts the major elements that describe a process.

A key assumption underlying the process model is that of fixed proportionality. Simply stated, this means that an increase in the output requires the same proportional increase in all of the inputs. Thus, doubling the output requires doubling all of the inputs. As shown in the iron and steel industry network, a given process may be operated in different operating modes (i.e., with different proportions of inputs). However, once these operating modes have been specified, it is assumed that they remain fixed, regardless of the production level.

DECISION PROCESS SIMULATIONS

In the process model framework two different types of decisions are being simulated. The first decision is the operation decision. This decision focuses on how to utilize the existing network with all process capacities fixed at

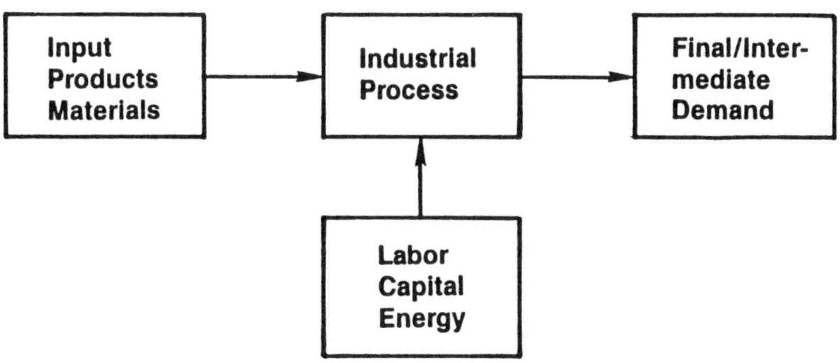

FIGURE 21-4
Process Characteristics

- **Process Descriptors**
  - **Current Capacity**
  - **Capacity Expansion Cost**
  - **Process Efficiency**
  - **Mix of Inputs**
  - **Other Operating Cost**

their current levels. The second decision is the investment decision. This decision focuses on how to expand or alter the network's capacities to satisfy growth in demand and to replace retired capacity.

The operating decision is based upon the assumption that the exogenously supplied demand must be satisfied by the industry. The utilization of the different processes is determined in such a way as to minimize operating cost. Two different computational procedures are used in the INDEPTH process model: a global optimization utilizing the linear programming procedure and a ranking procedure, according to cost. In this ranking procedure, competing processes are ranked from least to highest operative cost and the processes are utilized in this order until the product demand is satisfied. It should be noted that the importance of the operating decision depends to a large extent upon the level of excess capacity in an industry. Thus, if little or no excess capacity exists, all processes must be used to satisfy demand regardless of their operating cost. However, if there is substantial capacity, only the most cost-effective processes will be utilized.

The INDEPTH model treats the investment decision in two parts: (1) the magnitude of new capacity to be added and (2) the composition or mix of processes to be expanded. The magnitude of the investment is a function of economic obsolescence, retirement, and growth. Existing capacity is said to be economically obsolete if its operating cost is higher than the total cost of a new technology. In the model, all economically obsolete capacity is replaced. In addition, once existing capacity exceeds a specified retirement age it is also replaced. Finally, capacity may have to be added to accommodate growth in output in future time periods. In the case where an industry's output is expected to decline in the future, only that portion of economically obsolete or retired capacity that is needed to meet the reduced demand is replaced.

The composition of the needed capacity is determined on a least-cost basis. The cost used is a life-cycle cost including expected operating costs, capital expenditure, interest payments, and tax benefits.

LEVEL OF DETAIL

In the INDEPTH project, ten industries are being modeled at the process level of detail. Figure 21-5 lists the industries as well as the manner in which they are modeled. To illustrate the level of detail in which the industries are treated, the technologies incorporated in the iron and steel model are shown in Figure 21-6. It should be noted that each process is characterized by: capacity, operating cost, fuel costs, capital cost per unit of capacity, electricity requirements, process efficiency, capacity utilization, and the age distribution of the existing capacity. Competition takes place first among secondary processes and subsequently among primary processes.

PRELIMINARY RESULTS

To illustrate the model's capabilities, a number of preliminary outputs from the iron and steel industry are presented. Table 21-1 summarizes the base-case forecast. The electricity projections are based upon a national representation of the iron and steel industry and assume U.S. steel

	SIC	Model Type
Iron & Steel	3312	Tree
Chlor Alkali	2812	Tree
Textiles	2281	Tree
Metals Fabrication	3440	Tree
Petroleum	2911	L.P.
Automotive Stampings	3465	Tree
Cement	3241	Tree
Foundry	3321	Tree
Pulp and Paper	2600	L.P
Glass	3220	Tree

FIGURE 21-5
INDEPTH Process Industries

FIGURE 21-6
Technologies Embedded in Iron and Steel Model

Operation	Primary Processes	Secondary Processes
Sintering		
Coke Making	Wet Coke	
	Dry Coke	
	Form Coke	
Iron & Steel Making	Minimill	30% Direct Reduced
		0% Direct Reduced
	Integrated Basic Oxygen	Normal Blast Furnace
		High Temp/Pres.
	Integrated Basic Oxygen	Normal Blast Furnace
		High Temp/Pres.
	Integrated Open Hearth	Normal Blast Furnace
		High Temp/Pres.
	Integrated Electric Arc	Normal Blast Furnace—Hooded
		Normal Blast Furnace—Non Hooded
		High Temp/Pres. Hooded
		High Temp/Pres. Non Hooded
Casting	Continuous	
	Ingot	
Reheating	Pusher Tupe	
	Moving Beam	
	Induction	
Final Finishing	Billets	
	Slabs	
	Blooms	

TABLE 21-1
Model Projections for the Iron and Steel Industry (base case)

Year	Final Demand Total (000 Tons)	Actual (1000 MWh)	Electricity Projections (1000 MWh)	Pierce, Sparrow, and Pilati
1980	83323	52329*	51900	51670
1981	85107	--	52426	--
1982	86486	--	52858	--
1983	87777	--	53280	--
1984	89088	--	53720	--
1985	90420	--	54174	--
1986	91771	--	56844	--
1987	93143	--	60766	--
1988	94534	--	64788	--
1989	95945	--	68733	--
1990	97379	--	69508	--
1991	98834	--	70236	--
1992	100311	--	70719	--
1993	101810	--	71031	--
1994	103331	--	72461	--
1995	104875	--	73919	71511

*American Iron and Steel Institute Annual Report.

production rising from 83 million tons in 1980 to 105 million tons in 1995. This gives rise to a 42 percent increase in electricity consumption over the same time horizon.

Included in the table is also a comparison of the model results with those obtained by Pierce, Sparrow, and Pilati (1). In comparing the results it should be noted that most of the data used in the INDEPTH model were derived from the Pierce, Sparrow, and Pilati work. However, the detail and the structure of the INDEPTH model is significantly different from that used by Pierce et al. Thus the agreement of the two model results supports the lower level of detail of the INDEPTH model.

A convenient way of comparing electricity projections under different assumptions is to focus on electricity intensity expressed as kWh per ton of steel produced. Table 21-2 shows the sensitivity of electricity intensity to a number of inputs. It is interesting to note that the electricity intensity is quite sensitive to the growth in steel production. In the case where steel production declines two percentage points a year, there is virtually no change in the intensity. However, as production increases, new capacity, primarily electric arc furnaces, comes on line, increasing the electricity intensity from 673 kWh/ton to 744 kWh/ton.

TABLE 21-2
Electricity Intensity

	kWh/Ton of Steel	
Model	1980	1995
(Base case)	623	705
2% Decline in steel demand	623	618
5% Growth in steel demand	623	744
Technology choice Extremely cost sensitive	616	716
Technology choice Moderately cost sensitive	627	695

As pointed out above, the decision rule in selecting new technology is cost minimization. To test the sensitivity of model results to this decision rule two cases were analyzed. In the case "technology choice extremely cost sensitive," basically the least-cost alternative is implemented, while in the case "technology choice moderately cost sensitive," the least cost, as well as other more costly technologies, are implemented. The model results show that the cost sensitivity has only a minor impact on the electricity intensity. One important reason for this is that there currently is sufficient production capacity in the industry. Thus, substantial capacity additions are only needed under the high-growth case.

USE OF PROCESS MODELS IN UTILITY FORECASTING

Effective forecasting at the utility service-area level requires a combination of techniques. The large data requirements of process models make them appropriate only for those industries that are of special interest to a utility. This special interest may arise due to the size of the industry, the volatility of energy consumption, or the high potential for change in electricity requirements. For industries that are not of special interest, due to stable electricity consumption patterns and size, econometric models are much more appropriate. A comparison of important features of econometric and process models is shown in Table 21-3. As shown in this table, the process models have a much broader application than just forecasting. The detailed information produced by these models is useful in

TABLE 21-3
Comparison of Econometric and Process Model Features

Econometric Model	Process Model
Applicable for all sectors	Applicable only for major industries in service area
Focus on aggregate relationships	Focus on investment in industrial processes
Aggregate data	Detailed data
Historical behavioral response	Behavioral response predictable
Interest in equilibrium	Interested in path to equilibrium
Primary application: Forecasting	Primary application: Forecasting Strategic planning Marketing/DSM Electrification

strategic planning, marketing, and the analysis of demand-side alternatives.

In summary, process models are appropriate tools for forecasting industrial electricity requirements at the service-area level. They are data-intensive and thus require a substantial amount of effort for implementation. However, it should be noted that much of the information needed by process models is "market" information, which utilities should collect to obtain a better understanding of their customers' use of energy, regardless of the type of forecasting model used.

REFERENCE

1. Brookhaven National Laboratory. Industrial Process Models of Electricity Demand, Volume 3: The Iron and Steel Industry. EPRI EA-3507. Palo Alto, CA: Electric Power Research Institute, May 1984.

QUESTION-AND-ANSWER SESSION FOLLOWING PRESENTATION

MR. ADAM KAHANE, Lawrence Berkeley Laboratory: I have two questions that I think are important in understanding how to interpret the results. The first is: Does this cost minimization criterion for choosing between alternative ways of making steel account for the past? In other words, if you tried this process, starting in 1970, would you get out the choice of processes that we have actually seen over that decade?

MR. BROEHL: You bring up a very interesting and very important point, and that is the validation of models, particularly process models. Up to this point, we have not had the opportunity to do an extensive amount of work in the area of validation. One of the problems is that, not only do you need information on actual costs, and other variables, over the past, but you also need projections of these variables and new technologies that were made at that particular time. Thus the validation is a fairly extensive process.

MR. TOM SPARROW, Purdue University: We tried the validation at Brookhaven. You tell me what your price expectations are and I'll tell you what the model will do. You have to build into it a trajectory, and if, in 1970, you believed electricity prices were not going to go up, the model would show all sorts of electrification.

MR. KAHANE: What if you put in the actual prices that occurred over the decade? I am just wondering whether the cost minimization is a good assumption.

MR. BROEHL: Well, I think to some extent it is a good assumption. It is the same assumption that we use in the INDEPTH econometric model. The production function assumes that an industry will operate in such a way that it minimizes cost and will select capital, labor, energy, and materials in such a way as to minimize cost. So, it is consistent, in that respect, to the econometric model using the KLEM specification.

MR. SAM SUGIYAMA, Bonneville Power Administration: Tom Sparrow is helping us develop the process model for pulp and paper in the BPA region. One of the fortunate aspects of having the Brookhaven model is that they developed a model for us in 1982. We have 1980 data, and part of what we are going to do is use the model from 1980 and see how it tracks what our estimate of 1985 is. We are hoping to run it annually, but we haven't gotten to that stage yet.

MR. PAUL WERBOS, U.S. Department of Energy: Are you planning on putting in the actual values?

MR. SUGIYAMA: I think we do; we calibrated the model in 1980.

MR. BOB CILIANO, Dun and Bradstreet: I don't want to put myself into the role of an interpreter, but I think that Tom Sparrow's comment, which is really very important, may have gone over some people's heads. If you want to validate a model like this, you do not use the actual prices that occurred during the period 1970 and 1980; you almost have to go back and search the literature and see what the common wisdom was in 1970 with regard to the price trajectory for the '70s. In other words, in terms of anticipation, find out what decision-makers thought the future looked like in 1970, put in that price trajectory, and then see how the model results compare to actuals.

MR. BROEHL: That refers not only to prices but also to the technologies and their operating characteristics. All of this places some very difficult requirements on validating process models. On the other hand, we have to recognize that we must be looking forward. We have to analyze, for example, effects of electrotechnologies that we see on the horizon on a utility's service area. A utility has to take a look at the effect of these technologies not at the national industry level, but rather at the service-area level. In other words, what is the impact on the utility's customers?

MR. WERBOS: It is hard for me to resist a discussion of process versus econometric models. I may have been a little bit too strong before. There is a clear application for these process models, such as ISTUM II or INDEPTH, in evaluating the future of specific technologies. Some of the problems with the existing process models are the sources of their lists of technologies. If you get the list from Amory Lovins, you are going to get one kind of future; if you get it from Herron, you may get a different kind of future. I hope the Energy Modeling Forum will be able to get an electrotechnology list and some other lists so that some comparisons can be made.

It is clear that EPRI has a need to worry about electrotechnologies, and you can't avoid process models in that context. I still think you need to calibrate them a little bit and you need to somehow try to account for those other factors when you are forecasting. There is a difference between forecasting a technology penetration and getting a baseline projection.

At DOE, we have done some experiments on price expectations. Our policy office did a study of decision-makers' price expectations in the last ten years. Basically the study concluded that if you took the last three months in the last year and did a kind of a minimization process and extrapolated that out five years, you could match decision-makers' expectations obtained from a survey quite closely.

I think it is safe to say that, in the period we just came out of, that is the period before OPEC started having problems, people's price expectations were very high. Thus, forecasts from process models we used show all kinds of conservation. If they had used more realistic price expectations, they still would have gotten a great deal of conservation, more than what we observed.

MR. BROEHL: Let me make one final comment on the validation. The reason the validation question has not been fully addressed is that it is a Phase III INDEPTH activity. The first two phases of the project focused basically on developing the models and showing how they work. The third phase focuses on implementing these models for a number of case-study utilities and showing how these models work in a utility planning environment. As part of that activity, we will be doing a very extensive validation exercise at the service-area level.

MR. AHMAD FARUQUI, EPRI: Thank you. This is really a discussion that could continue for at least several hours.

I would like to add, as a concluding note to the discussion, that, recognizing that process models have their strengths and weaknesses just as econometric models and end-use equipment models do, the approach we have taken in the INDEPTH project is a portfolio approach. Thus, we are complementing the strengths of one approach with the strengths of the other, rather than comparing the weaknesses of one with the weaknesses of the other. So, there is an attempt to use a system of approaches here, rather than rely on just a single approach in isolation. Hopefully, this portfolio approach will give us a better chance of producing a good forecast, rather than relying on just a single approach.

22

INDEPTH CASE STUDY: PULP AND PAPER PROCESS MODEL

Carie E. Lee and Samuel O. Sugiyama
Bonneville Power Administration

INTRODUCTION

The Bonneville Power Administration (BPA) is a test utility for the Electric Power Research Institute (EPRI) Industrial End-Use Planning Methodology (INDEPTH) Project. The primary objective of this project is to design and demonstrate a methodology for forecasting and shaping industrial electricity use, which will be adaptable to individual utility service areas. A system of econometric and more detailed process models for all SIC groups is being built for use in forecasting electricity demand and performing analyses related to marketing and conservation. The process models cover industries at the four-digit SIC level in greater detail. These models are designed to analyze the dominant, electricity-intensive industries in a particular service area, and to incorporate any firm-specific data that are available. BPA's participation as a test utility in the INDEPTH project enabled us to receive, with the contribution of staff time, process models of the pulp and paper (SIC 2611, 2621, 2631) and the chlor-alkali (SIC 2812) industries, and econometric models at the two-digit SIC level of detail. This paper describes BPA's experience in the implementation of the pulp and paper process model in our service area.

BPA FORECASTING FRAMEWORK

DESCRIPTION OF BPA'S INDUSTRIAL SECTOR

BPA's industrial sector consists of 19 Direct Service Industries (DSIs), all manufacturing firms other than the DSIs, and a nonmanufacturing large power and light component. The Direct Service Industries include ten primary aluminum reduction operations and nine other industrial customers who produce a variety of products such as nickel, titanium sponge, ferrosilicon, chlorine, caustic soda, and paper. For forecasting purposes, the DSIs are modeled on a firm-specific basis, as opposed to the rest of the industrial sector, which is forecast on a two- or three-digit SIC industry basis. Electricity sales to the total industrial sector is the largest component of total sales in the Pacific Northwest, accounting for more than 40 percent in 1980, as shown in Figure 22-1.

The non-DSI industrial sector is the primary beneficiary of the EPRI INDEPTH project. This sector is dominated by five major industries—paper and allied products (SIC 26), lumber and wood products (SIC 24), chemical and allied products (SIC 28), food and kindred products (SIC 20), and primary metals, excluding aluminum (SIC 33). Figure 22-2 displays the industrial sector sales breakdown.

BPA'S LONG-TERM FORECASTING PROCESS

BPA's load forecasting models are an integral part of a system of computer models, which include an electricity

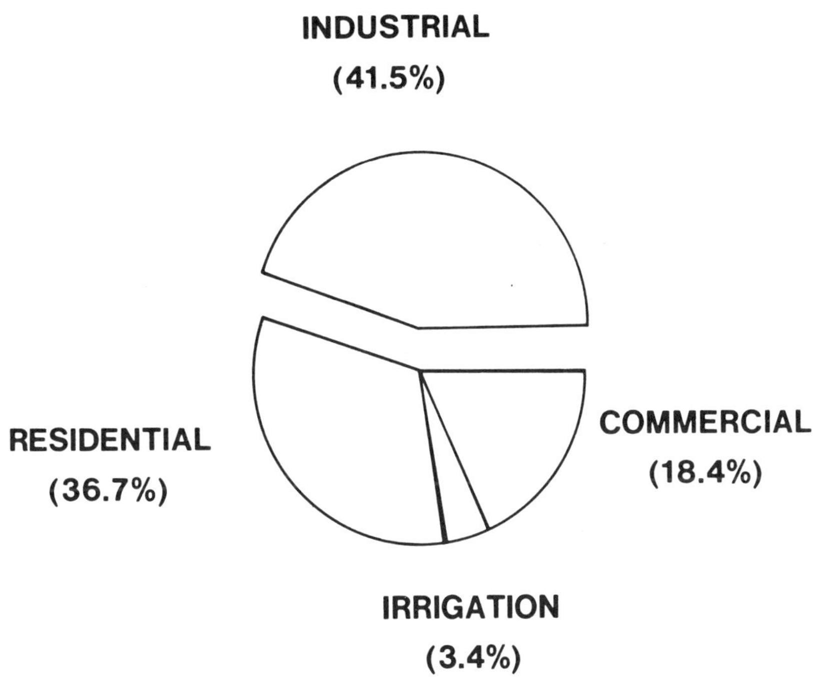

FIGURE 22-1
Pacific Northwest Regional Electricity Sales in 1980

Source: Bonneville Power Administration, Industrial Forecasting Section.

supply pricing model, a conservation supply curve model, and a linear programming model that determines the least-cost mix of resources that will satisfy the forecasted load. A diagram of this system of models is shown in Figure 22-3. This system requires, as a starting point, a basic set of economic driver variables, including electricity prices. The load forecasting models then iterate with the other models in the system until a condition of convergence is achieved with respect to electricity prices; i.e., the difference between the quantity of electricity supplied to quantity demanded in the final year of the forecast varies by less than 1 percent from iteration to iteration. The industrial forecasting models must fit into this system as much as possible to eliminate the need for manually iterating to achieve equilibrium.

At present, the EPRI INDEPTH model developed for the pulp and paper industry does not exactly fit these specifications, primarily because the model cannot produce a year-

FIGURE 22-2
Electricity Sales to the Pacific Northwest by Industrial Sector in 1980

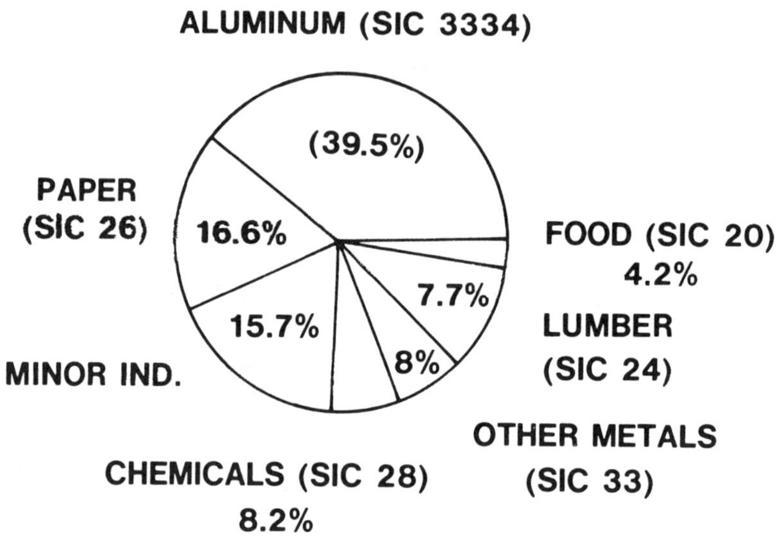

Source: Bonneville Power Administration, Industrial Forecasting Section.

to-year forecast without an analyst operating and handling the input and output for each year. A proposed solution to this problem is that a set of demand curves for electricity be constructed, containing the electricity price and quantity relationships indicated by running the full pulp and paper model using a variety of electricity price scenarios. This reduced-form version of the detailed pulp and paper process model could then be integrated into BPA's load forecasting system.

CURRENT FORECASTING MODEL

The current models used for forecasting electric energy consumption by BPA's non-DSI industries are econometric. The models for each industry consist of a translog fuel submodel with an ad hoc total energy specification. That is, the model forecasts total energy demand by industry, and then distributes this total to various energy types via forecasted fuel-cost shares that depend on relative fuel prices. In addition, a partial adjustment process is assumed, where the response of fuel consumption to changes in

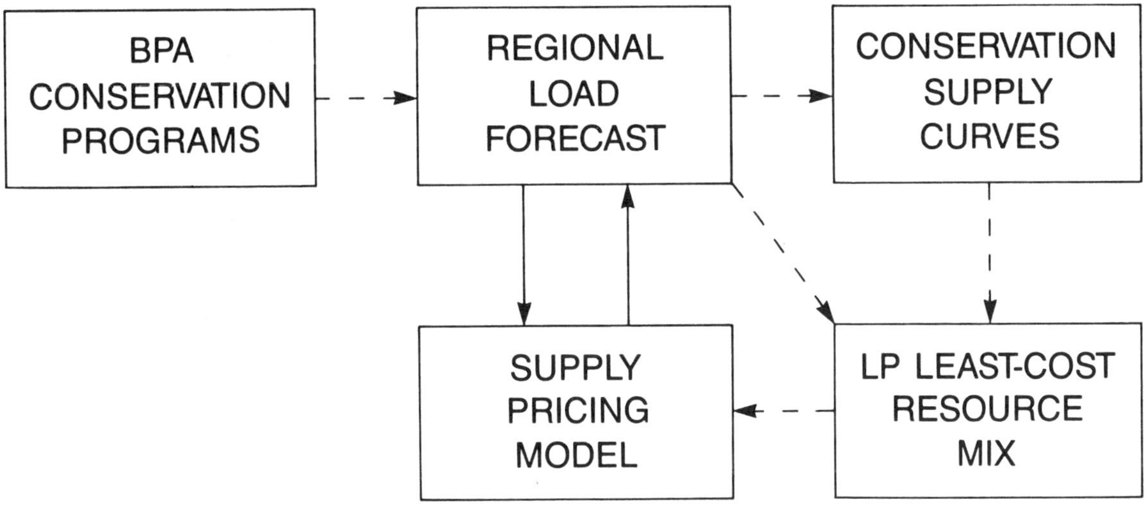

Source: Bonneville Power Administration, Industrial Forecasting Section.

FIGURE 22-3
BPA's Long-Term Forecasting Process

relative fuel prices is lagged, since investment in new equipment or conversions of old equipment do not occur immediately. The coefficients of the variables in these equations were estimated using national rather than regional data.

The above description clearly indicates that there is substantial room for improvement in modeling and analyzing the non-DSI industrial sector. The INDEPTH econometric specifications, although not yet received, appear to offer a consistently estimated set of models for BPA using all inputs in the production process and a more explicit recognition of the short-run and long-run situations. We are optimistic that the results will be significantly superior to our current specifications.

THE NEED FOR PROCESS MODELS AT BPA

Several events have occurred during the past five years that have increased the need for process models at BPA. This set of circumstances was not unique to BPA, and was similar to that which other U.S. utilities faced. First, the Regional Power Act passed in 1980 specified, among other things, that BPA was responsible for developing a strategy for conservation. Questions arose regarding the electricity conservation potential for each industry and, further, the programs that BPA could sponsor to capture those savings. Second, a leveling of the growth in electricity consumption by industries was taking place. Part of the decline in growth was attributable to rapidly increasing power rates, part to structural shifts to less energy-intensive sectors within each two-digit SIC industry, and part to the economic recession that severely impacted the major industries in our service area. It was necessary to analyze the importance of each factor in order to properly assess future electricity consumption. Third, the effects of the economic recession on Pacific Northwest industries were more than short-lived. Analysts were forecasting declining growth rates for industrial electricity sales, a trend also occurring in the residential and commercial sectors. Our analysis indicated that BPA was no longer in a deficit situation in the near term, and that a substantial amount of surplus power was available for marketing. Questions arose regarding the amount of electricity that would be consumed by specific industries if power rates were lower, and regarding BPA's ability to restart closed plants or attract new industry by offering a lower power rate. To provide answers to these and other questions raised by management, our staff needed to analyze industries at a more disaggregated level. Process modeling seemed to be the natural direction to take.

IMPLEMENTATION OF THE INDEPTH PULP AND PAPER PROCESS MODEL AT BPA

MODEL DESCRIPTION

The model installed on BPA's computer system was a single-region version of the national four-region model that the researchers at Brookhaven National Laboratory built under contract to the U.S. Department of Energy and the Electric Power Research Institute. This linear programming model forecasts fuels and electric energy demands under the assumption that firms strive to minimize their costs of production. Given a forecast of the demand for various paper and paperboard products and a forecast of fuel and electricity prices, the model chooses the combination of pulping and papermaking processes and energy-using equipment that will satisfy the end-product demand at the lowest possible cost. Figure 22-4 displays the structure of the model and the possible paths the model can choose in processing the raw material to manufacture the final end-products. The model contains 166 processes and 87 constraints. A detailed description of the Brookhaven National Laboratory's pulp and paper process model can be found in (1).

MODEL INSTALLATION REQUIREMENTS

Although BPA contracted with Brookhaven in 1981 to develop a single-region model for our forecasting purposes, the

FIGURE 22-4
BPA's Pulp and Paper Process Model Structure

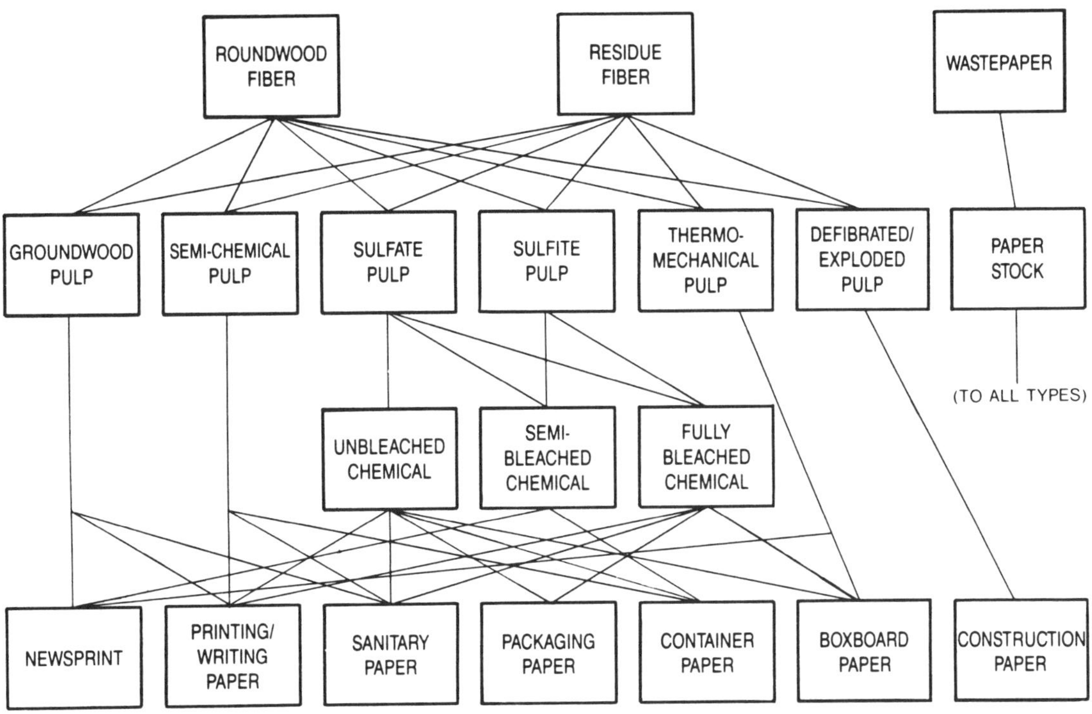

Source: Bonneville Power Administration, Industrial Forecasting Section.

model was not installed on BPA's computer system. Size restrictions with respect to the computer system and excessive staff support requirements prevented BPA from seriously considering in-house use of the model. Therefore, we relied upon our consultants at Brookhaven to run the model to produce long-term energy forecasts of consumption by the pulp and paper industry for our service area. Participation in the INDEPTH project presented us with an opportunity to have a process model installed at BPA.

There were several requirements BPA fulfilled to make model installation successful. At a minimum, we had to have a linear programming solution software package resident on our computer system. We discovered that we had the Mathematical Programming System Extended/370 (MPSX/370) package on our IBM mainframe. In addition, we needed a computer specialist conversant with the linear programming package and an economist who was familiar with the pulp and paper industry and who would later take responsibility for running the model.

There were two other factors that were not necessary, but that BPA felt would significantly enhance the use of the process model. First, we wanted the model to be readily incorporated into our overall forecasting framework. Second, we felt that it was important to discuss model inputs, structure, and results with regional industry representatives because periodic discussions of this nature would ensure that the model would better depict our regional situation. BPA did have the opportunity to meet twice with industry representatives through the Northwest Pulp and Paper Association; and we have established a working relationship with EKONO, an engineering consulting firm that is very familiar with the regional pulp and paper industry.

The model installation process began with a week-long site visit by INDEPTH project consultant, Dr. Tom Sparrow. Dr. Sparrow first installed a simple example to demonstrate the fundamentals of LP modeling and to allow us to exercise our LP software package. We then began construction of a static version of the pulp and paper process model on a piece-by-piece basis, with a discussion occurring first on paper, and subsequent transference of the model segment to the computer system. By the end of the week, we had a working model and our staff was comfortable in operating it. A second site visit by Dr. Sparrow was necessary in order to enlarge the existing model with additional process options and to make the model dynamic. The dynamic model allowed expansion of existing pulping and papermaking capacities and energy-using equipment capacities. What resulted from this experience was the installation of a static and a dynamic version of the pulp and paper process model on our computer system. Because these models were built using a "hands-on" approach, our analysts were able to run the model and to make appropriate changes with little difficulty.

DATA REQUIRED FOR PROCESS MODELS

The data that are required to run a process model are much more detailed and more difficult to obtain than the usual fuels and electricity price and quantity data required for econometric model estimation. Additional information on production, production capacity, the specific uses of fuel and electricity in the manufacturing process, and probable energy conservation measures or energy-saving processes need to be collected. Although most process models have nominal data sets available, some service area-specific data are necessary to make model acquisition worthwhile.

The collection of data necessary to "regionalize" the EPRI INDEPTH pulp and paper process model was not an insignificant task. We had a slight advantage, however, because our previous experience with the Brookhaven National Laboratory's pulp and paper process model provided us with a regional data set. Because the EPRI INDEPTH model was similar in structure to the BNL model, the data sets could be used, but not without consideration of future work in updating and revising the data.

The data required for the pulp and paper process model include the following:

- Base-year production capacities for pulping and papermaking activities.
- Base-year energy equipment capacities.
- Raw material input and output relationships (e.g., amount of wood fiber required per ton of kraft pulp).

- Energy input and output relationships (e.g., amount of natural gas required to raise 1,000 pounds of steam).
- Energy requirements per unit of output.
- Production costs (other than energy) per unit of output.
- Unit capacity expansion costs.

DATA SOURCES

There are 21 pulp and paper plants that account for 100 percent of regional electricity use. The primary source of data was EKONO, a local engineering consulting firm that has an established working relationship with the pulp and paper plants in the region. The consulting firm has the ability to survey the plants through the Northwest Pulp and Paper Association (NWPPA) and to provide results on an aggregate basis, thereby avoiding antitrust violations and disclosure of individual plant data by the survey participants. Also, if collected data are in need of further interpretation, the consultant is available to resolve those problems. BPA does, on occasion, meet with representatives of the NWPPA to ask general questions about the industry and to solicit comments related to our long-term electricity forecasts for their industry. During the last few years, the NWPPA has become increasingly active and knowledgeable in regional energy matters. They have changed their attitude with respect to supplying data to BPA. In the past, the industry feared that the data would result in adverse effects, such as mandatory targets for electricity conservation by all plants. The industry has subsequently learned that the more BPA understands, the better and more reasonable policy formulation will be.

In addition, there have been studies conducted in our region by agencies requiring answers to questions about industrial energy consumption. These studies have been sponsored by various state energy departments and regional planning entities. The data related to these studies have significantly contributed to our bank of information on regional industries.

DATA COLLECTION RECOMMENDATIONS

There have been numerous lessons learned with our experience in process modeling and related data collection. First and foremost, make certain you are knowledgeable about the specific industry before approaching industry officials. This approach will generate more cooperation as industry officials will then feel more confident that you will use their data properly. If possible, a reference data set for mill personnel to comment on is preferable to handing out a lengthy questionnaire. National "average" data are easily obtained through federal government publications, and specific case-study data can be found by perusing industrial trade periodicals. Second, remember that most firms are sensitive to releasing data since they may jeopardize their competitive position in the market. In addition, industry officials may be concerned that the data will be used to effect policies that are to their disadvantage.

Third, it is necessary at times to have third-party involvement in data compilation, as in the case of our consulting firm, in order to avoid antitrust violations. Fourth, exhaust the existing sources of information first before handing out survey questionnaires. Call your local industry trade associations to determine whether the data you need are already collected in some way. Look through industry directories such as Lockwood's Directory (**2**) for the pulp and paper industry, the Directory of Forest Products Industry (**3**) for the lumber and wood products industry, or the National Chlorine Institute's North American Chlor-Alkali Industry Plants and Production Data Book (**4**) for the chlor-alkali industry. These directories usually have production capacity and other general industry information available as a starting point. The various trade journals and publications contain the latest information on potential developments in energy-saving technologies or processes. It is also possible to find articles that give prototype specifications for plants employing a new process versus a plant employing a conventional process.

PRELIMINARY MODEL RESULTS

A demand curve showing the amounts of electricity that would be consumed by the pulp and paper industry at various electricity price levels for a given level of paper output demand is displayed in Figure 22-5. The model used to produce this demand curve represents the industry as it existed in 1980. The notes alongside each "step" of the demand curve explain the process and energy equipment use changes that occur as the price of electricity rises, i.e., as we move from right to left on the demand curve. For

FIGURE 22-5
Relationships of the Price and Quantity of Electricity Purchased by the Pacific Northwest Pulp and Paper Industry

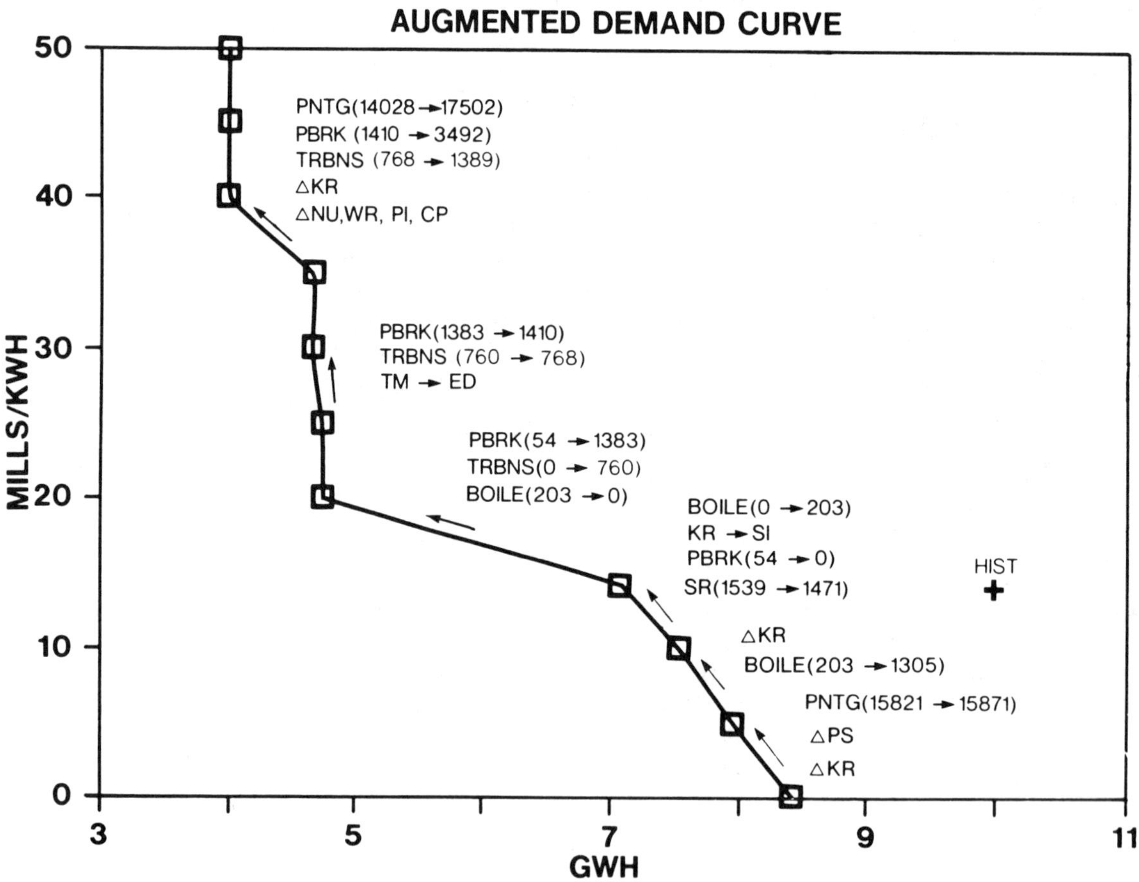

Source: Bonneville Power Administration, Industrial Forecasting Section.

example, as the price of electricity increases from 15 to 20 mills per kWh, the following changes occur to produce a reduction in the amount of electricity consumed: (1) bark purchases (PBRK) increase from 54 tons to 1,383 tons, (2) cogenerated electricity from steam turbines (TRBNS) increases from zero to 760 MWh, and (3) electric boiler activity (BOILE) decreases from 203 MWh to zero. In this particular step, only energy purchases and the use of energy equipment changed, and no process changes occurred. In contrast, the "step" below contained a process change—a substitution of sulphite pulping for kraft pulping (KR → S1).

This demand curve is labeled "augmented" because it was produced by running a version of the model that contained additional pulping and papermaking process options relative to the original model. For example, the original model contained one way to produce kraft pulp. The augmented model contains 49 ways to product kraft pulp.

These and other process options are defined in a report titled "Industrial Energy Productivity Project" (5). We have asked our consultant, EKONO, to review these options and comment on the probability of their use in our region. This is especially important, since the use of these options could result in significant energy savings, as evidenced by the comparison of the demand curves for runs produced by the augmented versus the unaugmented model in Figure 22-6. Note the position of the augmented demand curve to the left of the unaugmented demand curve. The lack of process options available in the unaugmented model is evident, in that a great portion of the demand curve is almost vertical, showing that the industry is unable to respond as easily to changes in electricity price, output demand remaining constant. The unaugmented demand curve may serve us well for analysis in the near term, while the augmented demand curve may be appropriate for longer-term analysis, provided there is a judicious selection of options to be included in the model. Historical consumption of electricity is marked with a "+".

Presently, our model results are slightly different from the actual electricity consumed in 1980, our base year. We are working on changing the model structure to correct this, and are also awaiting the delivery of data and information from our consultant to incorporate into the model.

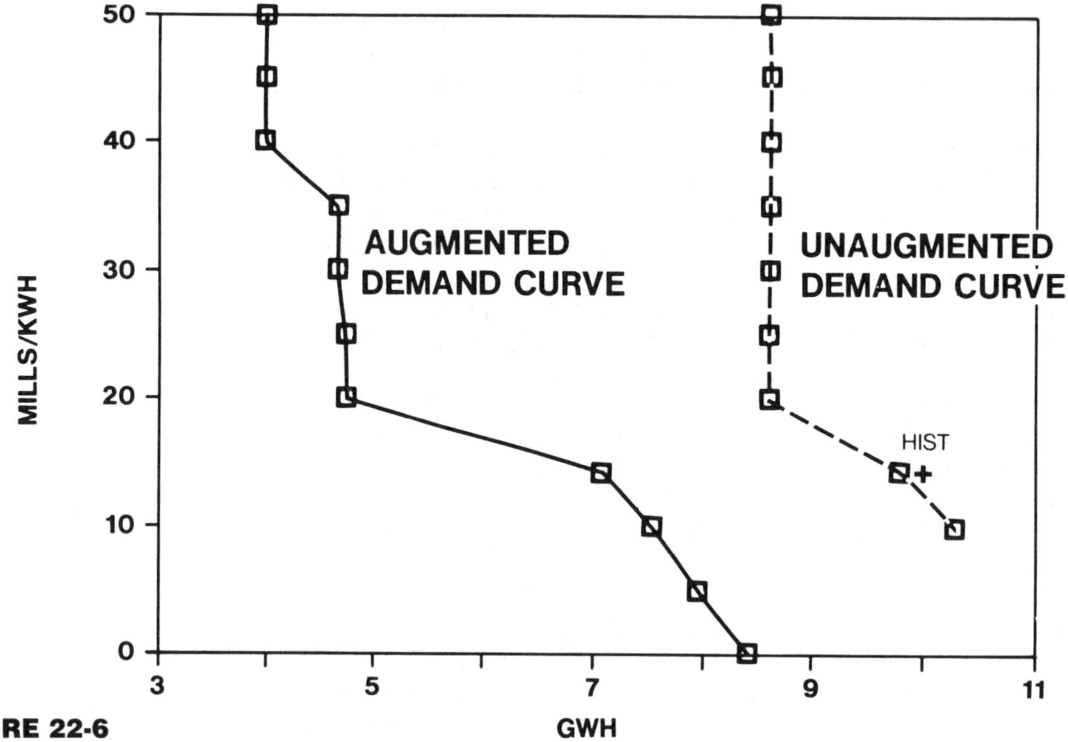

FIGURE 22-6
Results of the Augmented Versus Unaugmented Versions of the BPA Pulp and Paper Process Models

Source: Bonneville Power Administration, Industrial Forecasting Section.

CONCLUSIONS

Process models are becoming necessary tools for forecasting and analyzing electricity loads of BPA's large and electricity-intensive industrial customers. By now, many are aware of the obvious disadvantages of using process models. More staff time is required for the operation and maintenance of the models, the data requirements are large, and periodic consulting fees may need to be incurred. Furthermore, the model may be too large and cumbersome to fit into our forecasting framework, as in the case of the pulp and paper process model. The EPRI INDEPTH project, however, has provided the means by which utilities can experiment with process modeling, with the benefit of avoiding the expensive fees that consultants charge for model building alone. With the allocation of one industry analyst and one computer specialist, a utility can at least tap into the system of models being developed in the INDEPTH project, and then decide whether further investment in staff time and money will be worthwhile.

Based on our experience, we have found that there are several advantages to pursuing the process modeling effort. This effort encourages significant understanding of the industry being modeled. Even if the models are not put to use immediately, the very exercise of collecting the data to support the model is extremely beneficial. The analysts' understanding of how energy is used in the manufacturing process is enhanced, contact with industry is initiated, and the analyst will become familiar with the data sources available. Furthermore, a detailed analysis of electricity-intensive industries via the use of process models will enable BPA to address important questions that econometric models only poorly address at best. The INDEPTH project will produce for BPA a superior set of models capable of addressing questions related to marketing, conservation, and structural change.

REFERENCES

1. Brookhaven National Laboratory. Industrial Process Models of Electricity Demand, Volume 2: The Pulp and Paper Industry. EPRI EA-3507. Palo Alto, CA: Electric Power Research Institute, May 1984.

2. Lockwood's Directory of the Paper Allied Trade, 109th ed. New York: Vance Publishing Corporation, 1985.

3. Directory of the Forest Products Industry, 1985 ed. San Francisco, CA: Miller Freeman Publications, 1985.

4. North American Chlor-Alkali Industry Plants and Production Data Book. Chlorine Institute Pamphlet 10. January 1985.

5. Energy and Environmental Analysis, Inc. Industrial Energy Productivity Project, Final Report, The Pulp and Paper Industry. USDOE Publication No. CS/40151-1. February 1983.

QUESTION-AND-ANSWER SESSION FOLLOWING PRESENTATION

MR. ALTAF TAUFIQUE, Gulf State Utilities Company: In terms of structural changes, are you seeing any effect of imported pulp on your pulp and paper industry?

MR. SUGIYAMA: Significantly. Since 1979-1980, when the forest products industry was at its historical peak in our area, imports have declined significantly. According to Ms. Carie Lee, they will probably not reach those levels, in our long-term forecast anyway, until the late 1990s. So there is a significant impact.

MR. TAUFIQUE: Essentially where are the imports coming from in the pulp industry?

MR. SUGIYAMA: I think one of the major competitors is Canada.

MR. TAUFIQUE: In our area we have seen that people who are selling pulp are now going into papermaking because of the conflict in the pulp industry, and they are providing for a lot more flexibility in the papermaking area than what they had before. Is that what you have seen in your area, too?

MR. SUGIYAMA: We have integrated pulp mills. They go all the way from pulping to the final papermaking process. That is my understanding, at any rate.

MR. EDWARD FISCHLER, Georgia Power Company: If you had two different pulp and papermaking processes in your service area, such as a TMP that produced newsprint, a kraft that produced finished products.

MR. SUGIYAMA: Which we do.

MR. FISCHLER: Which you do, does that require essentially two different model calibrations, or can both industries be evaluated within the same modeling requirements?

MR. SUGIYAMA: We're planning to evaluate the process as a whole. We are going to look at output of the model in a given year. We are not trying to calibrate on a given plant;

rather we are going to calibrate for the particular four-digit SICs as a whole.

We will be looking at, as an end result, the amount coming out—I think we'll have that capability—to examine the amount of the kraft processing that is done. I'm not sure if we have all the data, but, if we do, we will certainly examine it.

MR. FISCHLER: You commented on the advantages and disadvantages of using process models as opposed to econometric models. Yesterday, I think Mr. John Broehl commented on some additional capabilities one has as a result of using a process model form. But, with the very narrow context of forecasting future load and energy requirements within a given industry, in your opinion—this probably has no really empirical data set to fall back on—but, in your opinion, does the additional gain in forecast accuracy that you might achieve, going to a process model form—is that justified in light of the additional resources that it takes to build a model of this type? Or do you get that benefit purely from being able to evaluate other kinds of questions?

MR. SUGIYAMA: You might be able to capture technology-switching with an econometric model, but you have to really understand the industry extremely well.

As a side note, I am a believer in an econometric approach, but I think the process models have something to offer. For those large industries—especially these smoke-stack industries, which the large industries usually are—I think there is a significant risk of losing them over the long run unless there are some significant technological developments.

Because of that, I think the process model gives us some indication, via the cost minimization objective criteria, as to when, approximately, that could come on line, and I think that is useful information. As far as levels, based on what I heard yesterday, that is something we need to be very careful about. It seems that the process model has a tendency to cogenerate extensively. I recall from our Brookhaven effort that we had to put some constraints on the cogeneration to get what we felt was a more reasonable forecast. But that is a way to do it.

An important additional response to that, in terms of constraining the model is this: We want to run it in a constrained way and in an unconstrained way, so we don't allow ourselves to be blind-sided.

MR. BOB CILIANO, Dun and Bradstreet: We were told, in some earlier presentations, that one of the main objectives of INDEPTH was to have a stable of models that would be highly portable across utilities. And I could think of many, many, many utilities that have a very important paper and pulp component in their industrial base that would be either unprepared or incapable of replicating the type of extensive effort that you have done in this area.

To what extent will your experience at BPA and your continued working with Tom and others at Battelle make this job easier for other utilities? What is it that they can save that you had to go through in this initial effort?

MR. SUGIYAMA: They need to be familiar, roughly, with the different kinds of pulping that are available. They need to get that information somehow. They can get some information from looking at Lockwood's directory. There are various other sources of information that are readily available. I also understand that Dun and Bradstreet has done considerable work on a data base.

MR. CILIANO: We do have a considerable amount of plant-level detail. But, I think, in looking at your data requirements, we have even come to the conclusion that, while we may have a relatively large percentage of that information, in many cases a lot of these utilities may have to do some limited-scale custom survey in order to get the other costs and parameters that would go into a model such as this.

So, it seems to be a major commitment that a utility has to make, and part of it has to be motivated by bridge-building to the industry and developing a rapport with these plants and industries, as much as it is to getting more accurate load forecasts.

MR. SUGIYAMA: That is true. One way around that, however—and we have done it with our econometric models—is to accept the national nominal data base, subject to trimming down in terms of the number of options available. You need to know at least whether you have a kraft process, how many kraft processes, sulfite processes, semi-chemical, and so on. If you can at least get to that level of detail, then use the national nominal data base to calibrate on the known output numbers. You can do that and perhaps generate some reasonable results. Also, the analysts need to know the amount of cogeneration possible. It they are not willing to put in that much effort, and that industry is large, then I would say something is wrong with their allocation of analytical effort.

23

STRUCTURAL CHANGE AND REGIONAL GROWTH

Thomas W. Moore
Tampa Electric Company

INTRODUCTION

Economists and forecasters have closely monitored the structural change that has occurred in our society during the past several decades as our nation evolves from an industrial to a service-oriented economy. The impact of this structural change can be viewed at the national level in the movement to a less unionized labor force and the expansion of industries geared toward data acquisition and processing. But this shift has also had a profound effect within individual regions and industries. The purpose of this discussion is to examine the impact of this change on an electric utility in Florida.

While Florida is often viewed as an area which is less industrialized than the rest of the country, there are geographic pockets within the state where there is a considerable amount of industrial activity. The area around Tampa, which is located in the west central portion of the state, is one of these. This greater intensity of manufacturing activity can be shown by examining the portion of electric utility energy sales that are industrial in nature for Tampa Electric Company, which serves this area. For 1984, Tampa Electric's industrial sales contribution was 34 percent, slightly below the 37 percent average of U.S. utilities but more than twice the 15 percent average for the state of Florida.

Within the Tampa Electric service area, there are several major changes occurring that relate to the manufacturing sector. The first, mentioned above, is the dynamic shift between manufacturing and services. The second is the transition within the industrial sector from heavy or energy-intensive production processes to lighter or less energy-intensive manufacturing. Finally there is the movement in the industrial sector toward self-generation by individual companies of their electricity requirements. Together, these forces are having a notable effect on industrial energy sales.

THE SHIFT FROM MANUFACTURING TO SERVICES

Table 23-1 examines the movement toward a service or commercial base as it shows the growth rates of the major sales categories at Tampa Electric during the past decade. Over this time span, industrial sales advanced at the slowest rate of the major groups. In fact, its pace of growth was well below the other classes, with commercial consumption expanding at a rate more than three times the industrial sector.

Table 23-2 views this shift from another perspective. Between 1970 and 1984, Tampa Electric's slower industrial energy growth resulted in an erosion of that sector's contribution to total sales from 45 percent to 34 percent. At the same time, the commercial category's share of company sales rose from 17 percent to 25 percent.

While Tables 23-1 and 23-2 examine the industrial to

TABLE 23-1
Energy Sales Growth 1974 to 1984

Sales Categories	Percent Growth	Average Annual Percent Change
Residential	42	3.5
Commercial	71	5.5
Industrial	22	2.0
Public authorities	60	4.8
Total sales	41	3.5

Source: Tampa Electric Company.

TABLE 23-2
Contribution to Total Energy Sales (percent)

	1970	1980	1984
Industrial	45	41	34
Commercial	17	20	25

Source: Tampa Electric Company.

commercial transition by looking at energy sales, it can also be viewed by analyzing local employment data. Table 23-3 contains a comparison of manufacturing and commercial employment growth in Hillsborough County, Florida, which represents a significant portion of Tampa Electric's service area. Similar to the sales data, commercial employment has expanded at an extraordinary pace since 1970, while the increase in manufacturing has been considerably slower.

Particularly noteworthy from Tables 23-2 and 23-3 has been the acceleration in the industrial to commercial shift between 1980 and 1984. This escalation in the industrial erosion can be attributed to several factors. First, there were two national recessions during this period, the second

TABLE 23-3
Hillsborough County Employment (thousands)

	1970	1980	1984	Average Annual Percent Change 1970-1984	1970-1980	1980-1984
Manufacturing	31.6	37.8	39.8	1.7	1.8	1.3
Commercial[a]	108.8	178.9	230.8	5.5	5.1	6.6

Source: U.S. Department of Commerce, Bureau of Economic Analysis and Department of Labor and Employment Security, State of Florida.

[a]Commercial category includes: services, trade, finance/insurance/real estate, and transportation/communications/public utilities.

of which was quite serious. Since manufacturing is considerably more sensitive to business cycles than the service sector, these downturns had a more substantial adverse impact on industrial energy sales than commercial. Also contributing to this trend has been the strong value of the dollar versus foreign currencies. This, as we know, hurts American manufacturing activity, which competes heavily with foreign producers both in United States and overseas markets. However, the service industry, for the most part, competes in local markets which are largely insulated from foreign competition. The combination of these two factors resulted in Tampa Electric's industrial sales falling 12 percent between 1980 and 1984, while commercial sales expanded a whopping 31 percent.

These developments had a particularly strong impact on the phosphate industry, which comprises 75 percent of Tampa Electric's industrial energy sales. Phosphate is a key ingredient in the production of agricultural fertilizers which are sold both in domestic and foreign markets. During the early 1980s, the sharp jump in the relative value of the dollar, the weak national economy, and the government's payment-in-kind (PIK) program, which paid U.S. farmers to set aside land from production, combined to dramatically reduce phosphate demand. In fact, phosphate production fell 30 percent between 1981 and 1982. While the industry's level of activity has improved over the past two years, the strong dollar continues to cause severe strains in the phosphate sector.

In summary, there are several important points related to the manufacturing/services shift in the Tampa Electric service area. The first is that these phenomena can be traced back at least over the past decade. Secondly, the transition has accelerated since 1980 due to several national economic downturns and an overvalued dollar.

THE TRANSITION FROM HEAVY TO LIGHT MANUFACTURING

Looking within the service-area industrial sector, there is occurring an alteration in the mix of manufacturing from high energy-intensive production, what is often called smokestack industry, to lower energy-intensive production, often given the tag of high tech. The decline of heavy industry is the result of intense foreign competition evolving from the strong dollar, a more youthful overseas capital stock, and the wage differential between foreign and U.S. labor markets. Evidence of the inroads being made by foreign industrial competition in the Tampa area is presented in Table 23-4, which shows sharp increases in imports of cement, steel building materials, machinery, and processed food products between 1980 and 1984. At the same time, Florida, like many other regions in the U.S., has a rapidly developing electronic component production industry. The primary geographic area for this expansion is a band across the center of the state which includes the Tampa region. Growth in this industry has helped to offset the erosion in heavy manufacturing.

This change at the local level can best be noted by examining employment trends. Table 23-5 contains a comparison of employment growth in SIC 36, which represents light industry, and SIC 32 and 33, which are examples of the heavy manufacturing categories. Between 1974 and 1984, employment in the SIC 36 group increased over eightfold due to substantial growth in communication equipment firms (SIC 336) and electronic component companies (SIC 367). Over this same 10-year period, the basic industry category lost workers, dropping nearly 300 jobs between 1980 and 1984. Since average electric usage per employee in the SIC 32-33 category is approximately 6 times greater than in SIC 36, it would require roughly an 1800 employee

TABLE 23-4
Foreign Imports—Tampa Port Authority (thousands of tons)

	1980	1981	1982	1983	1984
Building Materials	--	0.1	0.1	0.3	2.1
Cement	23.2	20.9	23.9	40.7	201.7
Steel	195.7	221.0	391.2	252.4	307.3
Machinery	--	0.1	2.0	0.8	3.7
Citrus concentrate	23.1	78.0	101.4	104.6	160.5

Source: Cargo Report, Tampa Port Authority.

TABLE 23-5
Comparison of Hillsborough County Employment Growth in SIC 36, and SIC 32 and 33

	1974	1980	1984	Percent Change 1974-1984
Electrical and electronic machinery, equipment and supplies (SIC 36)	509	2913	4198	725
Stone, clay, glass, concrete (SIC 32) Primary metals (SIC (33)	3817	3991	3694	-3

Source: Department of Labor and Employment Security, State of Florida.

increase in SIC 36 to compensate for the lost energy associated with the 300 person decline in SIC 32-33. The actual increase in SIC 36 employment in this period, however, was only 1285, so that the energy related to this growth was not sufficient to fully overcome the lost sales in the basic group.

The implication of this shift from heavier to lighter industry is that the substitutability related to this phenomenon requires a considerable amount of lighter industry growth to compensate sales lost in the energy-intensive sector. In the above example, this offset is not fully occurring, which provides a drag on industrial sales.

COGENERATION

Also important is the move within heavy industry toward self-generation or cogeneration. In the phosphate sector of our service area, there is a significant trend toward utilizing their waste heat as an energy source, which has the effect of removing the energy sales of these companies from our system. By 1990, Tampa Electric industrial customers are expected to be cogenerating about 1000 GWh which represents more than 25 percent of this sector's present energy sales.

FUTURE GROWTH TRENDS AND THEIR IMPLICATIONS

The anticipated impact of the above trends is illustrated in Table 23-6, which contains the Tampa Electric sales forecast for the major categories over the coming decade. Because of the shift from manufacturing to nonmanufacturing, the stagnation in heavy industry, and the trend toward cogeneration, the company's industrial sales are not expected to rise appreciably during the next ten years. In many areas of the country, this might have an adverse impact on utility revenues. However, Tampa is a fast-growing region fueled by an expanding population and related economic advances. Therefore, increases in commercial, residential, and government energy consumption resulting from a growing local economy will provide the major thrust behind sales gains for the company.

However, this shift away from the industrial sector does have an impact on the structure and efficiency of the company's electrical system. First, this transition represents a movement from non-weather-sensitive to weather-sensitive load as a percent of total system demand. Since Tampa Electric's system peak is in the winter and is needle

TABLE 23-6
Forecasted Energy Sales Growth 1984 to 1994 (percent)

Sales Category	Percent Growth	Average Annual Percent Change
Residential	28	2.5
Commercial	49	4.1
Industrial	1	0.1
Public authorities	30	2.7
Total sales	25	2.2

Source: Tampa Electric Company.

shaped, this shift would have the long-run effect of further accentuating the company's steeply shaped load profile. Also related to this problem is the implication that cogeneration and the slower growth of industrial sales can be expected to have a downward influence on the company's load factor. That is, the lower load factor residential/commercial/governmental group will be growing faster than the higher load factor industrial group.

To offset these trends, Tampa Electric has embarked on an aggressive conservation and load management program designed to reduce peak demand and improve the efficiency of the electrical system. In 1994, these programs are expected to reduce growth by around 500 MW or about half of the projected peak demand increase. This should successfully counterbalance the adverse influences of the structural change.

FORECASTING CONSIDERATIONS

From a utility-forecasting perspective, the best way to examine the possible impact of the trends that have been discussed is to utilize a more detailed forecasting methodology. At Tampa Electric, the nonphosphate industrial category is separated into four subcomponents grouped by energy intensiveness per employee. With this approach, the effect on total industrial sales of the faster-growing light manufacturing sector versus the slower growth in heavy manufacturing can be more efficiently analyzed.

A regression methodology can also provide insights into industrial energy trends and is usually in the form of industrial sales as a function of the relative price of electricity plus a national economic variable such as the industrial production index. In our service area, this relationship could not be disaggregated any further than at the total industrial level. Still, the industrial production index tracked Tampa Electric's industrial energy sales fairly closely during the 1970s, despite differences between local and national manufacturing mixes, because most industries moved more or less in tandem over this period. However, the strong relationship between our industrial sales and the industrial production index has deteriorated in recent years. This appears to be due to the widening growth dispersions among categories in the national industrial sector, as some industries are expanding rapidly while others are in decline. Therefore, we are finding that the industrial production index is becoming less useful as a surrogate for our service-area manufacturing activity. A good example of this development occurred in 1984, when the industrial production index increased 11.5 percent while manufacturing energy sales in Tampa grew less than 1 percent. Historically, the relationship of growth had been around 1 to 1.

In summation, the major economic changes occurring in the Tampa region include a shift in the share of energy from industrial to commercial (and also residential/governmental). Also important is the acceleration in this transition between 1980 to 1984, which appears to have been the result of the strong dollar and the weak economy of this period. There is also an internal change within manufacturing taking place in the form of a movement away from the heavy manufacturing processes. These developments, as well as the growing popularity of cogeneration, provide strong evidence that commercial, residential, and public authority sales growth should continue to exceed the industrial sector over the coming decade. While these trends could have future adverse implications for the company's load profile and load factor, Tampa Electric's conservation and load management programs should offset these influences and maintain the system's efficiency.

QUESTION-AND-ANSWER SESSION FOLLOWING PRESENTATION

MR. AHMAD FARUQUI, EPRI: You indicated that considerable structural change has been occurring in your service area, like from commercial to industrial, and, within the industrial, from heavy to light industry. Can you share some insights on how, as far as the future is concerned, you would forecast the structural change itself? Would it be based on some qualitative insights, or would it be perhaps the results of some models that you might be developing to explicitly analyze that, or maybe some other method?

MR. MOORE: Well, the end-use model is useful in that it basically defines and breaks into detail these categories within the industrial sector, so that you can see the heavy and lighter manufacturing disparities that are going on and forecast them that way.

What I was trying to explain was that the regression approach, at least in this form, didn't allow us to examine and separate these changes as neatly. But we expect that light industry will continue to be the major growth point within the industrial sector, and that heavy industry will continue to—as we've seen from the data—decline to some extent into the future. And that is the primary underlying assumption.

As far as work in the area of model building, we are in the process of reexamining our regression models. We are looking at quarterly models. Like Mr. Pradeep Gupta, we are interested in trying to build the value of the dollar or the strength of the dollar into our models, as an explanatory variable, and we are trying to disaggregate to the different pieces.

Unlike national data, when you try to build models at the regional level, data and data integrity are sometimes a problem, as I think most people out there would agree. So, it makes it hard sometimes to disaggregate the models as far as we would like.

But we are looking at other variables to try to explain these phenomena. Certainly, the price of electricity will continue to be a variable in our future regression models, but we will look toward the dollar as possibly being another explanatory variable.

MR. BOB CILIANO, Dun and Bradstreet: I am curious to see how integrated your forecasting models are among the sectors. For example, many of your diagrams related electricity use as a measure of intensity per employee.

It seems to me that, if you look at total multiplier effects, you may be better off to have Martin-Marietta building Pershing missiles in Orlando than someone making cement in your service territory, simply because, per unit of output, the number of employees that then become additional residential customers, and require so much in terms of services in commercial sector consumption, may more than offset that amount lost in some of your basic industries.

Are your models integrated so that the assumptions are directly transferable among the sectors?

MR. MOORE: Yes. And that really relates to our basic models. What I showed you here were the models that we use to develop the peak demand and energy forecast. Then, there is another set of models that is used to develop the assumptions which go into these models, which are assumptions related to employment, customer and population growth, and income growth. And these have the ties that you were talking about.

For example, we model migration as a function of the relative economic situation in our service area versus the rest of the United States. So, we have got migration in there. In addition, we have got commercial and industrial employment tied to local population trends. For example, we found that commercial employment can be explained by population and customer growth, and industrial employment through other factors related not just to the national economy but to the local economy. But, to answer your question: Yes, we have tied that all together.

One of the things that is important to look at, to our way of thinking, is the employment to population ratio. If you want to take it to its greatest detail, employment is compared to work-age population times the labor force participation rate to make sure the model is providing reasonable outputs. In other words, you don't want to be forecasting more jobs than population in the future, or you don't want an employment growth rate which is unreasonably high or low.

So, all of these ties are considered in our first model, which is an econometric model of about 20 equations which forecasts service area population and employment.

MR. CILIANO: So, I guess what could happen is: You could reach the consequence where, maybe, in your strategic interests, ignoring for the moment differences in load shapes and things of this sort, you actually encourage the demise of your industrial load simply because, by attracting less electric-intensive industry, you may overall increase your total electric sales when you consider these multiplier effects. Is that what is happening, as light industry and less energy-intensive industry move into your area?

MR. MOORE: Certainly they are happening at the same time. I'm not sure if the relationship is as neat as you have made it, but they are definitely happening. You have more population growth than you would have otherwise because lighter industry is adding more employees in our service area than are being lost through the decline in heavy industry.

MR. PRADEEP GUPTA, EPRI: One comment and one question. The comment is: In your presentation, at the beginning, you said that competitiveness does not relieve the impact of services. I would like to just mention that, on the national account level, services trade is actually keeping our situation a little bit bearable. We have a rather significant surplus, in terms of our trade with the other trading partners in the services area, and they are significant. Although it may not impact the service area of your utility significantly, on the national level it is significant.

My question to you regards the structure of your industrial load in future forecasting. One of the phenomena which I'm reading more and more about is the interutility competition for new growth in the industrial location, for example, two utilities might be competing for the same new customer. Do you see any such activity or any such concerns in your forecasting model important enough to be accounted for?

MR. MOORE: Well, I'm not sure if the utilities are often competing, but certainly the elements within the utility service area are, that is to say, the Chamber of Commerce and similar agencies. They are very aggressive about going around and spreading the word of how fantastic Tampa is versus Miami, and, of course, vice-versa. So, we see some of that competition outside the utility arena. Government officials of our state are not only going out to other parts of the country but also overseas to try to get firms to come into the state.

Yet, on the other hand, we have some conservation goals that we need to adhere to. Our state has conservation goals, so we have to balance both of these things.

There is some competition among utilities in selling power to another utility where they have a shortfall and you have an excess. There is plenty of that. But I do not know of any competition among utilities to bring new firms into their service areas.

I apologize for my comment on the commercial sector, because I meant to say, for the most part, it is insulated, but not completely. Certainly there is foreign competition in that.

MR. DAVID GOLDFARB, Georgia Power Company: You talked a little bit about some of the conservation programs that you were instituting in your service territory. Could you explain how large the impact is going to be in the future for your company; and, second, what, if any, evaluation methodology have you introduced to track the impact of such programs?

MR. MOORE: I think I had stated we have expectations for 500 MW of savings in 1994 from conservation and load management. While this figure doesn't mean anything by itself, compared to the potential growth of 1000 MW that we are anticipating, it means that potential growth would be reduced by about half.

Measuring the impact is difficult because in certain cally measured. Measuring conservation savings is even more difficult because you need to try to estimate how many controls are out there, and assign a savings value for each. Therefore, these savings do not represent a number that shows up on any one meter like peak demand or energy sales might; it is a number that is arrived at through some estimates and calculations.

Of course, we are concerned because we use the data in our forecast, and, therefore, we want both the estimates as well as their forecasts to be as accurate as possible. The state of the art in that area is improving but it still has a way to go. But we are, like most utilities, when we can, putting in better equipment to try to monitor exactly what that effect is.

Appendix A

BIOSKETCHES

JOHN A. ALIC

Mr. Alic has been a Project Director with the Office of Technology Assessment (OTA) of the U.S. Congress since 1979. Currently in the Industry, Technology, and Employment Program, he was responsible for two OTA studies: U.S. Industrial Competitiveness: A Comparison of Steel, Electronics, and Automobiles, published in 1981, and International Competitiveness in Electronics, published in 1983. Mr. Alic has also contributed to OTA assessments dealing with automobile fuel economy, and, most recently, structural unemployment and retraining of displaced workers. He is presently directing a study titled International Competitiveness in the Service Industries.

LAUREL ANDREWS

Ms. Andrews is a Senior Economist at Synergic Resources Corporation (SRC) in Seattle, Washington. She joined SRC in 1984, and specializes in demand forecasting and analysis of electricity conservation. Prior to joining SRC Ms. Andrews was an Economist in the Load Forecasting Group at Seattle City Light. Her responsibilities included the commercial short-range forecast and conducting a commercial building survey. Ms. Andrews also worked as an Energy Economist for the City of Seattle Office of Policy and Evaluation in 1980. Ms. Andrews holds an M.A. and "All But Dissertation" in Economics from University of California, Los Angeles, which she completed in 1979. She is a member of the International Association of Energy Economists.

BARRY BLUESTONE

Dr. Bluestone is Professor of Economics and Senior Research Associate at the Social Welfare Research Institute at Boston College. He attended the University of Michigan where he received his Ph.D. in Economics in 1974. As an Economist, Dr. Bluestone has specialized in the areas of labor studies, regional economic development, and industrial policy. At the Social Welfare Research Institute, Dr. Bluestone is currently working on a number of major research projects. One of these involves the development of a large-scale computerized microsimulation model capable of analyzing the consequences of a wide array of government tax, transfer, and expenditure policies. Part of his work involves consulting with trade unions and various government agencies. He was Executive Adviser to the Governor's Commission on the Future of Mature Industries in Massachusetts. Dr. Bluestone has testified before a number of congressional committees and lectures regularly before university, labor, community, and business groups. He also contributes regularly to academic, as well as popular journals, and is the author of five books.

JOHN H. BROEHL

Mr. Broehl is a Research Leader in the Applied and Technical Economics Department at Battelle Columbus Division. During his 18 years at Battelle, Mr. Broehl has been developing and applying quantitative decision analysis techniques in the area of energy economics. Since 1975 he has been primarily involved in utility planning projects, including load forecasting, load shape analysis, and demand-side management. He is one of the authors of the Edison Electric Institute/Electric Power Research Institute (EEI/EPRI) Primer on Demand-Side Management and a key member of the EPRI Demand-Side Management Project. Mr. Broehl holds an M.S. in Mathematics from Michigan State University and an M.B.A. from The Ohio State University. He is the author of numerous reports and papers in the area of utility planning and forecasting.

JAMES F. COLLINS

Mr. Collins is Executive Vice President and Treasurer of American Iron and Steel Institute. He previously served as Senior Vice President and Vice President of Public Affairs. From 1962 to 1967 he was employed by the U.S. Department of Commerce as Deputy Assistant Secretary of Commerce for Business Policy and he later became the Acting Assistant Secretary. Mr. Collins holds an A.B. in Economics from Brown University and has done graduate work in International Economics at Harvard and Columbia.

JOSEPH W. DUNCAN

Dr. Duncan joined The Dun & Bradstreet Corporation in January 1982, and serves as the Corporate Economist and Chief Statistician. In this role he works with the company's extensive information resources to develop new economics-related products and services to analyze economic trends and to evaluate the economic impact of government policies and business practices. He was previously Chief Statistician for the Office of Information and Regulatory Affairs in the Office of Management and Budget where he was responsible for national statistical policy. He also served as the U.S. Representative to the United Nations Statistical Commission, of which he was chairman in 1981. He is past President and Chairman of the National Economists Club, a fellow of the American Statistical Association, and an elected member of the International Statistical Institute.

Dr. Duncan received his B.S.M.E. from Case Institute of Technology, an M.B.A. from Harvard Graduate School of Business Administration, and a Ph.D. in Economics from The Ohio State University. He has lectured at a number of universities.

AHMAD FARUQUI

Dr. Faruqui manages the End-Use Assessment and Forecasting Subprogram at the Electric Power Research Institute (EPRI) in Palo Alto, California. He joined the Institute in 1979 as a Project Manager in the Rate Design Study Program, moving to the Demand and Conservation Program in 1980. In 1984, he was appointed Senior Project Manager responsible for demand-side management research. For the two years prior to joining EPRI, Dr. Faruqui was a Consultant to the California Energy Commission in Sacramento, California. In this position, he helped develop a statewide model for forecasting industrial energy demand. From 1975 to 1977 he taught Economics at the University of California, Davis, and in 1974, he was a Lecturer at the University of Karachi, Pakistan. Dr. Faruqui graduated from the University of Karachi, Pakistan, in 1973 and holds M.A. and Ph.D. degrees in Economics from the University of California, Davis. Dr. Faruqui is a member of the American Economics Association and has contributed 3 books and more than 25 papers to the professional literature on energy forecasting, market planning, demand-side management, electricity pricing, and economic development.

STANLEY J. FELDMAN

Dr. Feldman has been with Data Resources, Inc. (DRI) since 1980 and is Vice President of Interindustry and Regional Services. In this capacity he oversees a large staff of professional economists who specialize in detailed industry and regional analysis and forecasting, and in evaluating public policy issues that range from defense expenditure impacts to the economics of import restrictions. Before coming to DRI, Dr. Feldman was a Senior Economist with Prudential Insurance Company specializing in forecasting and analyzing financial markets and the macroeconomy. He also held a position as an Economist with the Federal Reserve Bank of New York. Dr. Feldman received a B.A. in Economics from Hunter College, City University of New York, and a Ph.D. from New York University. Dr. Feldman has prepared research studies, policy evaluations, and testimony for various public and private organizations.

PRADEEP GUPTA

Dr. Gupta is the Director of the Energy Analysis Department for the Electric Power Research Institute (EPRI) in Palo Alto, California. He was formerly the Manager of the EPRI Demand and Conservation Analysis Program. Before joining EPRI in 1980, Dr. Gupta spent eight years with the Southern California Edison Company in Rosemead, California, first as Supervisor of Load Forecasting and later as Supervisor of Load Management Planning. Previously, Dr. Gupta worked for three years at Systems Control, Inc., Palo Alto, as a Consultant on utility planning problems. He is an Advisor to the Alliance to Save Energy and The John A. Hartford Foundation. Originally from Delhi, India, Dr. Gupta received his M.S. and Ph.D. degrees in Electrical Engineering from Purdue University in 1966 and 1969, respectively. He received a B.S. degree from the Indian Institute of Technology in 1964. Dr. Gupta is a member of IEEE and the International Association of Energy Economists, and he is a Registered Professional Engineer in California.

BRUCE G. HUMPHREY

Dr. Humphrey joined Edison Electric Institute (EEI) in 1982 and serves as Director of Economics. He provides expertise on the macroeconomic environment in which utilities operate, microeconomic topics specific to electric utilities, and methodologies for forecasting, modeling, and analysis. Recent work has focused on business cycles, economic trends, and structural change as they relate to electricity. Before joining EEI, Dr. Humphrey served as an Economist in the Office of Regulatory Policy at the U.S. Department of Commerce, where he conducted economic analysis of regulatory policy, including assessment of its impact on industries, the economy, and international trade. Previously, Dr. Humphrey performed economic analysis of energy supply and demand, econometric modeling, and energy forecasting for the Commerce Department's Office of Energy Programs. He received his M.A. and Ph.D. in Economics from Tufts University, Boston.

HILLARD G. HUNTINGTON

Dr. Huntington is a Senior Research Associate in the Department of Engineering-Economic Systems and the Director of the Energy Modeling Forum (EMF) Affiliate Program at Stanford University. At EMF he has conducted studies to improve the use and usefulness of energy models for business planning and policy analysis. His principal research interests have focused on natural gas markets and policy, energy demand, and energy-economy interactions. Prior to coming to Stanford in 1980, he was the Director of the Washington Energy Office for Data Resources, Inc. Other positions he has held include: Senior Economist, Data Resources, Inc.; Staff Economist, Federal Energy Administration; Visiting Research Associate, Institute for Development Studies, University of Nairobi, Kenya; and Assistant to the Operations Engineer, Public Utilities Authority, Monrovia, Liberia. He received his B.S. in 1967 from Cornell University and his Ph.D. in Economics in 1974 from the State University of New York.

WILLIAM R. HUSS

Mr. Huss is the Manager of Forecasting and Planning at Battelle Columbus Division, where he has led a number of projects in planning and forecasting for the electric and gas utility industries. He is currently involved in a study for the Electric Power Research Institute (EPRI) which is developing a methodology for forecasting and simulating electrical energy, peak demand, and patterns of use by service area and establishment in the manufacturing sector. Mr. Huss also has been a key participant in transferring Battelle's SHAPES end-use load forecasting package to the personal computer. He has also led a project for a large gas utility to forecast the amount of fuel switching (from gas to oil) which will occur in the industrial sector given various future price scenarios. Mr. Huss holds a B.A., summa cum laude, in Mathematics and Physics from Gettysburg College, an M.A. in Public Administration, and an M.S. in Industrial and Systems Engineering from The Ohio State University. He is completing a Ph.D. in Forecasting and Decision Analysis from The Ohio State University.

KURT E. KARL

Mr. Karl, Director of Long-Term Forecasting Services at Wharton Econometric Forecasting Associates, is responsible for business unit management, project research, and production of forecast publications. His expertise in the area of macroeconomic analysis includes the summary of the long-term macroeconomic forecast, productivity, employment, prices and wages, and the housing sector. Before joining Wharton, Mr. Karl was Chief Statistician for the Government of Swaziland and Research Associate at Birbeck College, University of London. He received an M.Sc. in Economics from the London School of Economics. His numerous studies of the U.S. economy include an analysis of the possible effects of an international financial crisis on the U.S. economy, and various fiscal and monetary policy analyses.

STANLEY LANCEY

Dr. Lancey joined the Economics Department of the American Paper Institute in 1977 and presently holds the position of Director of Quantitative Analysis. His responsibilities include the preparation of market outlook reports concerning various grades of paper. He holds a B.S. in Mathematics from the City College of New York and a Ph.D. in Economics from the University of Pennsylvania.

ROBERT Z. LAWRENCE

Dr. Lawrence is a Senior Fellow in the Economic Studies Program at the Brookings Institution, specializing in international economics. He received his B.A. in Economics from the University of Witwatersrand, South Africa, in

1970, and earned his M.A. in International Relations in 1973, and a Ph.D. in Economics in 1978 from Yale University. Dr. Lawrence has been an Instructor at Yale University and a Professorial Lecturer at the Johns Hopkins School of Advanced International Studies. He has served as a Consultant to the Federal Reserve Bank of New York and to the World Bank, and as a member of the U.S. Congress Joint Economic Committee working on the Special Study on Economic Change. His many publications on domestic and international economic problems include contributions to the Brookings Papers on Economic Activity; Primary Commodity Markets and the New Inflation (co-author); and Can America Compete?, 1984.

ROBERT C. MARLAY

Dr. Marlay is Special Assistant to the Deputy Assistant Secretary for Conservation in the U.S. Department of Energy, where he is responsible for policy coordination and multiyear R&D planning. His primary focus is on advanced, highly efficient, energy end-use technologies. He also is involved in implementing congressionally mandated, national energy conservation programs. His areas of expertise include energy technology, strategic planning, and policy analysis. Most recently Dr. Marlay's research focuses on industrial energy productivity. His work is published in several reports, journals, and books. Dr. Marlay earned his Ph.D. from the Massachusetts Institute of Technology where he studied Nuclear Physics, Thermodynamics, and Reactor Engineering. He also holds an M.S. in Engineering Systems Analysis and another M.S. in Urban Planning, both from MIT. His undergraduate training was in Civil Engineering at Duke University. He is a Registered Professional Engineer in the District of Columbia.

TOM MOORE

Mr. Moore is Manager of Economic and Load Forecasting at Tampa Electric Company. His responsibilities include supervision of economic research and development of the peak demand/energy sales forecast for the company. Mr. Moore has an M.A. in Economics from Lehigh University, and an M.B.A. from Rutgers University. He also teaches graduate courses in Macroeconomics and Forecasting at the University of Tampa. Mr. Moore served on the Governor's Economic Advisory Committee, State of Florida, from 1979 to 1984.

WILLIAM J. SHEPPARD

Dr. Sheppard is Senior Chemical Market Analyst in the Advanced Materials Department of Battelle Columbus Division. An Economic Analyst, he has extensive experience in working with the planning, policy, and economic aspects of chemicals production and marketing. He also has experience in energy economics, including fossil fuel availability, energy storage, and renewable energy forms. He has conducted in-depth research into determining the economic impacts of regulations and other government policies on the energy and chemical industries, their markets, and their products. Dr. Sheppard has a Ph.D. and M.A. in Organic Chemistry from Harvard University. He has a B.A. in Chemistry from Oberlin College.

ALBERT J. SOBEY

Mr. Sobey is Senior Director of Energy and Advanced Product Economics at General Motors Corporation. He is also a member of the Economic Staff Policy Committee and the GM Corporation New Business Advisory Committee. In 1962 he directed the analysis of future automotive and freight transportation systems for both the design and research staffs of the GM Technical Center. In 1972 he joined Booz, Allen & Hamilton as a Consultant on transportation and energy. He returned to GM as Manager of Division Planning for the Transportation Systems Division when it was formed in 1974. At TSD he directed studies of advanced personal and freight transportation systems, and the economics of energy. Mr. Sobey has a degree in Mechanical Engineering and has taken advance courses at Purdue University, Butler, and Oakland (Michigan). He has published and spoken extensively on transportation and technical subjects, has authored a book on aircraft engine controls, and has been granted more than 20 patents. He holds professional engineering licenses in three states and Mr. Sobey participates in many national technical and professional activities.

WILLIAM L. STEWART

Mr. Stewart is Deputy Director of the Division of Science Resources Studies and Head of the R&D Economic Studies Section of the National Science Foundation. He directs a program of surveys and economic analysis of the supply and utilization of financial and manpower resources for science and technology in government, industry, universities and colleges, and other nonprofit organizations. He earned a Bachelor's Degree and has done graduate work in Economics at the American University in Washington, D.C. Mr. Stewart attended Stanford University on a NIPA Fellowship when he received a certificate in systematic analysis. Mr. Stewart is the U.S. Delegate to OECD and UNESCO meetings of experts on R&D statistics. He is a member of the American Economic Association and the American Association for the Advancement of Science.

SAM O. SUGIYAMA

Dr. Sugiyama is Chief of the Industrial Forecasting Section at Bonneville Power Administration. Employed at BPA since 1982 as an Industry Economist, he is currently responsible for model development, analysis, and forecasts of industrial sector electricity demands in the Pacific Northwest. He received a Ph.D. in Economics from the University of Utah in 1981 and a B.A. in Economics and Mathematics from Western Washington University in 1971. Dr. Sugiyama is presently an Adjunct Professor in the Masters of Public Administration program at Lewis & Clark College. He has also served as a Lecturer in Economics at Portland State University, Research Assistant Professor of Economics at the University of Utah, and Operations Research Analyst at Utah Power & Light Co.

PAUL WERBOS

Dr. Werbos is currently the Lead Analyst for the Energy Information Administration (EIA) in industrial transportation demand. In the past, when EIA projected to the year 2000 and beyond, he had lead repsonsibility for evaluating these forecasts and comparing them with other long-term projections, and for evaluating models of oil and gas supply. He has served on three interagency committees, involving a reassessment of the Global 2000 data and models, emissions assumptions for acid rain, and data revision policy. He holds four degrees from Harvard and the London School of Economics, in Economics, International Political Economy, Mathematical Physics, and Statistical Mathematics. He has published papers and government reports in subjects ranging from future electricity demand and a comparison of global models through to basic physics, mathematical models, and philosophy. Before working for the government, he taught and worked on research projects.

Appendix B

SEMINAR REGISTRATION LIST

John Adams
Supervisor
Load Forecasting
New York Power Pool
3890 Carmen Road
Schenectady, NY 12303
(518) 381-2139

Bill Baldwin
Engineer
Public Service Company of Oklahoma
212 East 6th Street
Tulsa, OK 74119
(918) 599-2271

James Nick Bayne
Senior Economic Development
Representative
Carolina Power & Light Company
P.O. Box 1551
Raleigh, NC 27602
(919) 836-6653

J. Alan Beamon
Economist
Energy Information Administration
Department of Energy
1000 Independence Avenue
Mail Stop 2F-021
Washington, DC 20585
(202) 252-4434

Dan Birch
Economist
Tennessee Valley Authority
MR5S46D-C
1101 Market Street
Chattanooga, TN 37402-2801
(615) 751-2574

Donald Bonney
Senior Energy Economist
Pacific Gas & Electric Company
77 Beale Street
Room 1183
San Francisco, CA 94106
(415) 972-2376

R. Larry Brantley
Corporate Planner
Georgia Power Company
333 Piedmont Avenue
23rd Floor
Atlanta, GA 30302
(404) 526-7218

Harry Brown
Dun & Bradstreet Technical
Economic Services
20 University Road
Charles Square
Cambridge, MA 02138
(617) 547-6000

Donald Bules
Energy Consultant
Energy Services Department
Pacific Gas & Electric Company
Suite A1268
333 Market Street
San Francisco, CA 94106
(415) 972-4238

Larry Butler
Director
Forecasting
Southern California Edison Company
Room 464
2244 Walnut Grove Avenue
Rosemead, CA 91770
(818) 302-2165

Robert Camfield
Economist
Georgia Power Company
333 Piedmont Avenue
Atlanta, GA 30308
(404) 526-7419

Bob Ciliano
Director
Market & Product Planning
Dun & Bradstreet Technical
Economic Services
20 University Road
Charles Square
Cambridge, MA 02138
(617) 547-6000

Connie Clonch
Energy Economist
Pacific Gas & Electric Company
77 Beale Street
Room 1183
San Francisco, CA 94106
(415) 972-2376

William Crane
Forecast Analyst
Alabama Power Company
600 North 18th Street
Birmingham, AL 35203
(205) 250-2396

Reed C. Davis
Supervisor
Budget & Forecasting
Utah Power & Light Company
1407 West North Temple
Salt Lake City, UT 84116
(801) 535-4139

Gene Dyar
Forecaster
Duke Power Company
400 South Church
Charlotte, NC 28242
(704) 373-8616

Kenneth Embry
Senior Engineer
Texas Utilities Electric Company
2001 Bryan Tower #1830
Dallas, TX 75201
(214) 653-4834

Ralph Ferraro
Program Manager
Industrial Department
Electric Power Research Institute
3412 Hillview Avenue
P.O. Box 10412
Palo Alto, CA 94303
(415) 855-2557

Edward Fischler
Forecast Manager
Georgia Power Company
P.O. Box 4545
Atlanta, GA 30302
(404) 526-7413

Joseph Frueh
Economist
Tennessee Valley Authority
MR5S43 D-C
1101 Market Street
Chattanooga, TN 37402-2801
(615) 751-2578

Paul A. Gnadt
Program Manager
Martin Marietta Energy Systems
Building 5500, MS-218
Oak Ridge National Laboratory
P.O. Box X
Oak Ridge, TN 37831
(615) 574-0266

David Goldfarb
Economic Analyst
Georgia Power Company
P.O. Box 4545
Atlanta, GA 30302
(404) 526-7413

Tom Grahame
U.S. Department of Energy
1000 Independence Avenue, S.W.
Washington, DC 20585
(202) 252-5735

John Grocki
Forecast Analyst
Public Service Electric & Gas
411 B
80 Park Plaza
Newark, NJ 07101
(201) 430-6685

Charles L. Hazlett
Manager
Market Development Department
Northern Indiana Public Service Company
5265 Hohman Avenue
Hammond, IN 46320
(219) 853-5904

E. L. Hillsman
Martin Marietta Energy Systems
P.O. Box X
Oak Ridge, TN 37831
(615) 473-5938

Derek H. Howell
Analyst
Economic & Load Forecasting
Northeast Utilities Service Company
107 Selden Street
Berlin, CT 06037
(203) 665-5048

William Idzerda
Energy Research Analyst
Pacific Gas & Electric Company
Room 1119
77 Beale Street
San Francisco, CA 94106
(415) 972-7534

Adam Kahane
Consultant
Lawrence Berkeley Laboratory
University of California
90-3125
Berkeley, CA 94720
(415) 972-7425

Jane Kilby
Director
Business Development
Central & South West Services,
Incorporated
Suite 2500
2121 San Jacinto
Dallas, TX 75266
(214) 754-1166

Donald F. King
Economic Analyst
Philadelphia Electric Company
2301 Market Street
Philadelphia, PA 19101
(215) 841-5635

Shannon M. Larson
Research Analyst
Boston Edison Company
800 Boylston Street P303
Boston, MA 02199
(617) 424-2946

Earl Lewis
Economic Research Associate
Baltimore Gas & Electric Company
Charles Center
P.O. Box 1475
Baltimore, MD 21203
(301) 234-5129

Dilip Limaye
President
Synergic Resources Corporation
Suite 1050
555 City Line Avenue
Bala Cynwyd, PA 19004-3406
(215) 667-2160

Frank Lin
Senior Analyst
New England Power Service
25 Research Drive
Westboro, MA 01581
(617) 366-9011 X2762

John Locher
Senior Engineer
Detroit Edison
2000 Second Avenue
Detroit, MI 48226
(313) 237-8839

Phil Loy
Research Assistant
Pennsylvania Power & Light
Two North Ninth Street
Allentown, PA 18101
(215) 770-6834

Wayne Lucas
Director
Engineering Special Projects
Kentucky Utilities
1 Quality Street
Lexington, KY 40507
(606) 255-1461

Edward Lynch
Forecast Specialist
Public Service Electric & Gas
80 Park Plaza
Newark, NJ 07101
(201) 430-6687

William Marks
Economist
Southern Company Services
211 Lakeshore I
P.O. Box 2625
Birmingham, AL 35202
(205) 877-7204

Mark Mayo
Senior Economist
Central & South West Services,
Incorporated
P.O. Box 660164
Dallas, TX 75266-0164
(214) 754-1149

David Meyer
U.S. Department of Energy
1000 Independence Avenue, S.W.
Washington, DC 20585
(202) 252-5735

Terry Morlan
Manager
Demand Forecasting
Northwest Power Planning Council
850 Southwest Broadway
Suite 1100
Portland, OR 97205
(503) 222-5161

Shishir Mukherjee
Senior Project Manager
Electric Power Research Institute
3412 Hillview Avenue
P.O. Box 10412
Palo Alto, CA 94303
(415) 855-2621

Fred Norrell
Economist
Alabama Power Company
P.O. Box 2641
Birmingham, AL 35291
(205) 250-4405

Gay Pasley
Supervisor
System Forecasting & Analysis
Baltimore Gas & Electric Company
Room 900
P.O. Box 1475
Baltimore, MD 21203
(301) 234-5531

Hayden E. Perkey
Baltimore Gas & Electric Company
P.O. Box 1475
Baltimore, MD 21203
(301) 234-5732

William Poe
Director
Economic Development
Carolina Power & Light Company
P.O. Box 1551
Raleigh, NC 27602
(919) 836-6653

Nick Poulios
Senior Economist
General Public Utilities
100 Interpace Parkway
Parsippany, NJ 07054
(201) 263-6277

Kathryn Price
Supervisor
Analyst-Forecasting
Pennsylvania Power & Light Company
2 North Ninth Street
Allentown, PA 18101
(215) 770-5034

T. Ross Reeve
Research Analyst
Utah Power & Light Company
1407 West North Temple
P.O. Box 899
Salt Lake City, UT 84110
(801) 535-4025

Stephen Ross
Staff Planner
Houston Lighting & Power Company
611 Walker Street
Houston, TX 77001
(713) 228-9211 X3845

Paul Rossi
Economic Analyst
San Diego Gas & Electric
P.O. Box 1831
San Diego, CA 92112
(619) 696-4026

Sam Schurr
Deputy Director
Energy Study Center
Electric Power Research Institute
3412 Hillview Avenue
P.O. Box 10412
Palo Alto, CA 94303
(415) 855-2909

Michael Sedmak
Associate
Booz, Allen & Hamilton Incorporated
4330 East West Highway
Bethesda, MD 20814
(301) 951-2051

Thomas Sheahen
Executive Director
Energy Research Advisory Board
U.S. Department of Energy
1000 Independence Avenue, S.W.
Washington, DC 20585
(202) 252-5444

Ray Squitieri
Senior Planning Economist
Electric Power Research Institute
3412 Hillview Avenue
P.O. Box 10412
Palo Alto, CA 94303
(415) 855-2329

Raymond E. Sund
Director
Research & Development
Toldeo Edison Company
300 Madison Avenue
Toldeo, OH 43652
(419) 249-5118

Blair Swezey
Energy Economist
Electric Power Research Institute
3412 Hillview Avenue
P.O. Box 10412
Palo Alto, CA 94303
(415) 855-2840

Stanley Szwed
Supervisor
Forecasts & Special Studies
The Cleveland Electric Illuminating
Company (CEI)
55 Public Square
Room 311B
Cleveland, OH 44113
(216) 622-9800 X3516

Richard Tate
Load Research Analyst II
Georgia Power Company
21st Floor
333 Piedmont Avenue
Atlanta, GA 30308
(404) 526-7398

Altaf Taufique
Senior Economist
Corporate Planning
Gulf States Utilities Company
P.O. Box 2951
Beaumont, TX 77704
(409) 838-6631

Timothy Viezer
Junior Forecast Analyst
The Cleveland Electric Illuminating
Company (CEI)
55 Public Square
Room 311B
Cleveland, OH 44113
(216) 622-9800 X3504

Joe Wharton
Project Manager
Demand & Conservation Program
Electric Power Research Institute
3412 Hillview Avenue
P.O. Box 10412
Palo Alto, CA 94303
(415) 855-2924

SPEAKERS

John A. Alic
Project Director
Office of Technology Assessment
600 Pennsylvania Avenue S.E.
Washington, DC 20510
(202) 226-2012

Laurel Andrews
Synergic Resources Corporation
Suite 820
Fourth and Pike Building
Seattle, WA 98101
(206) 624-8508

Barry Bluestone
Professor of Economics
Boston College
515 McGuinn Hall
Chestnut Hill, MA 02167
(617) 552-8000

John H. Broehl
Research Leader
Battelle Columbus Division
505 King Avenue
Columbus, OH 43201-2693
(614) 424-4325

James F. Collins
Executive Vice President
American Iron & Steel Institute
100 Sixteenth Street N.W.
Washington, DC 20036
(202) 452-7100

Joseph W. Duncan
Corporate Economist
& Chief Statistician
The Dun & Bradstreet Corporation
299 Park Avenue
New York, NY 10171
(212) 593-6947

Ahmad Faruqui
Manager, End-Use Assessment and Forecasting
Electric Power Research Institute
3412 Hillview Avenue
P.O. Box 10412
Palo Alto, CA 94303
(415) 855-2630

Stanley J. Feldman
Vice President
Interindustry Services
Data Resources Incorporated
29 Hartwell Avenue
Lexington, MA 02173
(617) 861-0165

Pradeep Gupta
Director
Energy Analysis Department
Electric Power Research Institute
3412 Hillview Avenue
P.O. Box 10412
Palo Alto, CA 94303
(415) 855-2143

Bruce G. Humphrey
Director
Economics
Edison Electric Institute
1111 19th Street N.W.
Washington, DC 20036
(202) 828-7456

Hillard G. Huntington
Senior Research Associate
Energy Modelling Forum
Stanford University
Stanford, CA 94305
(415) 497-1050

William R. Huss
Principal Research Scientist
Battelle Columbus Division
505 King Avenue
Columbus, OH 43201-2693
(614) 424-4607

Kurt E. Karl
Director
Long Term Forecasting Services
Wharton Econometric Forecasting
Associates
3624 Science Center
Philadelphia, PA 19104
(215) 386-9000

Stanley Lancey
Director
Quantitative Analysis
American Paper Institute
260 Madison Avenue
New York, NY 10016
(212) 340-0600

Robert Z. Lawrence
Senior Fellow
The Brookings Institution
1775 Massachusetts Avenue N.W.
Washington, DC 20036
(202) 797-6000

Robert C. Marlay
Special Assistant to the Deputy
Assistant Secretary for Conservation
U.S. Department of Energy
Forrestal Building
Mail Stop CE-10
Washington, DC 20585
(202) 252-9232

Tom Moore
Manager
Economic & Load Forecasting
Tampa Electric Company
P.O. Box 111
Tampa, FL 33601
(813) 228-4462

William J. Sheppard
Senior Research Scientist
Battelle Columbus Division
505 King Avenue
Columbus, OH 43201-2693
(614) 424-4755

Albert J. Sobey
Director
Energy & Advanced Product Economics
General Motors Corporation
Room 12168
3044 West Grand Boulevard
Detroit, MI 48202
(313) 556-4360

William L. Stewart
Acting Division Director
Division of Science Resources Studies
National Science Foundation
1800 G Street N.W.
Washington, DC 20550
(202) 634-4625

Sam O. Sugiyama
Chief
Industrial Forecasting Section
Bonneville Power Administration
P.O. Box 3621
Portland, OR 96208
(503) 230-3692